Logic Colloquium 2004

LECTURE NOTES IN LOGIC

A Publication of
The Association for Symbolic Logic

This series serves researchers, teachers, and students in the field of symbolic logic, broadly interpreted. The aim of the series is to bring publications to the logic community with the least possible delay and to provide rapid dissemination of the latest research. Scientific quality is the overriding criterion by which submissions are evaluated.

See end of book for a list of the books in the series. More information can be found at http://www.aslonline.org/books-lnl.html.

LECTURE NOTES IN LOGIC 29

Logic Colloquium 2004

Proceedings of the Annual European Summer Meeting of the Association for Symbolic Logic, Held in Torino, Italy, July 25–31, 2004

Edited by

ALESSANDRO ANDRETTA
Università degli Studi di Torino

KEITH KEARNES
University of Colorado, Boulder

DOMENICO ZAMBELLA
Università degli Studi di Torino

ASSOCIATION FOR SYMBOLIC LOGIC

CAMBRIDGE
UNIVERSITY PRESS

CAMBRIDGE UNIVERSITY PRESS
Cambridge, New York, Melbourne, Madrid, Cape Town, Singapore, São Paulo, Delhi

Cambridge University Press
32 Avenue of the Americas, New York, NY 10013-2473, USA

www.cambridge.org
Information on this title: www.cambridge.org/9780521884242

Association for Symbolic Logic
David Marker, Publisher
Department of Mathematics, Statistics, and Computer Science (M/C249)
University of Illinois at Chicago
851 S. Morgan St.
Chicago, IL 60607, USA
http://www.aslonline.org

First published 2008

Printed in the United States of America

This book was typeset in LaTeX by the African Group for the ASL Typesetting Office, from electronic files produced by the authors, using the ASL document class asl.cls. The font is Monotype Times Roman.

A catalog record for this publication is available from the British Library.

Library of Congress Control Number: 87659142

ISBN 978-0-521-88424-2 hardback

CONTENTS

INTRODUCTION

The Logic Colloquium 2004, the European Summer Meeting of the Association for Symbolic Logic, was held in Torino, Italy, 25–31 July 2004.

With more than 280 participants from all over the world, ten one-hour plenary lectures, four tutorials (three hours each), a panel discussion, three special sessions (with 20 speakers), and more than 200 contributed talks, the Logic Colloquium 2004 managed to present a broad spectrum of ideas, results, techniques, and problems in current research in logic.

The Program Committee consisted of: Toshiyasu Arai, Warren Goldfarb, Greg Hjorth, Dick de Jongh, Steffen Lempp, Gabriele Lolli, David Marker, Tony Martin, Ralph McKenzie, Wolfram Pohlers (Chair), Wilfried Sieg, Andrea Sorbi, and Alex Wilkie.

The Local Organizing Committee consisted of: Alessandro Andretta (Chair), Stefano Berardi, Riccardo Camerlo, Ugo de'Liguoro, Mariangiola Dezani, Alberto Marcone, Nicola Olivetti, and Domenico Zambella.

The meeting was organized by the logicians at the Dipartimento di Matematica and Dipartimento di Informatica of the University of Torino and was held at the Centro Congressi dell'Unione Industriale.

Sponsors. We would like to express our sincerest thanks to the following institutions, which, with their generous financial support, made this meeting possible. First and foremost, the Association for Symbolic Logic, which gave us the opportunity to organize the meeting and supported many students and post-docs with travel grants. A special mention to the Fondazione San Paolo, which supported, by means of a generous grant, a large number of young PhD students and researchers from developing countries. These proceedings would have not been possible without a grant from the Fondazione Cassa di Risparmio di Torino. In no particular order, we would like to mention: the Associazione Italiana di Logica e Applicazioni; the Istituto Nazionale di Alta Matematica–G.N.S.A.G.A.; the Departments of Mathematics and of Computer Science and the Facoltà di Scienze Matematiche, Fisiche e Naturali of the University of Torino; Assessorato alla Cultura della Città di Torino; Unione Industriale; Torino Internazionale; Oxford University Press; A.K.Peters Ltd.; and Rekordata srl.

The program of the meeting is listed on the following pages. All invited speakers were invited to submit a paper for this volume, but not all did. The editors would like to thank everyone who contributed to this volume. All submitted papers were refereed according to the usual high standards for publications in logic journals, and we would like to express our gratitude to the referees for their work.

The Editors
Alessandro Andretta
Keith Kearnes
Domenico Zambella

SPEAKERS AND TITLES

Tutorials

Alexander Kechris and Slawomir Solecki, *Logic and dynamical systems*
Thomas Scanlon, *Geometric stability theory*
Reed Solomon, *Computable algebra*
Ross Willard, *An overview of modern universal algebra*

Plenary talks

John T. Baldwin, *Quasiminimal sets and categoricity in abstract elementary classes*
Tomek Bartoszynski, *Measure and category from set-theoretic viewpoint*
Alessandro Berarducci, *Definable groups in o-minimal structures*
Alberto Marcone, *Logical aspects of wqo theory*
Grigori Mints, *Intuitionistic Frege systems are polynomially equivalent*
Michael Möllerfeld, *On the proof theory of* Π^1_2*–comprehension*
Frank Stephan, *Kolmogorov complexity*
Göran Sundholm, *The METAlinguistic turn*
Guglielmo Tamburrini, *Computation and the explanation of intelligent behaviours: ethologically motivated restart*
Matthew Valeriote, *The constraint satisfaction problem and universal algebra*
W. Hugh Woodin, *Extender models and the inner model problem for supercompact cardinals and beyond*

Panel discussion

Kant's legacy for the philosophy of logic
Panelists: Mirella Capozzi, Cesare Cozzo, Riccardo Pozzo
Moderator: Dag Prawitz

Special session in model theory

Organizers: D. Marker, A. Wilkie

Matthias Aschenbrenner, *Faithfully flat Lefschetz extensions*
Alexander Berenstein, *Probability spaces with a generic automorphism*
Raf Cluckers, *Ax-Kochen-Eršov principles for integrals over local fields*
Tamara Servi, *On the decidability of the real exponential field: candidates for a complete axiomatization*
Ivan Tomašić, *Probabilistic independence theorem in pseudofinite fields*
Monica VanDieren, *Shelah's categoricity conjecture for tame abstract elementary classes*
Xavier Vidaux, *The analogue of Büchi's problem in polynomial rings*
Karim Zahidi, *Elimination theory for addition and Frobenius in certain rings of positive characteristic*

Special session in universal algebra

Organizers: R. Willard, M. Valeriote, K. A. Kearnes, G. McNulty

Paolo Aglianò, *Basic-algebras: an overview*
Erhard Aichinger, *The variety of near-rings is generated by its finite members*
Steve Givant, *Interconnections between groups and algebras of relations*
Benoit Larose, *Some results on the complexity of strong colourings of hypergraphs with fixed group of symmetry*
Paolo Lipparini, *Tolerance intersection properties*
Miklós Maróti, *Finite basis problems and results for quasivarieties*
Vera Vértesi, *Checking equations in finite algebras*
Lászlo Zádori, *Bounded width problems and algebras*

Special session in proof theory

Organizers: T. Carlson, M. Rathjen, A. Weiermann

Andrea Cantini, *Classifications, operations and (constructive) sets*
Robert Lubarsky, *On the independence of variants of power set under IZF–power set*
Anton Setzer, *Universes in type theory: Mahlo and* Π_3*–reflection*
Alan Woods, *Probability limit laws for ordinal numbers*

ABSTRACT ELEMENTARY CLASSES: SOME ANSWERS, MORE QUESTIONS

JOHN T. BALDWIN

Abstract. We survey some of the recent work in the study of Abstract Elementary Classes focusing on the categoricity spectrum and the introduction of certain conditions (amalgamation, tameness, arbitrarily large models) which allow one to develop a workable theory. We repeat or raise for the first time a number of questions; many now seem to be accessible.

Much late 19th and early 20th century work in logic was in a 2nd order framework; infinitary logics in the modern sense were foreshadowed by Schroeder and Pierce before being formalized in modern terms in Poland during the late 20's. First order logic was only singled out as the 'natural' language to formalize mathematics as such authors as Tarski, Robinson, and Malcev developed the fundamental tools and applied model theory in the study of algebra. Serious work extending the model theory of the 50's to various infinitary logics blossomed during the 1960's and 70's with substantial work on logics such as $L_{\omega_1,\omega}$ and $L_{\omega_1,\omega}(Q)$. At the same time Shelah's work on stable theories completed the switch in focus in first order model theory from study of the logic to the study of complete first order theories As Shelah in [44, 46] sought to bring this same classification theory standpoint to infinitary logic, he introduced a total switch to a semantic standpoint. Instead of studying theories in a logic, one studies the class of models defined by a theory. He abstracted (pardon the pun) the essential features of the class of models of a first order theory partially ordered by the elementary submodel relation. An abstract elementary class AEC $(\boldsymbol{K}, \prec_{\boldsymbol{K}})$ is a class of models closed under isomorphism and partially ordered under $\prec_{\boldsymbol{K}}$, where $\prec_{\boldsymbol{K}}$ is required to refine the substructure relation, that is closed under unions and satisfies two additional conditions: if each element M_i of a chain satisfies $M_i \prec_{\boldsymbol{K}} M$ then $M_0 \prec_{\boldsymbol{K}} \bigcup_i M_i \prec_{\boldsymbol{K}} M$ and $M_0 \prec_{\boldsymbol{K}} M_2, M_1 \prec_{\boldsymbol{K}} M_2$ and $M_0 \subseteq M_1$ implies $M_0 \prec_{\boldsymbol{K}} M_1$ (coherence axiom). Further there is a Löwnenheim-Skolem number κ associated with $\boldsymbol{K}$ so that if $A \subseteq M \in \boldsymbol{K}$, there is an M_1 with $A \subset M_1 \prec_{\boldsymbol{K}} M$ and $|M_1| \leq |A| + \kappa$.

Partially supported by DMS -0500841

Logic Colloquium '04
Edited by A. Andretta, K. Kearnes, and D. Zambella
Lecture Notes in Logic, 29

In this paper we will review some of the reasons for considering AEC's, outline several major lines of study in the subject, and offer a series of problems whose solution would advance the various lines. The fundamental ideas discussed here are due to Shelah. However, we explore in some detail areas that have been developed in the very recent past by such authors as Grossberg, Hyttinen, Kolesnikov, Lessmann, VanDieren, and Villaveces; generally speaking these studies proceed by putting further model theoretic conditions on an AEC and we will expound some of these conditions. In the closing pages we give a short introduction to the mainline of Shelah's research [52, 53, 51, 50].

Our survey focuses primarily on problems closely related to categoricity. We have attempted to attribute both results and questions correctly. But many of the questions are just writing out what people in the area are thinking about. For expositional purposes, we frequently cite [1]; the default is that results in that monograph are not new although the proofs may be. I thank Tapani Hyttinen, the anonymous referee, and especially Rami Grossberg for useful comments on this article.

Increased interest in nonelementary classes arose recently for several reasons. First, the increased emphasis, signaled in [49, 54] and emphasized in [21], on hypotheses such as amalgamation or tameness as fruitful conditions to create a workable theory of AEC, has led to a number of new results. The need for studying AEC became more clear for two reasons. On the one hand the pursuit of specific problems in the first order setting has led to constructions which can no longer be formalized by first order means. On the other, the paradigm: study an interesting structure by studying its first order theory has broken down in some significant cases because the first order theory is not sufficiently nice.

The work of Kim and Pillay [31] showed that the essential distinction between stable and simple theories [45] lay in the fact that for a stable theory, Lascar strong type equals strong type. Strong types are first order objects; Lascar strong types are not. Analysis of this problem led to the introduction of hyperimaginaries and other properly infinitary objects and ultimately to compact abstract theories CATS [12]. In a slightly different direction, the 'Hrushovski construction' [28, 27] leads in nice cases (when the generic is ω-saturated) to the construction of first order theories with special properties. However, in certain notable cases, the best that has so far been found is a Robinson theory (in the search for a bad field [3, 5]) or even only a positive Robinson theory (in the search for a simple theory where strong type is not equal to Lascar strong type [40]). Despite the terminology, a (positive) Robinson theory, refers to the class of models of a first order theory which omit certain types; it can be described only in infinitary logic.

The first order theory of the field of complex numbers with exponentiation is intractable; the ring of integers and their order is first order definable. But Zilber suggested in a sequence of papers [60, 59, 57, 58] the notion of considering the $L_{\omega_1,\omega}(Q)$-theory of $(\mathbb{C}, +, \cdot, \exp)$. The intuition is that the essential wildness will be contained by forcing the kernel of the exponential map to always be exactly the standard integers. In his proof of categoricity for quasiminimal excellent classes Zilber discovered a special case of Shelah's notion of excellence that is easy to describe. He works in a context where there is a well-behaved notion of closure, cl which defines a combinatorial geometry. The aim is to show that if X is isomorphic to Y, then $\mathrm{cl}(X)$ is isomorphic to $\mathrm{cl}(Y)$. In general, this condition is non-trivial; it follows from excellence. In this context, excellence means that for every n, if $A = \{a_1, \ldots, a_n\}$ is an independent set then for any $a \in \mathrm{cl}(A)$ the type of a over $Z = \bigcup_{i<n} \mathrm{cl}(A - \{a_i\})$ is determined by the type of a over a finite subset of Z. Shelah works in a more general situation, where combinatorial geometry is replaced by a 'forking'-like notion. Consequently his notion is harder to describe and we omit the description here. Crucially, in both cases a condition (excellence) on countable models has important consequences (e.g. amalgamation) in all cardinalities.

Various other attempts to formalize analytic structures (notably Banach spaces [25, 26]) provide examples of 'homogeneous model theory' ([43, 13] and many more); Banach spaces are also an example of CATS [11]. Strictly speaking, the class of Banach spaces is not closed under unions of chains so doesn't form an AEC. But, Banach space model theory can be thought of as the study not of Banach spaces, but of structures whose completion is a Banach Space and this provides an interpretation of classes of Banach spaces as AEC's. Further mathematical examples include locally finite groups [19] and some aspects of compact complex manifolds (Although here, the first order theory is an attractive topic for model theorists (e.g. [39, 41]).).

Many, but not all, of these 'infinitary' formalizations can be captured in the framework of AEC's. (In particular, CATS are inherently different.) The work that I'll describe here has a complementary motivation. Stability theory provided a very strong tool to classify first order theories and then for extremely well-behaved theories (those below the 'main gap') to assign invariants to models of the theories. This insight of Shelah spread beyond stable theories with the realization that very different tools but some of the same heuristics allowed the study of o-minimal theories. By these techniques, o-minimality and stability, model theorists have learned much about the theories of both the real and the complex numbers and many other algebraic structures. But Shelah asks an in some ways more basic question. What are the properties of first order logic that make stability theory work? To what extent can we extend our results to wider classes, in particular to AEC?

Most known mathematical results are either *extremely cardinal dependent:* about finite or countable structures or at most structures of cardinality the continuum; *or completely cardinal independent:* about every structure satisfying certain properties. Already first order model theory has discovered problems that have an intimate relation between the cardinality of structures and algebraic properties of the structures:

1. Stability spectrum and counting models
2. A general theory of independence
3. Decomposition theorems for general models

There are structural algebraic, not merely combinatorial features, which are non-trivially cardinal dependent. (For example, the general theory of independence is intimately related with the class of cardinals in which the theory is stable and even for stable first order theories, stability in κ depends on the cofinality of κ.)

As usual a class of models $\boldsymbol{K}$ with a distinguished notion of submodel has *joint embedding property* (jep) if any two members of $\boldsymbol{K}$ have a common extension and $\boldsymbol{K}$ has the *amalgamation property* (ap) if any two extensions of a fixed model M have a common extension (over M).

If we were to take the fundamental analogy to be that an abstract elementary class represents a *complete* first order theory then we would add to the definition that the class $(\boldsymbol{K}, \prec_{\boldsymbol{K}})$ has the amalgamation and the joint embedding property. But completeness is a bit much to ask even in $L_{\omega_1,\omega}$. Here completeness (all models Karp equivalent) is not necessarily compatible with Löwenheim number ω. Some uncountable models do not have countable Karp equivalent submodels. The standard first order proof of the theorem, 'categoricity in power implies completeness' is a triviality but it assumes *both* the upwards and downwards Löwenheim-Skolem theorem for a set of sentences. Even for a sentence of $L_{\omega_1,\omega}$ in a countable language the reduction for an arbitrary categorical sentence ψ to one which is complete and has essentially the same spectrum is not at all trivial [44, 46]. It is substantially easier if ψ is assumed to have arbitrarily large models ([1] VII.2) than without that hypothesis ([1] VII.3). The difficult case is carried out in full in [44, 46]; the easier case is hinted at in [44, 46] but spelled out in the expository [3, 1]. In either case a notion of stability (counting the number of types) is used to obtain even the completeness result.

Moreover, unlike the first order case, completeness does not immediately yield the amalgamation property. The only known proof [46, 47] that a categorical sentence in $L_{\omega_1,\omega}$ has the amalgamation property invokes the weak continuum hypothesis and introduces the much more intricate notion of excellence. Moreover few models in every cardinal up to $\aleph_\omega$ is assumed; indeed, categoricity in every cardinal up to $\aleph_\omega$ is essential to get eventual categoricity [24, 1]. Similarly, although Zilber's quasiminimal excellent classes do have the amalgamation property the existing proof deduces the result from the proof of excellence, which has non-trivial algebraic content (e.g. [57]).

We will discuss first AEC with arbitrarily large models and then move to a harder case where that assumption is not made.

QUESTION 1. Must the class of models of a sentence in $L_{\omega_1,\omega}$ (or more generally an AEC) that has arbitrarily large models and is categorical in a sufficiently large cardinal have the amalgamation property (at least for sufficiently large models). This is an interesting question even assuming the weak GCH; the necessity of such an assumption presents a different set of problems.

Grossberg (e.g. [15]) has posed this question for AEC and for $L_{\omega_1,\omega}$. For sentences of $L_{\omega_1,\omega}$, Shelah's result reported above gives a partial answer modulo weak gch. He deduces excellence and thus amalgamation from categoricity up to $\aleph_\omega$. But although Grossberg's question is on the 'assume arbitrarily large models side', it is more demanding than Shelah's result in asking that categoricity in one cardinality suffice. Trying to obtain a proof (even for $L_{\omega_1,\omega}$) from the arbitrarily large model assumption without passing through excellence is a 'warm-up' strategy for the AEC version.

Shelah's presentation theorem is a crucial tool for the study of AEC. It asserts that every AEC $\boldsymbol{K}$ may be seen as the class of reducts of a collection of models defined by a first order theory (in a language of size $\mathrm{LS}(\boldsymbol{K})$) which omit a specified collection of (at most $2^{\mathrm{LS}(\boldsymbol{K})}$) types. Let us state a crucial corollary. Fix a vocabulary τ. For each pair of a first order theory and set of types Φ (in a vocabulary τ' extending τ), and each linear order I, $EM(I,\Phi)$ denotes the reduct to τ of the τ'-structure which satisfies Φ. The presentation theorem implies that for each $\boldsymbol{K}$, there is a Φ such that $EM(\_,\Phi)$ is a functor into $\boldsymbol{K}$ (which takes subordering to $\prec_{\boldsymbol{K}}$). A straightforward use of Ehrenfeucht-Mostowski models over indiscernibles yields: If $\boldsymbol{K}$ has a model of cardinality greater than $\beth_{(2^{\mathrm{LS}(\boldsymbol{K})})^+}$ then $\boldsymbol{K}$ has arbitrarily large models. In the vernacular, we say the Hanf number for AEC with vocabulary of size at most κ and Löwenheim-Skolem number at most κ is at most $H(\kappa) = \beth_{(2^\kappa)^+}$. We call this function H as we use it to compute Hanf numbers. It might be more appropriate to call it ER as it actually computes the bound for applying the Erdos-Rado theorem to obtain indiscernibles.

Many of the ideas expounded here were presaged in earlier work such as [37, 33] dealing with languages $L_{\kappa,\omega}$ with strong hypotheses (e.g. compactness, measurability) on the cardinal κ. The earliest result in this series was:

THEOREM 2. [37] *Let κ be strongly compact. If a sentence $\psi \in L_{\kappa,\omega}$ is categorical in $\lambda^+ > \mu_0 = \beth_{(2^\kappa)^+}$ then it is categorical in all cardinals greater than μ_0.*

In view of the set-theoretic requirements on the syntax of the underlying logic, we don't discuss this line but deal with the more general notion of AEC.

For most of the rest of this paper, we will assume $\boldsymbol{K}$ is an AEC with the amalgamation property. It is then trivial to reduce to the study of AEC with both the amalgamation and joint embedding properties. Under these hypotheses, when $\boldsymbol{K}$ has arbitrarily large models, we are able to work inside a monster model which behaves much like the first order situation but is weaker in a significant way. We have amalgamation only over submodels, thus the monster model is homogeneous only over submodels. The stronger condition, assuming that there is a 'monster model' that is homogeneous over sets, gives rise to the area known as homogeneous model theory. For the major literature in this area consult such authors as Hyttinen, Lessmann, and Shelah.

Working within a model-homogeneous 'monster model' (i.e. in an AEC with amalgamation), we define the Galois type of a over M to be the orbit of a under automorphisms of the monster which fix M. We write $\mathcal{S}(M)$ for the collection of Galois types over M. Then we can define a model M to be κ-saturated if every Galois type over a submodel of M with cardinality $< \kappa$ is realized in M. A somewhat more general definition (without assuming ap) occurs in [48, 49].

We begin by discussing classes which have arbitrarily large models. Invoking the presentation theorem, we are able to build Ehrenfeucht-Mostowski models over sequences of order indiscernibles. As Shelah remarks in the introduction to [52], this yields the non-definability of well-ordering and so gives us an approximation to compactness. Most of these notes concern this case and build on [49]. We return at the end to the much more difficult situation, where one attempts to find information about AEC simply from the information that it has one (or few models) in some specific cardinalities. We will sketch some of Shelah's extensive work on this subject; our emphasis on classes with arbitrarily large models represents the extent of our understanding, not importance.

Assuming $\boldsymbol{K}$ has arbitrarily large models, the proof that categoricity in λ implies stability in all cardinals smaller than λ has the same general form as in the first order case. But other arguments involving Galois types over models generated by order indiscernibles require significantly more complicated analysis of the linear orders than in the first order case. This is in interesting contrast with the Laskowski-Pillay study of 'gross-models' [34]; a model is gross if every infinite definable subset of it has full cardinality. Morley's theorem can be proved in this context using the normal first order notion of type. Thus, the categoricity implies stability is routine. Intriguingly, the Laskowski-Pillay work was inspired by investigations of Moosa on the first order theory of compact complex manifolds.

The fundamental test question for the study of AEC is:

CONJECTURE 3 (Shelah's categoricity conjecture). *There is a cardinal $\mu(\kappa)$ such that for all AEC with Löwenheim number at most κ, if $\boldsymbol{K}$ is categorical in some cardinal greater than $\mu(\kappa)$ then $\boldsymbol{K}$ is categorical in all $\lambda \geq \mu(\kappa)$.*

The best approximation to the categoricity conjecture takes $\mu(\kappa)$ as the 'second Hanf number': $H_2 = H(H(\mathrm{LS}(\boldsymbol{K})))$. The initial step in the analysis [49] (see also [1]) requires the lifting to this setting of a clever integration of Morley's omitting types theorem and Morley's two cardinal theorem.

THEOREM 4. [49] *Suppose $\boldsymbol{K}$ has the amalgamation property and arbitrarily large models. Suppose $\boldsymbol{K}$ is λ^+-categorical with $\lambda > H_2$. Then, $\boldsymbol{K}$ is H_2-categorical and indeed categorical on the interval $[H_2, \lambda^+]$.*

The proof requires using the omitting types theorem twice. The second time one names as many constants (H_1) as required for the first use. Categoricity on the interval is then proved by induction, making essential use of Theorem 7. Theorem 4 leads to a natural question.

QUESTION 5. Prove or disprove. Suppose $\boldsymbol{K}$ has the amalgamation property and arbitrarily large models. Suppose $\boldsymbol{K}$ is λ^+-categorical with $\lambda > H_1$. Then, $\boldsymbol{K}$ is H_1-categorical.

In order to understand further progress on the categoricity transfer problem, we introduce an important notion (first named in [21]; the cardinal parameters were added in [4]).

DEFINITION 6. The AEC $\boldsymbol{K}$ is (χ, λ)-(weakly) tame if for any (saturated) model M of cardinality λ, if $p, q \in \mathcal{S}(M)$ (the Galois types over M) are distinct then there is a submodel N of M with $N \leq \chi$ so that $p \restriction N \neq q \restriction N$.

Of course any first order theory is tame; i.e. $(\aleph_0, \infty)$-tame. And by [46, 47], it is consistent with ZFC that every categorical AEC defined by a sentence of $L_{\omega_1,\omega}$ is tame. But aside from the first order case (and homogeneous model theory where again every class is tame), there is no example where $(\aleph_0, \infty)$-tameness has been deduced from categoricity except as a corollary to the Morley theorem for the class. (E.g. Zilber's quasiminimal excellent classes and categorical classes in $L_{\omega_1,\omega}$ are each shown to be tame in [1]; but the result is not needed for the transfer of categoricity proof given but only an observation.)

Nontameness can arise in natural mathematical settings. An Abelian group is $\aleph_1$-free if every countable subgroup is free. An Abelian group H is *Whitehead* if every extension of Z by H is free. Shelah constructed (in ZFC) an Abelian group of cardinality $\aleph_1$ which is $\aleph_1$-free but not a Whitehead group. (See [14, Chapter VII.4].) Baldwin and Shelah [8] code this into an example of nontameness. Essentially a point codes an abelian group which is the right end of a short exact sequence; every countable approximation to the group splits but the whole group does not. Thus the AEC is not $(\aleph_0, \aleph_1)$-tame. Baldwin and Shelah [8] also show that nontameness is essentially a distinct phenomena from non-amalgamation by showing any AEC $\boldsymbol{K}$ which exemplifies a nontameness property (satisfying a mild condition) can be transformed to one which does satisfy amalgamation and still fails the tameness. But this transformation destroys categoricity and even stability. In my view, the most significant (nontrivial) sufficient condition for tameness is due to Shelah:

THEOREM 7. [49] *Suppose $\boldsymbol{K}$ has the amalgamation property and arbitrarily large models. Suppose $\boldsymbol{K}$ is λ-categorical with $\lambda > H_1$. For every κ with $H_1 \leq \kappa < \lambda$, $\boldsymbol{K}$ is (χ, κ)-weakly tame for some $\chi < H_1$.*

The combination of Shelah's downward categoricity argument and the tameness argument gives the result for 'tame' instead of 'weakly tame' if H_1 is allowed to grow to H_2. The argument for Theorem 7 in [49] is flawed. A short and correct argument due to Hyttinen, correcting and elaborating various exegises given separately by Baldwin and Shelah, appears in [1]. This result poses several questions.

QUESTION 8. Suppose $\boldsymbol{K}$ has the amalgamation property and arbitrarily large models. Suppose $\boldsymbol{K}$ is λ-categorical with $\lambda > H_1$.

1. Is there any way to reduce the upper bound on χ in Theorem 7 (or find a lower bound above $\mathrm{LS}(\boldsymbol{K})$)?
2. Is there any way to replace weakly tame by tame?
3. And most important, (compare 1.16 of [21]), can $\kappa = \lambda$ in Theorem 7?

A positive answer to Question 8.3 would yield a full solution of the categoricity problem for AEC with amalgamation and arbitrarily large models.

Is there any way to weaken the categoricity hypothesis in Theorem 7 to stability?

QUESTION 9. Suppose $\boldsymbol{K}$ has the amalgamation property and arbitrarily large models. Prove or disprove: If $\boldsymbol{K}$ is κ-stable with $\kappa > H_1$ then $\boldsymbol{K}$ is (weakly) (H_1, κ)-tame.

In the light of Theorem 4 and Theorem 7, it is interesting to examine Question 9 at the successor of the categoricity cardinal in the hypothesis of Theorem 4. How much would it help to know stability in λ^{++}?

Shelah speaks rather loosely of locality in various places. We have broken this notion into three precise concepts. Following [21], we have chosen tame as the name of one of these. We call the others locality and compactness. There is considerable to be learned about the relations among the parameterized versions of these notions; the following survey just touchs on some of the natural questions that arise. Essentially, they are a few of the many ways one might make specific the general question, 'Are there AEC which are eventually categorical without the many nice properties such as tameness, excellence, locality of the known examples?'

DEFINITION 10.

1. $\boldsymbol{K}$ has (κ, λ)-*local Galois types* if for every continuous increasing chain $M = \bigcup_{i<\kappa} M_i$ of members of $\boldsymbol{K}$ with $|M| = \lambda$ and for any $p, q \in \mathcal{S}(M)$: if $p \upharpoonright M_i = q \upharpoonright M_i$ for every i then $p = q$.
2. Galois types are (κ, λ)-*compact* in $\boldsymbol{K}$ if for every continuous increasing chain $M = \bigcup_{i<\kappa} M_i$ of members of $\boldsymbol{K}$ with $|M| = \lambda$ and every increasing chain $\{p_i : i < \kappa\}$ of members $\mathcal{S}(M_i)$ there is a $p \in \mathcal{S}(M)$ with $p \upharpoonright M_i = p_i$ for every i.

The proof of Theorem 7 is very much about tameness rather than locality.

QUESTION 11. Is there any way to replace (weakly) tame by local in Theorem 7?

The constructions in [8] that create amalgamation destroy categoricity; can this be avoided? More precisely,

QUESTION 12. Find an AEC (in a countable language) which is categorical in all uncountable powers, has the amalgamation property, and which is not $(\aleph_0, \infty)$-tame (or $(\aleph_0, \infty)$-local).

Grossberg and Kolesnikov [17] recently completed an important analysis of the relationship between excellence and tameness. They work in classes which are posited to have an independence relation analogous to forking in the first order case. They show in particular that if the class satisfies the extension property for independence, the appropriate version of stationarity and the forking has $< \lambda$-character then (λ, λ^+)-tameness and (λ, λ^+)-locality follow. Further, they show that if $\boldsymbol{K}$ is χ-excellent (under an extended definition for this context), then $\boldsymbol{K}$ is (χ, ∞)-tame. There is no simple test, such as failure of the order property in the first order case, to generate a wealth of examples of classes satisfying these hypotheses. It is not even clear that eventual categoricity yields the properties. But as with tameness, these hypotheses provide a platform on which to develop a stability theory for AEC.

A positive answer to either Question 8.1 or Question 11 would seem to require essentially new methods. The distinction between syntactic (given by a set of formulas in some logic) and semantic or Galois types (given by the ability to amalgamate embeddings or as orbits in a suitably homogeneous model) leads to a quest for further examples.

QUESTION 13. What are some AEC's which are not basically given syntactically? Which of the many examples of extended logics in [10] give rise to AEC's?

A few examples appear in [15, 7, 1], but there should be many more. Zilber's work on excellent classes raises several issues here [60, 59]. He phrases his work for certain models (those satisfying the countable closure condition) in a class defined in $L_{\omega_1,\omega}$. So the class could be described in $L_{\omega_1,\omega}(Q)$; but such a formulation is of no value for the proof. The hardest part of the argument, the verification of excellence, is in the standard vein of algebraic model theory. But here infinitary conditions are being interwoven with not only algebraic but analytic arguments. Zilber's model theoretic perspective produces an intriguing group of conjectures about the complex numbers. In particular, even a very simple case of showing the complex exponential field is 'strongly exponentially closed' in the sense of [59], has only been answered using Schanuel's conjecture and Hadamard factorization [38].

In another direction, one might try to weaken the categoricity assumption for proving tameness. The following version doesn't shed much light since we don't have any clear way at hand to verify it (aside from categoricity). Shelah called this notion rigid. I discuss this notion because a number of central steps in the analysis of categoricity in [49], existence of non-splitting extensions, (H_1, λ)-tameness, and unions of saturated models are saturated are fundamentally about AEC which are epi.

DEFINITION 14. The AEC $\boldsymbol{K}$ is *epi* if there is an EM-template Φ such that the functor $EM(\_, \Phi)$ is an *epi*morphism from linear orders onto the models of $\boldsymbol{K}$.

For example, the core of the proof of Theorem 7 shows:

COROLLARY 15. *If $\boldsymbol{K}$ is epi then $\boldsymbol{K}$ is (H_1, ∞)-tame.*

Categoricity is used in the proof of Theorem 7 to get that the AEC is epi. (Of course this terminology isn't used.)

Grossberg and VanDieren [20] strengthen the hypothesis of Theorem 4 by adding (μ, ∞)-tameness for some $\mu < \lambda$ with powerful results.

THEOREM 16. [21] *Suppose $\boldsymbol{K}$ has the amalgamation property and arbitrarily large models. Suppose $\boldsymbol{K}$ is λ and λ^+ categorical for some $\lambda > \mathrm{LS}(\boldsymbol{K})$ and is (μ, ∞)-tame for some $\mu < \lambda$. Then $\boldsymbol{K}$ is categorical in all cardinals above λ.*

There are a number of variations on this result and on the elimination of the (categorical in λ)-hypothesis [36, 20, 7, 23, 55] to get 'upwards categoricity from a single cardinal'. We don't go into this further here except to remark that any full proof from these hypotheses involves an intensive investigation of EM models to show that a union of a short chain of saturated models is saturated [49, 1]. Natural extensions, which remain open as far as I know, are to replace categoricity in a single successor cardinal by categoricity in a regular or an arbitrary cardinal; a different idea is needed to replace the role of two cardinal models.

The stability spectrum theorem is fundamental for the study of first order theories; it is the essence of the classification of theories. But no similar result is known for general abstract elementary classes. The stability spectrum of an AEC $\boldsymbol{K}$ is the function from cardinals to cardinals which gives the supremum of the cardinals of the number of Galois types over a model in $\boldsymbol{K}$ of fixed cardinality.

QUESTION 17. Is the stability spectra of an abstract elementary classes (even in a countable language with $\mathrm{LS}(\boldsymbol{K}) = \omega$) one of a finite set of functions? Does ω-stable imply stable in all cardinals?

Baldwin, Kueker, and VanDieren [6] give a positive response to the last question but only under the extremely strong hypotheses of both $(\aleph_0, \infty)$-tameness and $(\aleph_0, \infty)$-locality. Grossberg and VanDieren earlier noted in [21] (they state stronger hypotheses):

THEOREM 18. *Let $\mu \geq \mathrm{LS}(\boldsymbol{K})$. If $\boldsymbol{K}$ is (μ, ∞)-tame and μ-stable then $\boldsymbol{K}$ is stable in all κ with $\kappa^\mu = \kappa$.*

This follows by the standard argument after you have shown every Galois type does not μ-split over a set of size μ and (using tameness) that Galois types (over appropriate base models) have unique non-splitting extensions. In the first order case, the converse to Theorem 18 requires the definition of $\kappa(T)$ and a simple argument [2] uses a considerable amount of the forking calculus. A natural question is

QUESTION 19. What is the 'correct' notion of superstablity for AEC?

There are a number of suggestions [49, 20, 15] revolving around variants in defining a concept analogous to the $\kappa(T)$ in first order logic: the length of a string of 'forking' extensions. Further complications arise from finding an appropriate notion of dependence. Below we discuss another candidate for superstability from [52, 53] where there is no assumption that $\boldsymbol{K}$ has arbitrarily large models. We ask naively

QUESTION 20. Is there a μ so that stability for all $\lambda > \mu$ is a robust concept for AEC? How does it interact with the purely structural notion: Every union of a chain of saturated members of $\boldsymbol{K}$ is saturated.

While it is straightforward in regular cardinals κ to show stability in κ implies the existence of a saturated model in κ, considerations like those above would be necessary to extend this to any κ in which $\boldsymbol{K}$ is stable. Note that the converse (non-stable in λ implies no saturated model in λ) is a rather technical argument in the first order case.

The positive results on the categoricity spectrum for classes with arbitrarily large models depend at least indirectly on subtle applications of EM-models [4]. The work of [21, 22, 20, 23, 30] proceeds in quite a different direction. Largely eschewing the use of EM-models the authors try to identify construction and 'stability theoretic conditions' that allow one to carry out more refined versions of the first order analysis (aiming towards geometric stability) in suitably restricted AEC.

Grossberg and VanDieren originated this trend in their analysis of the stability spectrum for tame AEC in [21]; it continued in further work on the stability spectrum [6] and the analysis of categoricity in [22, 20, 7] and under even stronger hypotheses in [30, 29]. This kind of work suggests several directions of inquiry.

Some of the crucial notions in this development are limit models, a new notion of strong type, towers of models and the means for analyzing them, variations on splitting (with assorted cardinal parameters). Determining the interrelations of these notions provides a fertile field of study. The notion of limit model is essential for studying structures with cardinality $\mathrm{LS}(\boldsymbol{K})$ because under the usual notion of Galois type (the domain of a type must be a model), the concept of a saturated model in cardinality $\mathrm{LS}(\boldsymbol{K})$ is vacuous.

The work of [30] is novel as it introduces a notion of type defined by mappings (i.e. a kind of Galois type) but considers types over arbitrary subsets. Key to this is the observation that the proof of transitivity in establishing an equivalence relation by $\mathrm{tp}_A(a) = \mathrm{tp}_B(b)$ if there is an automorphism of the universe taking a to b uses amalgamation not of the domain models (in the sense of Shelah's definition [46, 49, 15]) but rather amalgamation of the models in which the types are realized. They work in the general context of AEC with arbitrarily large models and amalgamation and make the additional assumptions that $\boldsymbol{K}$ has a prime model and extend amalgamation to disjoint amalgamation. Making effective use of this notion of Galois type requires the condition we'll call H-local: if $A \subset B$ and for every finite $\boldsymbol{a}$ in A, $\mathrm{tp}_A(\boldsymbol{a}) = \mathrm{tp}_B(\boldsymbol{a})$, then $A \prec_{\boldsymbol{K}} B$. They further prove that for tame AEC satisfying their general conditions, their notion of weak type coincides with Galois types. (In general $\boldsymbol{a}$ and $\boldsymbol{b}$ have the same Galois type implies they have same weak type implies they have the same syntactic type.)

One can ask for mathematical examples to justify the study of particular families of hypotheses. For example,

QUESTION 21. Is there an example of an AEC $\boldsymbol{K}$ with $\mathrm{LS}(\boldsymbol{K}) = \omega$ which is $(\aleph_0, \infty)$-tame and is not defined by a sentence in $L_{\omega_1,\omega}(Q)$?

Excellence is another condition to impose; in [16], the class $\boldsymbol{K}$ is defined in $L_{\omega_1,\omega}$. Excellence requires a notion of independence; essentially excellence consists in requiring the existence of 'prime models' over independent n-cubes. See [3] for an intuitive introduction. Grossberg and Hart [16] prove a 'main gap' theorem in their context. It would be interesting to try to replace ω-stable (part of the definition of excellence in this context) by stable. Grossberg, Kolesnikov and Lessmann [17, 32, 18] deal with AEC that are equipped with an independence notion. Thus, this work is in the tradition of [48, 9] and has some similarity to Shelah's [53, 51] study of frames.

Much of the work described here has been under the hypothesis of amalgamation. Naturally, the original more general question remains.

QUESTION 22. Explore AEC which have arbitrarily large models but without assuming the amalgamation property.

Shelah and Villaveces [54] and VanDieren [56] weaken 'amalgamation property' to 'no maximal models'. With the use of the Devlin-Shelah diamond they are able to prove the existence of a 'dense' family of amalgamation bases and carry over much of the analysis. In [56], under further model theoretic hypotheses, the uniqueness of limit models is established. With the increased understanding of the categoricity transfer problem for classes with amalgamation, the following is much more accessible.

QUESTION 23. Prove Theorem 4 and Theorem 16, but weakening the hypothesis of amalgamation property to no maximal models. To get the full information we have for classes with amalgamation, prove (an appropriate variant) of Theorem 7 under the weaker hypothesis.

The study of infinitary logic often appeared to have a heavy dependence on axiomatic set theory. This was perhaps exacerbated by studies (e.g. [37, 33]) of logics $L_{\kappa,\omega}$ or $L_{\kappa,\kappa}$ where κ was a large (compact or measurable) cardinal. But some extensions beyond ZFC are needed for the harder theme in studying AEC; *we no longer assume that* $\boldsymbol{K}$ *has arbitrarily large models.* Certain landmark results depend on set theory. In particular, Shelah originally assumed $\diamond$ to prove:

THEOREM 24. [44] *If a sentence of* $L_{\omega_1,\omega}(Q)$ *is* $\aleph_1$*-categorical then it has a model of power* $\aleph_2$.

But, in [48] and as expounded in [15, 1] the result can be given an extremely beautiful proof in ZFC; it is not quite possible to describe the result just in the language of AEC's [42, 1]. In contrast two other results:

THEOREM 25. *Suppose* $2^{\aleph_0} < 2^{\aleph_1}$.

1. *If the AEC* $\boldsymbol{K}$ *is* $\aleph_1$*-categorical then* $\boldsymbol{K}$ *is* ω*-stable.*
2. *Suppose* $2^{\lambda} < 2^{\lambda^+}$. *If the AEC* $\boldsymbol{K}$ *is* λ*-categorical, with* $\lambda \geq \mathrm{LS}(\boldsymbol{K})$ *but fails the amalgamation property in* λ *then it has* 2^{λ^+} *models of cardinality* λ^+.

actually require the set theoretic hypothesis. This necessity is outlined in [48], and more clearly in the revised version of that paper [42]; see [15] for a good account of the positive Theorem 25 2). Complete proofs of both results including finding counterexamples in $L_{\omega_1,\omega}$ showing the necessity of weak CH appear in [1]. The most striking result which has no apparent upward Löwenheim-Skolem assumption is Shelah's proof of the appropriate version of the Morley conjecture for $L_{\omega_1,\omega}$:

THEOREM 26. [46, 47] *Assume for each* $n < \omega$, $2^{\aleph_n} < 2^{\aleph_{n+1}}$. *If* ψ *is a sentence in* $L_{\omega_1,\omega}$ *in a countable vocabulary that is categorical in all cardinals less than* $\aleph_\omega$ *the* ψ *defines an excellent class that is categorical in all cardinals.*

The only full accounts of this is [46, 47, 1]; there are several accounts of the deduction of categoricity from excellence (e.g. [35, 1]. The excellence is actually derived from the hypothesis that there are less than the maximum number of models in each cardinal below $\aleph_n$ (there is a further set theoretic hypothesis hidden here; 'maximum' is a little more complicated than usual; see [46, 47]).

The $L_{\omega_1,\omega}$ work depends heavily the assumption that $\mathrm{LS}(\boldsymbol{K}) = \omega$ and for this reason it does not apply nor extend in a straightforward manner to $L_{\omega_1,\omega}(Q)$. More generally, to extend this kind of result to $\boldsymbol{K}$ with $\mathrm{LS}(\boldsymbol{K}) > \omega$ requires different methods. Shelah has a number of works in this area, which have not yet been published. Assuming, $2^{\lambda} < 2^{\lambda^+} < 2^{\lambda^{++}}$, Shelah asserts in the introduction to [52] that categoricity of an AEC in three successive cardinals implies the existence of a model in the next cardinal. Further, he asserts that categoricity in the third cardinal can be replaced by 'fewer than the maximal number of models'. Further work approaches the goal of showing that for an arbitrary AEC, categoricity on a sufficiently long interval of cardinals implies the existence of arbitrarily large models.

Shelah's work [53, 50, 51] introduces the notion of a *frame* and the stronger notion of a good frame. He regards this as a notion of 'superstability' for this context. In short, a frame describes more extensive conditions on the models in an AEC $\boldsymbol{K}$ of cardinality κ which suffice to move upward (say from categoricity in κ to existence or uniqueness of models in larger powers). In the other direction, the existence of a frame in cardinality κ is derived from categoricity in cardinals above κ; a major theme is to reduce the number of cardinals above κ in which one must make the categoricity hypothesis.

In one direction one can hope to generalize to AEC the fundamental results of first order stability theory; to classify AEC by some kind of stability notion, develop a robust notion of independence, and compute the possible spectra of an AEC under natural conditions.

A greater challenge is to relate this general study more directly to problems from mainstream mathematics. Broadly speaking, in the first order case, it was found that the studying the first order theory of a structure, the reals, the complexes, and others allowed one to get serious information about the structure. If the most optimistic scenario concerning Zilber's conjectures worked out, there would be evidence for the $L_{\omega_1,\omega}(Q)$-theory playing a similar role. But there is no candidate at present for associating a more general AEC with a structure. It seems more likely to me that the significance of these ideas will rest more in their role of enabling us to understand that Cantor's paradise is not merely combinatorial; rather future generations will understand a rich world of complex mathematical structures of unlimited cardinality. In particular, the fact that combinatorial principles (Devlin-Shelah diamond) which are derived from the weak GCH, allow the development of a smooth model theory may eventually be viewed as evidence for the naturality of the weak GCH. Further, as we begin to consider in depth structures with cardinality beyond the continuum, we may focus even more on considering naturally defined classes of structures rather than individual structures.

REFERENCES

[1] J. T. Baldwin, *Categoricity*, `www.math.uic.edu/~jbaldwin`.

[2] ———, ***Fundamentals of Stability Theory***, Springer-Verlag, Berlin, 1988.

[3] ———, *Notes on quasiminimality and excellence*, ***The Bulletin of Symbolic Logic***, vol. 10 (2004), no. 3, pp. 334–366.

[4] ———, *Ehrenfeucht-Mostowski models in abstract elementary classes*, ***Logic and its Applications*** (Yi Zhang, editor), Contemporary Mathematics, vol. 380, American Mathematical Society, Providence, 2005, pp. 1–15.

[5] J. T. BALDWIN and K. HOLLAND, *Constructing ω-stable structures: Rank 2 fields*, ***The Journal of Symbolic Logic***, vol. 65 (2000), no. 1, pp. 371–391.

[6] J. T. BALDWIN, D. W. KUEKER, and M. VANDIEREN, *Upward stability transfer for tame abstract elementary classes*, ***Notre Dame Journal of Formal Logic***, vol. 47 (2006), no. 2, pp. 291–298.

[7] J. T. BALDWIN and O. LESSMANN, *Upward categoricity of very tame abstract elementary classes with amalgamation*, to appear in APAL.

[8] J. T. BALDWIN and S. SHELAH, *Examples of non-locality*, submitted.

[9] ———, *The primal framework. I*, ***Annals of Pure and Applied Logic***, vol. 46 (1990), no. 3, pp. 235–264.

[10] J. Barwise and S. Feferman (editors), ***Model-Theoretic Logics***, Springer-Verlag, New York, 1985.

[11] I. BEN-YAACOV, *Positive model theory and compact abstract theories*, ***Journal of Mathematical Logic***, vol. 3 (2003), no. 1, pp. 85–118.

[12] ———, *Simplicity in compact abstract theories*, ***Journal of Mathematical Logic***, vol. 3 (2003), no. 2, pp. 163–191.

[13] S. BUECHLER and O. LESSMANN, *Simple homogeneous models*, ***Journal of the American Mathematical Society***, vol. 16 (2003), no. 1, pp. 91–121.

[14] P. EKLOF and ALAN MEKLER, ***Almost Free Modules: Set Theoretic Methods***, North-Holland, Amsterdam, 1990.

[15] R. GROSSBERG, *Classification theory for abstract elementary classes*, ***Logic and Algebra*** (Yi Zhang, editor), Contemporary Mathematics, vol. 302, American Mathematical Society, Providence, 2002, pp. 165–204.

[16] R. GROSSBERG and BRADD HART, *The classification theory of excellent classes*, ***The Journal of Symbolic Logic***, vol. 54 (1989), no. 4, pp. 1359–1381.

[17] R. GROSSBERG and ALEXEI KOLESNIKOV, *Excellent abstract elementary classes are tame*, preprint.

[18] R. GROSSBERG and OLIVIER LESSMANN, *Abstract decomposition theorem and applications*, ***Logic and its Applications***, Contemporary Mathematics, vol. 380, American Mathematical Society, Providence, 2005, pp. 73–108.

[19] R. GROSSBERG and S. SHELAH, *On universal locally finite groups*, ***Israel Journal of Mathematics***, vol. 44 (1983), no. 4, pp. 289–302.

[20] R. GROSSBERG and M. VANDIEREN, *Shelah's categoricity conjecture from a successor for tame abstract elementary classes*, ***The Journal of Symbolic Logic***, vol. 71 (2006), no. 2, pp. 553–568.

[21] ———, *Galois-stability for tame abstract elementary classes*, ***Journal of Mathematical Logic***, vol. 6 (2006), no. 1, pp. 25–48.

[22] ———, *Shelah's categoricity conjecture from a successor for tame abstract elementary classes*, ***The Journal of Symbolic Logic***, vol. 71 (2006), no. 2, pp. 553–568.

[23] R. GROSSBERG, M. VANDIEREN, and ANDRÉS VILLAVECES, *Limit models in stable classes with amalgamation*, preprint.

[24] BRADD HART and S. SHELAH, *Categoricity over P for first order T or categoricity for $\phi \in l_{\omega_1\omega}$ can stop at $\aleph_k$ while holding for $\aleph_0, \ldots, \aleph_{k-1}$*, ***Israel Journal of Mathematics***, vol. 70 (1990), no. 2, pp. 219–235.

[25] W. HENSON, *Countable homogeneous relational structures and $\aleph_0$-categorical theories*, ***The Journal of Symbolic Logic***, vol. 37 (1972), pp. 494–500.

[26] W. HENSON and JOSÉ IOVINO, *Ultraproducts in analysis*, ***Analysis and Logic***, London Mathematical Society Lecture Notes, vol. 262, Cambridge University Press, Cambridge, 2002, pp. 1–110.

[27] E. Hrushovski, *Strongly minimal expansions of algebraically closed fields*, ***Israel Journal of Mathematics***, vol. 79 (1992), no. 2-3, pp. 129–151.

[28] ———, *A new strongly minimal set*, ***Annals of Pure and Applied Logic***, vol. 62 (1993), no. 2, pp. 147–166.

[29] T. Hyttinen, *Uncountably categorical local tame abstract elementary classes with disjoint amalgamation*, ***Archive for Mathematical Logic***, vol. 45 (2006), no. 1, pp. 63–73.

[30] T. Hyttinen and M. Viljanen, *Independence in local aec: Part i*, preprint.

[31] Byunghan Kim and Anand Pillay, *Simple theories*, ***Annals of Pure and Applied Logic***, vol. 88 (1997), no. 2-3, pp. 149–164.

[32] Alexei Kolesnikov, *Dependence relations in non-elementary classes*, ***Logic and its Applications***, Contemporary Mathematics, vol. 380, American Mathematical Society, Providence, 2005, pp. 203–230.

[33] Oren Kolman and S. Shelah, *Categoricity of theories in $L_{\kappa\omega}$, when κ is a measurable cardinal. I*, ***Fundamenta Mathematicae***, vol. 151 (1996), no. 3, pp. 209–240.

[34] C. Laskowski and A. Pillay, *Morley's theorem for gross models*, preprint.

[35] Olivier Lessmann, *Categoricity and U-rank in excellent classes*, ***The Journal of Symbolic Logic***, vol. 68 (2003), no. 4, pp. 1317–1336.

[36] ———, *Upward categoricity from a successor cardinal for tame abstract classes with amalgamation*, ***The Journal of Symbolic Logic***, vol. 70 (2005), no. 2, pp. 639–660.

[37] Michael Makkai and S. Shelah, *Categoricity of theories in $L_{\kappa\omega}$, with κ a compact cardinal*, ***Annals of Pure and Applied Logic***, vol. 47 (1990), no. 1, pp. 41–97.

[38] D. Marker, *A remark on Zilber's pseudo-exponentiation*, preprint.

[39] R. Moosa, *On saturation and the model theory of compact Kähler manifolds*, ***Journal für die Reine und Angewandte Mathematik***, vol. 586 (2005), pp. 1–20.

[40] M. Pourmahdian and F. Wagner, *A simple positive Robinson theory with $Lstp \neq Stp$*, ***Model Theory Confernce in Tehran*** (Mojtaba Moniri, editor), 2003.

[41] Dale Radin, *A definability result for compact complex spaces*, ***The Journal of Symbolic Logic***, vol. 69 (2004), no. 1, pp. 241–254.

[42] S. Shelah, *Abstract elementary classes near $\aleph_1$ sh88r*, revision of Classification of nonelementary classes II, Abstract elementary classes; consult Shelah for copy.

[43] ———, *Finite diagrams stable in power*, ***Annals of Mathematical Logic***, vol. 2 (1970), no. 1, pp. 69–118.

[44] ———, *Categoricity in $\aleph_1$ of sentences in $L_{\omega_1,\omega}(Q)$*, ***Israel Journal of Mathematics***, vol. 20 (1975), no. 2, pp. 127–148.

[45] ———, *Simple unstable theories*, ***Annals of Mathematical Logic***, vol. 19 (1980), no. 3, pp. 177–203.

[46] ———, *Classification theory for nonelementary classes. I. The number of uncountable models of $\psi \in L_{\omega_1,\omega}$. Part A*, ***Israel Journal of Mathematics***, vol. 46 (1983), no. 3, pp. 212–240, paper 87a.

[47] ———, *Classification theory for nonelementary classes. I. The number of uncountable models of $\psi \in L_{\omega_1,\omega}$. Part B*, ***Israel Journal of Mathematics***, vol. 46 (1983), no. 4, pp. 241–273, paper 87b.

[48] ———, *Classification of nonelementary classes. II. Abstract elementary classes*, ***Classification Theory (Chicago, IL, 1985)*** (J. T. Baldwin, editor), Lecture Notes in Mathematics, vol. 1292, Springer, Berlin, 1987, paper 88: Proceedings of the USA–Israel Conference on Classification Theory, Chicago, December 1985, pp. 419–497.

[49] ———, *Categoricity for abstract classes with amalgamation*, ***Annals of Pure and Applied Logic***, vol. 98 (1999), no. 1-3, pp. 261–294, paper 394. Consult Shelah for post-publication revisions.

[50] ———, *On categorical a.e.c.*, paper 734 in preparation, 200?

[51] ———, *Toward classification theory of good λ frames and abstract elementary classes*, preprint 705, 200?

[52] ———, *Categoricity of an abstract elementary class in two successive cardinals*, ***Israel Journal of Mathematics***, vol. 126 (2001), pp. 29–128.

[53] ———, *Categoricity of abstract elementary classes: going up inductive step*, preprint 600, 200x.

[54] S. Shelah and Andrés Villaveces, *Toward categoricity for classes with no maximal models*, ***Annals of Pure and Applied Logic***, vol. 97 (1999), no. 1-3, pp. 1–25, revised version: `www.math.rutgers.edu/~shelah`.

[55] M. VanDieren, *Categoricity and no Vaughtian pairs*, preprint.

[56] ———, *Categoricity in abstract elementary classes with no maximal models*, ***Annals of Pure and Applied Logic***, vol. 141 (2006), no. 1-2, pp. 108–147.

[57] B. I. Zilber, *Covers of the multiplicative group of an algebraically closed field of characteristic* 0, to appear in JLMS, 2000.

[58] ———, *Raising to powers in algebraically closed fields*, ***Journal of Mathematical Logic***, vol. 3 (2003), no. 2, pp. 217–238.

[59] ———, *Pseudo-exponentiation on algebraically closed fields of characteristic* 0, ***Annals of Pure and Applied Logic***, vol. 132 (2005), no. 1, pp. 67–95.

[60] ———, *A categoricity theorem for quasi-minimal excellent classes*, ***Logic and its Applications***, Contemporary Mathematics, vol. 380, American Mathematical Society, Providence, 2005, pp. 297–306.

DEPARTMENT OF MATHEMATICS
STATISTICS AND COMPUTER SCIENCE
UNIVERSITY OF ILLINOIS AT CHICAGO

ON THE DENSITY OF HAUSDORFF ULTRAFILTERS

TOMEK BARTOSZYNSKI AND SAHARON SHELAH

Abstract. An ultrafilter U is Hausdorff if for any two functions $f, g \in \omega^\omega$, $f(U) = g(U)$ iff $f{\restriction}X = g{\restriction}X$ for some $X \in U$. We will show that the statement that Hausdorff ultrafilters are dense in the Rudin-Keisler order is independent of ZFC.

§1. Introduction. For $f \in \omega^\omega$ and an ultrafilter U on ω define $f(U) = \{X \subseteq \omega : f^{-1}(X) \in U\}$, and for $f, g \in \omega^\omega$ we say that $f = g \mod U$ if there is $X \in U$ such that $f(n) = g(n)$ for $n \in X$.

We say that U is Hausdorff if for any two functions $f, g \in \omega^\omega$, if $f(U) = g(U)$ then $f = g \mod U$.

Let FtO be the collection of all finite-to-one functions $f \in \omega^\omega$. Recall that an ultrafilter U is a p-point if for every function $f \in \omega^\omega$ either there is n such that $f^{-1}(\{n\}) \in U$ or there exists $g \in$ FtO such that $f = g \mod U$. Similarly, U is Ramsey if for every function $f \in \omega^\omega$ either there is n such that $f^{-1}(\{n\}) \in U$ or there exists a one-to-one function $g \in \omega^\omega$ such that $f = g \mod U$.

In this paper we will assume that all ultrafilters U, and their images $f(U)$ are non-principal.

It is worth mentioning that the following appears as an exercise in [7]. If $f(U) = U$ then $f = id \mod U$. Therefore, if U is not Hausdorff, then this is witnessed by two functions, both not one-to-one mod U. It follows from it that Ramsey ultrafilters are Hausdorff.

The notion of a Hausdorff ultrafilters was reintroduced and studied by Mauro Di Nasso, Marco Forti and others in a sequence of papers [3, 4, 5, 6] in context of topological extensions. They used the name Hausdorff because Hausdorff ultrafilters are precisely those ultrafilters whose ultrapowers

2000 *Mathematics Subject Classification.* 03.

Key words and phrases. Hausdorff, ultrafilters, consistency.

The first author was partially supported by NSF grant DMS 0200671 and by KBN grant 5 P03A 037 20.

The second author was partially supported by United States - Israel Binational Science Foundation. Publication 826.

Logic Colloquium '04
Edited by A. Andretta, K. Kearnes, and D. Zambella
Lecture Notes in Logic, 29

equipped with the standard topology are Hausdorff topological spaces. In this paper we will show that it is consistent that for every ultrafilter U there exists a function $f \in \omega^\omega$ such that $f(U)$ is Hausdorff. A counterexample to this theorem is an ultrafilter called strongly non-Hausdorff.

DEFINITION 1. An ultrafilter U is strongly non-Hausdorff if for every $f \in \omega^\omega$, $f(U)$ is either a trivial ultrafilter or $f(U)$ is not Hausdorff.

We will prove the following two theorems:

THEOREM 2. *Assume* CH. *There exists a strongly non-Hausdorff p-point.*

THEOREM 3. *It is consistent that there are no strongly non-Hausdorff ultrafilters.*

§2. A construction of a non-Hausdorff ultrafilter. Let $I \subseteq \omega$ be a finite set and let $\Delta = \{(n,n) : n \in \omega\}$. Denote by $[I]^2 = (I \times I) \setminus \Delta$. For a set $X \subseteq [I]^2$ define

$$\|X\|_I = \min \left\{ k : \exists \{A_i, B_i : i \leq k\}\, \forall i < k\ A_i \cap B_i = \emptyset \text{ and } X \subseteq \bigcup_{i \leq k} A_i \times B_i \right\}.$$

We will drop the subscript I if it is clear from the context what it is.

LEMMA 4. (1) $\|[I]^2\|_I \longrightarrow \infty$ *as* $|I| \to \infty$.

(2) $\|X \cup Y\|_I \leq \|X\|_I + \|Y\|_I$,

(3) *if* $Z \subseteq I$ *and* $X \subseteq [I]^2$, $\|X\|_I > 2$, *then either* $\|[Z]^2 \cap X\|_I \geq \|X\|_I/2 - 1$ *or* $\|[I \setminus Z]^2 \cap X\|_I \geq \|X\|_I/2 - 1$.

PROOF. If (1) fails then there is $k \in \omega$ and sets $\{A^n_j, B^n_j : n, j \leq k\}$ such that $A^n_j \cap B^n_j = \emptyset$ for $j \leq k$ and $[n]^2 = \bigcup_{j \leq k} A^n_j \times B^n_j$. By compactness we get sets $\{A^\omega_j, B^\omega_j : j \leq k\}$ such that $A^\omega_j \cap B^\omega_j = \emptyset$ for $j \leq k$ and $[\omega]^2 = \bigcup_{j \leq k} A^\omega_j \times B^\omega_j$, which is not possible.

A more direct argument shows that $\|[I]^2\|_I \geq |I| - 2$.

(2) is obvious.

(3) Note that

$$\|X\|_I \leq \|([Z]^2 \cup [I \setminus Z]^2 \cup (Z \times (I \setminus Z)) \cup ((I \setminus Z) \times Z)) \cap X\|_I \leq \|[Z]^2 \cap X\|_I + \|[I \setminus Z]^2 \cap X\|_I + 1 + 1.$$

Thus

$$\|[Z]^2 \cap X\|_I + \|[I \setminus Z]^2 \cap X\|_I \geq \|X\|_I - 2. \qquad \dashv$$

For $I \in [\omega]^{<\omega}$ let $\pi_0, \pi_1 : [I]^2 \longrightarrow I$ be projections onto first and second coordinate respectively.

LEMMA 5. *Suppose that* $X \subseteq [I]^2$, *and* $\|X\|_I > 1$. *Then* $\pi_0(X) \cap \pi_1(X) \neq \emptyset$.

PROOF. Since $X \subseteq \pi_0(X) \times \pi_1(X)$, it follows that if $\pi_0(X) \cap \pi_1(X) = \emptyset$ then $\|X\|_I \leq 1$. ⊣

Next we define functions $f^0, g^0 \in \mathsf{FtO}$ that will witness that ultrafilter V_0 that we are about to construct is not Hausdorff.

Let $\{I_k, J_k : k \in \omega\}$ be two sequences of disjoint consecutive intervals such that for $k \in \omega$,

(1) $\|[I_k]^2\|_{I_k} \geq 2^{2^k}$,
(2) $|J_k| = |[I_k]^2|$.

Bijection implicit in (2) allows us to define projections $\pi_0^k, \pi_1^k : J_k \longrightarrow I_k$. Let $f^0 = \bigcup_k \pi_0^k$ and $g^0 = \bigcup_k \pi_1^k$. Note that $f^0(x) \neq g^0(x)$ for any $x \in J_k = [I_k]^2$, $k \in \omega$.

As a warm-up let us use these definitions to show the following:

LEMMA 6. *Assume* CH. *There exists a p-point that is not Hausdorff.*

PROOF. We will need the following easy observation:

LEMMA 7. *If $f, g \in \mathsf{FtO}$ and U is an ultrafilter then the following conditions are equivalent*:

(1) $f(U) \neq g(U)$,
(2) $f[X] \cap g[X] = \emptyset$ *for some* $X \in U$.

We will build an ultrafilter V_0 on the set $\bigcup_k [I_k]^2$ which we identified with ω. Let $\{Z_\alpha : \alpha < \omega_1\}$ be enumeration of $[\omega]^\omega$.

We will build by induction a sequence $\{X_\alpha : \alpha < \omega_1\}$ so that

(1) $\forall \beta < \alpha \; X_\alpha \subseteq^\star X_\beta$,
(2) $X_{\alpha+1} \cap Z_\alpha = \emptyset$ or $X_{\alpha+1} \subseteq Z_\alpha$ for all α.
(3) for every $\alpha < \omega_1$, $f^0[X_\alpha] \cap g^0[X_\alpha] \neq \emptyset$.
(4) for every $\alpha < \omega_1$, $\limsup_k \|X_\alpha \cap J_k\|_{I_k} = \infty$.

Let $V_0 = \{X : \exists \alpha \; X_\alpha \subseteq^\star X\}$. Note that the conditions (1) and (2) guarantee that V_0 is a p-point, and lemma 7 and (3) implies that $f^0(V_0) = g^0(V_0)$. Finally, (4) is the requirement that (by lemma 5) implies (3).

SUCCESSOR STEP. Suppose that X_α is given. Find a strictly increasing sequence $\{l_k : k \in \omega\}$ such that the set $A = \{k : \|X_\alpha \cap J_k\|_{I_k} = l_k\}$ is infinite. Let $A_0 = \{k : \|X_\alpha \cap Z_\alpha \cap J_k\|_{I_k} \geq l_k/2 - 1\}$ and $A_1 = \{k : \|(X_\alpha \setminus Z_\alpha) \cap J_k\|_{I_k} \geq l_k/2 - 1\}$. By lemma 4(2), one of these sets, say A_0, is infinite. Let $X_{\alpha+1} = \bigcup_{k \in A_0} X_\alpha \cap Z_\alpha \cap J_k$. The other case is the same.

LIMIT STEP. Given $\{X_\beta : \beta < \alpha < \omega_1\}$ let $\{\beta_k : k \in \omega\}$ be an increasing sequence cofinal in α. By finite modifications we can assume that $X_{\beta_{k+1}} \subseteq X_{\beta_k}$ for all k. Build by recursion a strictly increasing sequence $\{u_k : k \in \omega\}$ such that

$$\forall k \, \forall j \leq k \, \exists i \in [u_k, u_{k+1}) \; \|X_{\beta_j} \cap J_i\|_{I_i} \geq k,$$

and let

$$X_\alpha = \bigcup_k \left(X_{\beta_k} \cap \bigcup_{i \in [u_k, u_{k+1})} J_i \right).$$

It is clear that X_α satisfies (1) and (4). ⊣

Observe that CH was only needed in the limit step. If we do not require that that V_0 is a p-point then we have the following:

THEOREM 8. *There exists an ultrafilter that is not Hausdorff.*

PROOF. As in lemma 6, we will build an ultrafilter on the set $\bigcup_k [I_k]^2$. Let

$$\mathcal{I} = \left\{ X \subseteq \bigcup_k [I_k]^2 : \limsup_k \|X \cap J_k\|_{I_k} < \infty \right\}.$$

Note that $\mathcal{I}$ is an ideal, and let U be any ultrafilter disjoint with $\mathcal{I}$. Functions f^0, g^0 witness that U is not Hausdorff. ⊣

These constructions are optimal. Suppose that $f_0, f_1 \in \omega^\omega$ witness that some ultrafilter U is not Hausdorff. Without loss of generality we can assume that $f_0(n) \neq f_1(n)$ for all n. Let $\mathcal{I}_{f_0,f_1}$ be the ideal generated by $[\omega]^{<\omega} \cup \{X : f_0[X] \cap f_1[X] = \emptyset\}$. Clearly $U \cap \mathcal{I}_{f_0,f_1} = \emptyset$. Moreover, if V is another ultrafilter which is disjoint with $\mathcal{I}_{f_0,f_1}$, then V is not Hausdorff as witnessed by the same functions. If U is a p-point then $\mathcal{I}_{f_0,f_1}$ is defined as $\mathcal{I}$ above. To see it, let us start with the following observation.

LEMMA 9. *Suppose that U is a p-point. The following conditions are equivalent.*

1. *U is not Hausdorff,*
2. *there exist functions $f_0, f_1 \in \omega^\omega$ and a sequence of disjoint intervals $\langle I_n : n \in \omega \rangle$ such that*
   (a) $\forall n\ f_0(n) \neq f_1(n)$,
   (b) $f_0(U) = f_1(U)$,
   (c) $\forall n\, \forall i = 0, 1\ f_i \restriction I_n : I_n \longrightarrow I_n$.

PROOF. One implication is obvious we will prove the other. Suppose that U is not Hausdorff and let functions $\bar{f}_i$, $i = 0, 1$ witness that. We can assume that sets $A^i_n = \bar{f}_i^{-1}(\{n\}) \notin U$ for all n and $i = 0, 1$. Moreover, by restriction to an element of U we can assume that $A^0_n \cap A^1_n = \emptyset$ for all n. Since U is a p-point there is a set $B \in U$ such that $B \cap A^i_n$ is finite for all n and i. Let $k_0 = 0$ and define $k_{n+1} = \max\{A^i_n \cap B : A^i_n \cap B \cap [0, k_n] \neq \emptyset,\ i = 0, 1\}$. Without loss of generality $C = \bigcup_{n \in \omega} [k_{2n}, k_{2n+1}) \in U$. Put $I_n = [k_{2n}, k_{2n+2})$

for $n \in \omega$ and let $A_i = \bigcup_n \{A^i_n \setminus k_{2n} : A^i_n \cap B \cap [k_{2n}, k_{2n+1}) \neq \emptyset\}$. Finally, put

$$f_0(n) = \begin{cases} \bar{f}_0(n) & \text{if } n \in A_0 \\ \min(I_k) & \text{if } n \in I_k \setminus A_0 \end{cases},$$

$$f_1(n) = \begin{cases} \bar{f}_1(n) & \text{if } n \in A_1 \\ \max(I_k) & \text{if } n \in I_k \setminus A_1 \end{cases}.$$

Clearly for $i = 0, 1$, $\bar{f}_i = f_i \mod U$, as exemplified by $B \cap C$, and f_0, f_1 have the other required properties as well. ⊣

Suppose that $f_0, f_1, \langle I_n : n \in \omega \rangle$ are as in Lemma 9. For a set $X \subseteq I_n$ define

$$\|X\|_n = \min\{k : X = X_1 \cup \dots \cup X_k \ \& \ \forall i \leq k \ f_0[X_i] \cap f_1[X_i] = \emptyset\}.$$

Then $\mathcal{I}_{f_0,f_1} = \{X \subseteq \omega : \exists k \ \forall n \ \|X \cap I_n\|_n \leq k\}$.

§3. A construction of a strongly non-Hausdorff ultrafilter under CH. Now we are ready to prove Theorem 2 and to construct a p-point ultrafilter U_0 whose all finite-to-one images are not Hausdorff.

Let $\langle h_\alpha, Z_\alpha : \alpha < \omega_1 \rangle$ be enumeration of $\mathsf{FtO} \times [\omega]^\omega$. We will construct a sequence $\langle X_\alpha, e_\alpha : \alpha < \omega_1 \rangle$ such that $U_0 = \{X \in [\omega]^\omega : \exists \alpha \ X_\alpha \subseteq^\star X\}$ is the ultrafilter that we are looking for. We will require that

(1) $\forall \beta < \alpha \ X_\alpha \subseteq^\star X_\beta$,
(2) for all α either $X_\alpha \subseteq Z_\alpha$ or $X_\alpha \cap Z_\alpha = \emptyset$,
(3) $f^0 \circ e_\alpha, g^0 \circ e_\alpha$ witness that $h_\alpha(U_0)$ is not Hausdorff.

As before, (1) and (2) guarantees that U_0 is a p-point, and (3) implies that U_0 is strongly non-Hausdorff.

Definition 10. A finite set $Y \subseteq \omega$ is a (n, β)-witness if there exists $k \in \omega$ such that $\|e_\beta \circ h_\beta[Y] \cap J_k\|_{I_k} \geq n$.

To satisfy (3), we demand that,

(4) $\forall \beta \leq \alpha < \omega_1 \ \limsup_k \|e_\beta \circ h_\beta[X_\alpha] \cap J_k\|_{I_k} = \infty$, or equivalently $\forall n \ \forall \beta \leq \alpha \ \exists Y \in [X_\alpha]^{<\omega} \ Y$ is a (n, β)-witness.

The last condition concerns the inductive requirement for the function e_α:

(5) for all $\beta < \alpha$,
$$\lim_{n \in \omega} \max\{k : (e_\beta \circ h_\beta)^{-1}(\{n\}) \cap X_\alpha \text{ contains a } (k, \beta)\text{-witness}\} = \infty.$$

Successor step. Suppose that $\{X_\beta : \beta < \alpha\}$ satisfying (1)-(5) are already defined and we want to define $X_{\alpha+1}$ satisfying (1)-(5).

Case 1. If for some $\beta \leq \alpha$, $\limsup_k \|e_\beta \circ h_\beta[Z_\alpha \cap X_\alpha] \cap J_k\|_{I_k} < \infty$ then let $X_{\alpha+1} = X_\alpha \setminus Z_\alpha$.

Case 2. If for some $\beta \leq \alpha$, $\limsup_k \|e_\beta \circ h_\beta[X_\alpha \setminus Z_\alpha] \cap J_k\|_{I_k} < \infty$ then let $X_{\alpha+1} = X_\alpha \cap Z_\alpha$.

In all other cases let $X_{\alpha+1} = X_\alpha \cap Z_\alpha$.

We have to check that cases 1 and 2 are exclusive. By the inductive hypothesis,

$$\forall \beta \le \alpha \ \limsup_k \left\| e_\beta \circ h_\beta[X_\alpha] \cap J_k \right\|_{I_k} = \infty.$$

Assume that for the two possible choices for $X_{\alpha+1}$: $A = X_\alpha \cap Z_\alpha$ and $B = X_\alpha \setminus Z_\alpha$ one of them is rejected. Let $\beta_0 \le \alpha$ be the minimal ordinal witnessing this. Without loss of generality we can assume that for some $\beta_0 \le \alpha$,

$$\limsup_k \left\| e_{\beta_0} \circ h_{\beta_0}[A] \cap J_k \right\|_{I_k} < \infty.$$

The following lemma will complete the construction:

Lemma 11. *For every $\beta \le \alpha$,*

$$\limsup_k \left\| e_\beta \circ h_\beta[B] \cap J_k \right\|_{I_k} = \infty.$$

Proof. By minimality of β_0, the statement is true for $\beta < \beta_0$. By the induction hypothesis it is also true for $\beta = \beta_0$. In particular, we must have $\beta_0 < \alpha$.

Suppose that the Lemma is false and let $\beta_1 > \beta_0$ be the first ordinal such that for some $N \in \omega$,

$$\limsup_k \left\| e_{\beta_1} \circ h_{\beta_1}[B] \cap J_k \right\|_{I_k} < N.$$

Since by the induction hypothesis, $X_\alpha = A \cup B$ contains a sequence of (k_n, β_1)-witnesses where $k_n \longrightarrow \infty$, it follows that A contains a sequence of $(k_n - N, \beta_1)$-witnesses. Since $e_{\beta_1} \circ h_{\beta_1}[A \cup B] = e_{\beta_1} \circ h_{\beta_1}[A] \cup e_{\beta_1} \circ h_{\beta_1}[B]$, it means that

$$e_{\beta_1} \circ h_{\beta_1}[A] \setminus e_{\beta_1} \circ h_{\beta_1}[B] \text{ is infinite.}$$

By the induction hypothesis, for all but finitely many $n \in e_{\beta_1} \circ h_{\beta_1}[A] \setminus e_{\beta_1} \circ h_{\beta_1}[B]$, $(h_{\beta_1} \circ e_{\beta_1})^{-1}(\{n\}) \cap A = (h_{\beta_1} \circ e_{\beta_1})^{-1}(\{n\}) \cap (A \cup B)$ contains an (l_n, β_0)-witness, where $l_n \longrightarrow \infty$. This means that $\limsup_k \left\| e_{\beta_0} \circ h_{\beta_0}[A] \cap J_k \right\|_{I_k} = \infty$, a contradiction. ⊣

Finally we define the function $e_{\alpha+1}$. Let $\{\gamma_k : k \in \omega\}$ be an enumeration of α.

Define strictly increasing sequence $\langle u_k : k \in \omega \rangle$ such that

(1) $\forall k \, \forall j \le k \, [u_k, u_{k+1}) \cap X_{\alpha+1}$ contains a (k, γ_j)-witness,

(2) $\forall k \, [u_k, u_{k+1}) \cap h_{\alpha+1}[X_{\alpha+1}] \ne \emptyset$.

Define $e_{\alpha+1}(l) = k \iff j \in [u_k, u_{k+1})$. It is easy to see that the inductive conditions are satisfied.

Limit step. Suppose that $\{X_\beta : \beta < \alpha\}$ are defined and α is a limit ordinal. Let $\{\beta_k : k \in \omega\}$ be an increasing sequence cofinal in α, and let $\{\gamma_k : k \in \omega\}$ be an enumeration of α such that $\gamma_j \le \beta_k$ for $j \le k$. By finite modifications we can assume that $X_{\beta_{k+1}} \subseteq X_{\beta_k}$ for all k. Build by recursion a strictly increasing sequence $\langle u_k : k \in \omega \rangle$ such that

$$\forall k\ \forall j \leq k\ [u_k, u_{k+1}) \cap X_{\beta_k} \text{ contains a } (k, \beta_j)\text{-witness},$$

and let

$$X_\alpha = \bigcup_k X_{\beta_k} \cap [u_k, u_{k+1}),$$

and $e_\alpha(l) = k \iff l \in [u_k, u_{k+1})$. It is clear that X_α satisfies (1)-(5).

§4. A model where there are no strongly non-Hausdorff ultrafilters. In the next two sections we will show that:

THEOREM 12. *It is consistent with* ZFC *that for every ultrafilter U there is $h \in$* FtO *such that $h(U)$ is a Hausdorff p-point. In particular, it is consistent that there are no strongly non-Hausdorff ultrafilters.*

In order to prove this theorem we will show that there exists a proper forcing notion $\mathbb{M}$ which has the following properties:

(1) If U is an ultrafilter in V then $V^{\mathbb{M}} \models \exists h \in \mathsf{FtO}\ h(U)$ is a p-point ultrafilter,
(2) If U is a p-point in V then $V^{\mathbb{M}} \models \exists h \in \mathsf{FtO}\ h(U)$ is a Hausdorff p-point,
(3) $\mathbb{M}$ preserves p-points,
(4) $\mathbb{M}$ preserves Hausdorff p-points.

We will show that this suffices for the proof. Let $\mathbb{M}_{\omega_2}$ be a countable support support iteration of $\mathbb{M}$. We will show that $V^{\mathbb{M}_{\omega_2}}$ is the model we are looking for. Suppose that $U \in V^{\mathbb{M}_{\omega_2}}$ is an ultrafilter. By the standard Skolem-Lowenheim argument we can find $\delta < \omega_2$ such that $V^{\mathbb{M}_\delta} \models U \cap V^{\mathbb{M}_\delta}$ is an ultrafilter. By (1) there is $h_1 \in V^{\mathbb{M}_{\delta+1}} \cap \mathsf{FtO}$ such that $V^{\mathbb{M}_{\delta+1}} \models\ h_1(U)$ is a p-point. By (2) there is $h_2 \in V^{\mathbb{M}_{\delta+2}} \cap \mathsf{FtO}$ such that $V^{\mathbb{M}_{\delta+2}} \models\ h_2(h_1(U))$ is a Hausdorff p-point. The rest follows from the following theorem that we will prove in the next section.

THEOREM 13. *Suppose that $\langle \mathcal{P}_\alpha, \mathcal{Q}_\alpha : \alpha < \delta \rangle$ is a countable support iteration of proper forcing notions such that $\Vdash_\alpha \mathcal{Q}_\alpha$ preserves Hausdorff p-points. Then $\mathcal{P}_\delta$ preserves Hausdorff p-points.*

We will show that the rational perfect set forcing $\mathbb{M}$ defined below has the required properties. The rational perfect forcing $\mathbb{M}$ is the following forcing notion:

$p \in \mathbb{M}$ if $p \subseteq \omega^{<\omega}$ is a perfect tree of finite increasing sequences of natural numbers and $\forall s \in p\ \exists t \in p\ (s \subseteq t\ \&\ |\mathsf{succ}_p(t)| = \aleph_0)$. For $p, q \in \mathbb{M}$, $p \geq q$ if $p \subseteq q$. Without loss of generality we can assume that $|\mathsf{succ}_p(s)| = 1$ or $|\mathsf{succ}_p(s)| = \aleph_0$ for all $p \in \mathbb{M}$ and $s \in p$. Conditions of this type form a dense subset of $\mathbb{M}$.

Let $\mathsf{split}(p) = \{s \in p : |\mathsf{succ}_p(s)| > 1\} = \bigcup_{n\in\omega} \mathsf{split}_n(p)$, where $\mathsf{split}_n(p) = \{s \in \mathsf{split}(p) : |\{t \subsetneq s : t \in \mathsf{split}(p)\}| = n\}$.

For $p, q \in \mathbb{M}$, $n \in \omega$, we let $p \geq_n q \iff p \geq q$ & $\forall i, j \leq n$ if s is j-th element of $\mathsf{split}_i(q)$ then $s \in \mathsf{split}_i(p)$.

We have to verify that $\mathbb{M}$ has property (2) and (4) above, property (1) was proved in [2] and property (3) in [8]. Property (2) follows from the following result of Dow that we present here for completeness.

LEMMA 14 (Dow). *Suppose that $U \in V$ is a p-point. Then $V^{\mathbb{M}} \models \exists h \in \omega^\omega\ h(U)$ is a Ramsey ultrafilter. In particular, $h(U)$ is Hausdorff in $V^{\mathbb{M}}$.*

PROOF. Define $\boldsymbol{h}(n) = k$ if $m_k \leq n < m_{k+1}$, where $\langle m_k : k \in \omega\rangle$ is the generic sequence added by the Miller real. Since Miller forcing preserves p-points $V^{\mathbb{M}} \models \boldsymbol{h}(U)$ is a p-point. Thus it remains to check that if $p \Vdash_{\mathbb{M}} \dot f \in$ FtO then there exists $q \geq p$ and $X \in U$ such that $q \Vdash_{\mathbb{M}} \dot f \restriction \boldsymbol{h}[X]$ is one-to-one or constant. By shrinking we can assume that for every $t \in \mathsf{split}(p)$ and every $n \in \mathsf{succ}_p(t)$, there exists $k_{t^\frown n} \in \omega$ such that $p_{t^\frown n} \Vdash \dot f(|t|) = k_{t^\frown n}$. In particular, $p_{t^\frown n} \Vdash \dot f(\boldsymbol{h}(j)) = k_{t^\frown n}$ for $t(|t|-1) \leq j < n$. By further shrinking, we can assume that for every $t \in \mathsf{split}(p)$ and $n, m \in \mathsf{split}(p(t))$, either $k_{t^\frown n} = k_{t^\frown m}$ or that $k_{t^\frown n} < k_{t^\frown m}$ if $n < m$. This defines a partition of $\mathsf{split}(p) = A \cup B$ where $A = \{t \in \mathsf{split}(p) : k_{t^\frown n} = k_{t^\frown m} = k_t$ for all $n, m \in \mathsf{succ}_p(t)\}$ and $B = \mathsf{split}(p) \setminus A$. By passing to a stronger condition we can assume that one of these sets is empty (see [8]). Specifically, one of the following cases holds:

(1) $\mathsf{split}(p) \subseteq A$ and all k_t are distinct for $t \in \mathsf{split}(p)$,
(2) $\mathsf{split}(p) \subseteq A$ and $k_t = \bar k$ for all $t \in \mathsf{split}(p)$,
(3) $\mathsf{split}(p) \cap A = \emptyset$ and all $k_{t^\frown n}$ are distinct for all $t \in \mathsf{split}(p)$ and $n \in \mathsf{succ}_p(t)$.

For $t \in \mathsf{split}_l(p)$ and $n \in \mathsf{succ}_p(t)$ let $I_{t^\frown n} = [n, s(|s|-1))$, where $t^\frown n \subseteq s$ and $s \in \mathsf{split}_{l+1}(p)$. By shrinking p we can find $X \in U$ such that for every $t \in \mathsf{split}(p)$, $X \cap \bigcup\{I_{t^\frown n} : n \in \mathsf{succ}_p(t)\} = \emptyset$. We proceed as follows, given $t \in \mathsf{split}(p)$ we first thin out $\mathsf{succ}_p(t)$ so that intervals $I_{t^\frown n}$ are pairwise disjoint. Next we thin out $\mathsf{succ}_p(t)$ so that $B_t = \bigcup\{I_{t^\frown n} : n \in \mathsf{succ}_p(t)\} \notin U$. Since U is a p-point there is $X \in U$ such that $X \setminus B_t$ is finite for each t. Finally we shrink the set $\mathsf{succ}_p(t)$ so that $X \cap B_t = \emptyset$ for all t. This is the condition q that we are looking for. Suppose that $n, m \in X$ and $r \geq q$ is such that $r \Vdash_{\mathbb{M}} \dot f(\boldsymbol{h}(n)) = \dot f(\boldsymbol{h}(m))$. Without loss of generality we can assume that $r = q_{\bar s}$ where $n, m < \bar s(|\bar s|-1)$. Let $t_0 \subseteq t_1 \subseteq \cdots \subseteq t_k \subseteq \bar s$ be the list of all splitting nodes of q shorter than $\bar s$. Since $n, m \notin \bigcup_{j\leq k} B_{t_j}$ it follows that $n, m \in \bigcup_{j<k}[\bar s(|t_j|-1), \bar s(|t_j|))$. It follows that for some $j \leq k$, $q_{\bar s} \Vdash_{\mathbb{M}} \dot f(\boldsymbol{h}(n)) = \dot f(\boldsymbol{h}(m)) = k_{t_j}$ (or $k_{t_j^\frown \bar s(|t_j|)}$ or $\bar k$). In the first two cases it follows that $n, m \in [\bar s(|t_j|-1), \bar s(|t_j|))$, so $q_{\bar s} \Vdash_{\mathbb{M}} \boldsymbol{h}(n) = \boldsymbol{h}(m) = |t_j| - 1$

(so $\dot{f} \circ \boldsymbol{h}$ is forced to be one-to-one on X) and in the second case it means that $q_{\bar{s}} \Vdash_{\mathbb{M}} \dot{f}(\boldsymbol{h}(n)) = \dot{f}(\boldsymbol{h}(m)) = \bar{k}$ (so $\dot{f} \circ \boldsymbol{h}$ is forced to be constant on X). ⊣

In the remainder of this section we will show that Miller forcing preserves Hausdorff p-points.

We will start with the following observation.

Lemma 15. *The following conditions are equivalent for an ultrafilter U:*

(1) *U is not Hausdorff,*
(2) *there are functions $f_0, f_1 \in \omega^\omega$ such that*
  (a) $f_0(U) = f_1(U)$,
  (b) $\forall n\ f_0(n) \neq f_1(n)$,
  (c) $\forall n\ f_0(n), f_1(n) \leq 2n + 2$.

Proof. We will show that (1) implies (2), the other implication is obvious.

Let $g_0, g_1 \in \omega^\omega$ be such that $g_0(U) = g_1(U)$ and $g_0 \neq g_1 \mod U$. Let $A = \{n : g_0(n) \neq g_1(n)\} \setminus \{0, 1\}$. By the assumption $A \in U$. Define $h \in \omega^\omega$ as follows: $\mathsf{dom}(h) = \mathsf{range}(g_0 \restriction A) \cup \mathsf{range}(g_1 \restriction A)$ and let $h(n) = \min\{(2k + i : i \in 2,\ g_i(k) = n\}$. Next put for $i = 0, 1$, $f_i(n) = h(g_i(n))$ for $n \in A$ and $f_i(n) = i$ if $n \notin A$. Note that $f_0(U) = h(g_0(U)) = h(g_1(U)) = f_1(U)$, the other two properties are equally easy to see. ⊣

Lemma 16. *Suppose that $U \in \boldsymbol{V}$ is a Hausdorff p-point. Then $\boldsymbol{V}^{\mathbb{M}} \models U$ is a Hausdorff p-point.*

Proof. Let $U \in \boldsymbol{V}$ be a Hausdorff p-point. First of all recall that U remains a p-point in $\boldsymbol{V}^{\mathbb{M}}$ [8, Proposition 4.2]. Thus we only have to show that U remains Hausdorff. Let $p \in \mathbb{M}$ and $\dot{f}_0, \dot{f}_1$ be such that

(1) $p \Vdash_{\mathbb{M}} \forall n\ \dot{f}_0(n) \neq \dot{f}_1(n)$,
(2) $p \Vdash_{\mathbb{M}} \forall n\ \dot{f}_0(n), \dot{f}_1(n) \leq 2n + 2$.

It suffices to find $q \geq p$ and $X \in U$ such that $q \Vdash_{\mathbb{M}} \dot{f}_0[X] \cap \dot{f}_1[X] = \emptyset$.

Define a game $G(U)$ played by players I and II. Player I on his n-th move plays a set $X_n \in U$ and player II responds with a finite set $a_n \subseteq X_n$. Together they construct a sequence $X_1, a_1, X_2, a_2, \dots$. Player I wins if $\bigcup_{n \in \omega} a_n \notin U$; otherwise player II wins. It is well known, [1], that player I has a winning strategy in $G(U)$ if and only if U is not a p-point. We will be simultaneously playing game $G(U)$ and constructing sequences of $\langle q_n : n \in \omega \rangle$, $\langle X_n : n \in \omega \rangle$ and $\langle a_n : n \in \omega \rangle$ such that

(1) $\langle X_n, a_n : n \in \omega \rangle$ is a play in $G(U)$,
(2) $q_0 = p$ and $q_{n+1} \geq_n q_n$,
(3) $q_{n+1} \Vdash_{\mathbb{M}} \dot{f}_0[\bigcup_{j \leq n} a_j] \cap \dot{f}_1[\bigcup_{j \leq n} a_j] = \emptyset$.

Since U is a p-point, player II does not have a winning strategy in $G(U)$. Therefore there exists a play for player II such that $X = \bigcup_n a_n \in U$. At the same time, if $q \geq q_n$ for all n then, by (3) above, $q \Vdash_{\mathbb{M}} \dot{f}_0[X] \cap \dot{f}_1[X] = \emptyset$.

Without loss of generality we can assume that for every $t \in \mathsf{split}(p)$ and $n \in \mathsf{succ}_p(t)$, there exists $(f_0^{t^\frown n}, f_1^{t^\frown n}) \in (\omega^{<\omega})^2$ such that $p_{t^\frown n} \Vdash_{\mathbb{M}} \forall i = 0, 1\ \dot{f}_i \restriction |t| + n = f_i^{t^\frown n} \restriction |t| + n$. For given t, sequences $\{(f_0^{t^\frown n}, f_1^{t^\frown n}) : n \in \omega\}$ form a finitely branching tree. Let (f_0^t, f_1^t) be one of its infinite branches. Now prune the tree so that (f_0^t, f_1^t) is the unique branch for every $t \in \mathsf{split}(p)$. Since these functions need not be finite-to-one for each t we have one of the following four cases: either there is $i \in \{0, 1\}$, $a_i^t \in \omega$ and $X_i^t \in U$ such that $f_i^t \restriction X_i^t$ is constant with value a_i^t, or for $i \in \{0, 1\}$ there is $X^t \in U$ such that $f_0^t, f_1^t \restriction X^t$ are both finite-to-one. By passing to a stronger condition we can assume that exactly one of these cases holds. Furthermore, if it is one of the first three cases then we can assume that for the relevant $i \in \{0, 1\}$ the values a_i^t are all the same or all distinct.

The construction follows the cases described above. Suppose that $\langle q_k, X_k, a_k : k \leq n \rangle$ are already constructed. Let $S_n \subseteq \mathsf{split}(q_n)$ be the set of nodes that need to be contained in $\mathsf{split}(q_{n+1})$.

Case 1 $f_0^t, f_1^t \restriction X^t$ are finite-to-one for $t \in \mathsf{split}(p)$.

Since U is Hausdorff and functions are finite-to-one, for each $t \in S$ there is $X_t \in U$ such that $f_0^t[X_t \cap X^t] \cap f_1^t[X_t \cap X^t] = \emptyset$, and $f_i^t[X_t \cap X^t] \cap f_{1-i}^t[\bigcup_{k \leq n} a_k] = \emptyset$ for $i = 0, 1$. Define $X_{n+1} = \bigcap_{t \in S} X_t \cap X^t$, this is the move played in the ultrafilter game. Player II responds with $a_{n+1} \subseteq X_{n+1}$. Finally, let q_{n+1} be obtained by keeping only thise nodes $s \in \mathsf{split}(q_n) \setminus \bigcup_{k \leq n} S_k$ such that $|s| + n \geq \max(a_{n+1})$ for $n \in \mathsf{succ}_{q_n}(s)$. It is easy to verify that q_{n+1}, X_{n+1} have the required properties.

Case 2 $f_0^t \restriction X^t$ is constant with value a_0^t and a_0^t are distinct for $t \in \mathsf{split}(p)$ and $f_1^t \restriction X^t$ is finite-to-one.

Assume inductively that for each $s \in \mathsf{split}(q_n) \setminus S_{n-1}$, $a_0^s \notin f_1^t[\bigcup_{k \leq n} a_k]$. Since U is Hausdorff, for each $t \in S$ there is $X_t \in U$ such that $f_0^t[X_t \cap X^t] \cap f_1^t[X_t \cap X^t] = \emptyset$, and $f_1^t[X_t \cap X^t] \cap f_0^t[\bigcup_{k \leq n} a_k] = \emptyset$.

Define $X_{n+1} = \bigcap_{t \in S} X_t \cap X^t$, this is the move played in the ultrafilter game. Player II responds with $a_{n+1} \subseteq X_{n+1}$. Let q'_{n+1} be obtained by keeping only thise nodes $s \in \mathsf{split}(q_n) \setminus \bigcup_{k \leq n} S_k$ such that $|s| + n \geq \max(a_{n+1})$ for $n \in \mathsf{succ}_{q_n}(s)$. Finally let q_{n+1} be the condition obtained from q'_{n+1} by removing all splitting nodes $s \notin \bigcup_{k \leq n} S_k$ with $a_0^s \in \bigcup_{t \in S_n} f_1^t[a_{n+1}]$. Again, q_{n+1}, X_{n+1} have the required properties.

Remaining cases are handled similarly. ⊣

It is possible to show directly that an iteration of Miller forcing of countable (and thus arbitrary) length preserves Hausdorff p-points as well. This can be done by introducing $\leq_{F,n}$ order on the iteration and modifying the proof of Lemma 16. Instead we will prove much more general Theorem 13.

§5. Preserving Hausdorff p-points. In this section we will prove a general theorem concerning the preservation of Hausdorff p-points. The theorem is a specific application of a general preservation theorem due to Shelah [9, Chapter XVIII].

Let $\boldsymbol{C} = \{(f_0, f_1) \in \omega^\omega \times \omega^\omega : \forall n\, f_0(n) \neq f_1(n)\}$. Consider the following relation $\sqsubseteq = \bigcup_n \sqsubseteq_n$ on $\boldsymbol{C} \times P(\omega)$ defined as

$$(f_0, f_1) \sqsubseteq_n X \iff f_0[X \setminus n] \cap f_1[X \setminus n] = \emptyset.$$

Note that if U is a Hausdorff filter then U is a $\sqsubseteq$-dominating family, that is, for every $(f_0, f_1) \in \boldsymbol{C}$ there is $X \in U$ such that $(f_0, f_1) \sqsubseteq X$.

Suppose that U is a family of subsets of ω. Let $\bar{U} = \{Y \subseteq \omega : \exists n\, \exists X \in U\, X \setminus n \subseteq Y\}$. For a Hausdorff p-point U, let $S \subseteq [U]^{\leq \aleph_0}$ be the stationary set consisting of sets of form $N \cap U$ where N is a countable elementary submodel of $\boldsymbol{H}(\chi)$. Let $\boldsymbol{g} = \{X_a : a \in S\}$, where $X_a \in U$ is such that $X_a \subseteq^\star X$ for $X \in a$.

Shelah's preservation theorem says:

THEOREM 17. *Suppose that* $(\sqsubseteq, S, \boldsymbol{g})$ *strongly covers and* $\langle \mathcal{P}_\alpha, \dot{\mathcal{Q}}_\alpha : \alpha < \delta\rangle$ *is a countable support iteration of forcing notions such that for every* $\alpha < \delta$, $\Vdash_\alpha \dot{\mathcal{Q}}_\alpha$ *preserves* $(\sqsubseteq, S, \boldsymbol{g})$. *Then* $\mathcal{P}_\delta$ *preserves* $(\sqsubseteq, S, \boldsymbol{g})$.

We will define all the relevant notions below and we will show that Theorem 13 follows from it.

DEFINITION 18. A countable model N is good if every $(f_0, f_1) \in N \cap \boldsymbol{C}$, $(f_0, f_1) \sqsubseteq X_{N \cap U}$. We say that $(\sqsubseteq, S, \boldsymbol{g})$ covers (in $\boldsymbol{V}$) if every countable model $N \preceq \boldsymbol{H}(\chi)$ containing $(\sqsubseteq, S, \boldsymbol{g})$ is good.

Note that this corresponds to the settings $\alpha^\star = \omega$, $\bar{R} = \sqsubseteq$ in [9, Definition 3.2 p. 889].

LEMMA 19. *Suppose that* $\mathcal{P}$ *is a proper forcing notion which preserves p-points. The following conditions are equivalent for a Hausdorff p-point* U *in* $\boldsymbol{V}$:

(1) $\bar{U}$ *is a Hausdorff p-point in* $\boldsymbol{V}^{\mathcal{P}}$,
(2) $(\sqsubseteq, S, \boldsymbol{g})$ *covers in* $\boldsymbol{V}^{\mathcal{P}}$.

PROOF. (1)$\rightarrow$(2) Suppose that $N \preceq \boldsymbol{H}(\chi)^{\boldsymbol{V}^{\mathcal{P}}}$ is a countable model containing $(\sqsubseteq, S, \boldsymbol{g})$. Let $(f_0, f_1) \in N \cap \boldsymbol{C}$. Since $\bar{U}$ is Hausdorff in $\boldsymbol{V}^{\mathcal{P}}$ there is $Y \in U$ such that $(f_0, f_1) \sqsubseteq Y$. Since N is an elementary submodel containing all relevant objects we can assume that $Y \in N$. Since $X_{U \cap N} \subseteq^\star Y$ we are done.

(2)$\rightarrow$(1) If $\bar{U}$ is not Hausdorff in $\boldsymbol{V}^{\mathcal{P}}$ then there are functions $f_0, f_1 \in \boldsymbol{V}^{\mathcal{P}} \cap \omega^\omega$ witnessing this. Since $\mathcal{P}$ preserves p-points, without loss of generality we can assume by Lemma 9, that $(f_0, f_1) \in \boldsymbol{C}$. Therefore $|f_0[X] \cap f_1[X]| = \aleph_0$ for all $X \in U$. It follows that if N contains f_0, f_1 then $(f_0, f_1) \not\sqsubseteq X_{N \cap U}$. ⊣

The definition below is a special case of the Definition 3.3 on page 899 in [9] (Possibility A, case $\oplus'_k$).

DEFINITION 20. We say that $(\sqsubseteq, S, \boldsymbol{g})$ strongly covers if

(1) if $(\sqsubseteq, S, \boldsymbol{g})$ covers in $\boldsymbol{V}$,
(2) for each n, relation $\sqsubseteq_n$ is a closed subset of $\boldsymbol{C} \times P(\omega)$,
(3) the following holds in every proper forcing extension $\boldsymbol{V}^{\mathcal{P}}$ in which $(\sqsubseteq, S, \boldsymbol{g})$ covers: IF $N \prec \boldsymbol{H}(\chi)$ is a countable model, $k \in \omega$ and $\langle (f_0^{n,j}, f_1^{n,j}) : j \leq k, n \in \omega \rangle \in N \cap \boldsymbol{C}$ such that
   (a) $\lim_n (f_0^{n,j}, f_1^{n,j}) = (f_0^j, f_1^j)$,
   (b) for each $j \leq k$, there is $n_j \in \omega$ such that $(f_0^j, f_1^j) \sqsubseteq_{n_j} X_{N \cap U}$,
   then there exists $n^\star$ such that

$$\forall j \leq k\, (f_0^{n^\star,j}, f_1^{n^\star,j}) \sqsubseteq_{n_j} X_{N \cap U}.$$

The next lemma shows that for the forcing notions preserving Hausdorff p-points both notions coincide.

LEMMA 21. *If* $(\sqsubseteq, S, \boldsymbol{g})$ *covers then* $(\sqsubseteq, S, \boldsymbol{g})$ *strongly covers.*

PROOF. Let $\mathcal{P}$ be a proper forcing notion that preserves Hausdorff p-points. Work in $\boldsymbol{V}^{\mathcal{P}}$. Suppose that $N \prec \boldsymbol{H}(\chi)^{\boldsymbol{V}^{\mathcal{P}}}$ is a countable model and let $k \in \omega$, $\langle (f_0^{n,j}, f_1^{n,j}) : j \leq k, n \in \omega \rangle, \langle (f_0^j, f_1^j) : j \leq k \rangle$, $\langle n_j : j \leq k \rangle$ be as required. We have to show that there exists $n^\star$ such that

$$\forall j \leq k\, (f_0^{n^\star,j}, f_1^{n^\star,j}) \sqsubseteq_{n_j} X_{N \cap U}.$$

Since U is a p-point in N for each $n \in \omega, j \leq k$ there is a set $A^{n,j} \in U \cap N$ such that $(f_0^{n,j}, f_1^{n,j}) \sqsubseteq_0 A^{n,j}$. Let $A \in N \cap U$ be such that $A \subseteq^\star A^{n,j}$ for $n \in \omega, j \leq k$. Work in N and define sequence $\langle m_l : l \in \omega \rangle$ such that

(1) $m_0 = 0, m_l < m_{l+1}$,
(2) $\forall n \geq m_{l+1}\, \forall j \leq k\, \forall i = 0, 1\; f_i^{n,j} \restriction m_l = f_i^j \restriction m_l$,
(3) $A \setminus m_{l+1} \subseteq \bigcap_{n \leq m_l, j \leq k} A^{n,j}$,
(4) $\forall i = 0, 1\, \forall n \leq m_l\, \forall j \leq k\, \forall u \leq m_l\; f_i^{n,j}(u) < m_{l+1}$,
(5) $\forall i = 0, 1\, \forall n \leq m_l\, \forall j \leq k\, \forall u \geq m_{l+1}\; f_i^{n,j}(u) \geq m_l$.

Without loss of generality the set $B = \bigcup_j [m_{3l+1}, m_{3l+2}) \in U \cap N$. Let $l^\star$ be such that $X_{N \cap U} \setminus m_{3l^\star} \subseteq A \cap B$ and put $n^\star = m_{3l^\star}$. We have to show that for all $j \leq k, (f_0^{n^\star,j}, f_1^{n^\star,j}) \sqsubseteq_{n_j} X_{N \cap U}$, that is $f_0^{n^\star,j}[X_{N \cap U} \setminus n_j] \cap f_1^{n^\star,j}[X_{N \cap U} \setminus n_j] = \emptyset$. Fix $j \leq k$ and suppose that $x, y \in X_{N \cap U} \setminus n_j$.

CASE 1 $x, y < m_{3l^\star}$.

In this case $x, y \leq m_{3l^\star - 1} = m_{3(l^\star - 1)+2}$. Since for $i = 0, 1$, $f_i^{m_{3l^\star},j} \restriction m_{3(l^\star-1)+2} = f_i^j \restriction m_{3(l^\star-1)+2}$ it follows that $f_0^{n^\star,j}(x) \neq f_1^{n^\star,j}(y)$ since $f_0^j(x) \neq f_1^j(y)$.

CASE 2 $x, y \geq m_{3l^\star}$. In this case $x, y \geq m_{3l^\star+1}$ and since $A \setminus m_{3l^\star+1} \subseteq A^{3l^\star,j}$ it follows that $x, y \in A^{3l^\star,j}$. Thus $f_0^{n^\star,j}(x) \neq f_1^{n^\star,j}(y)$ by the choice of $A^{n^\star,j}$.

CASE 3 $x < m_{3l^\star} \leq y$.

In this case $x < m_{3l^\star-1} < m_{3l^\star+1} \leq y$, and $f_0^{n^\star,j}(x) \neq f_1^{n^\star,j}(y)$ by the property of sequence $\langle m_l : l \in \omega \rangle$. ⊣

DEFINITION 22. We say that a forcing notion $\mathcal{P}$ is $(\sqsubseteq, S, \boldsymbol{g})$-preserving if whenever $N \prec \boldsymbol{H}(\chi)$ is a countable model containing $\mathcal{P}$ and $\sqsubseteq$ whenever $\langle p_n : n \in \omega \rangle \in N$ is an increasing sequence of conditions interpreting $\langle (\dot{f}_0^0, \dot{f}_1^0), \dots, (\dot{f}_0^k, \dot{f}_1^k) \rangle \in N$ as $\langle (f_0^0, f_1^0), \dots, (f_0^k, f_1^k) \rangle$ then there is an N-generic condition $q \geq p_0$ such that:

1. $q \Vdash_{\mathcal{P}} N[\dot{G}]$ is $(\sqsubseteq, S, \boldsymbol{g})$-good and
2. $\forall n \in \omega\, \forall j \leq k\; q \Vdash_{\mathcal{P}} ((f_0^j, f_1^j) \sqsubseteq_n X_{N \cap U} \rightarrow (\dot{f}_0^j, \dot{f}_1^j) \sqsubseteq_n X_{N \cap U})$.

LEMMA 23. *Suppose that $\mathcal{P}$ is a forcing notion which preserves p-points and U is a Hausdorff p-point ultrafilter in $\boldsymbol{V}$. The following conditions are equivalent:*

1. *$\bar{U}$ is a Hausdorff p-point in $\boldsymbol{V}^{\mathcal{P}}$,*
2. *$\mathcal{P}$ is $(\sqsubseteq, S, \boldsymbol{g})$-preserving.*

PROOF. (2)→(1) If $\mathcal{P}$ is $(\sqsubseteq, S, \boldsymbol{g})$-preserving then $(\sqsubseteq, S, \boldsymbol{g})$ covers in $\boldsymbol{V}^{\mathcal{P}}$. In particular, U is a Hausdorff ultrafilter in $\boldsymbol{V}^{\mathcal{P}}$.

(1)→(2) Suppose that $N \prec \boldsymbol{H}(\chi)$ is a countable model containing $\mathcal{P}$ and $\sqsubseteq$. Since U is Hausdorff, N is $(\sqsubseteq, S, \boldsymbol{g})$-good. Let $\langle p_n : n \in \omega \rangle \in N$ be an increasing sequence of conditions interpreting $\langle (\dot{f}_0^0, \dot{f}_1^0), \dots, (\dot{f}_0^k, \dot{f}_1^k) \rangle \in N$ as $\langle (f_0^0, f_1^0), \dots, (f_0^k, f_1^k) \rangle$. In other words, for every n,

1. $p_{n+1} \geq p_n$,
2. $p_n \Vdash \forall j \leq k\; \forall i = 0, 1\; \dot{f}_i^j \restriction n = f_i^j \restriction n$.

We have to show that there exists $q \geq p_0$ such that

$$q \Vdash_{\mathcal{P}} \forall n\, \forall j \leq k\; \big((f_0^j, f_1^j) \sqsubseteq_n X_{N \cap U} \rightarrow (\dot{f}_0^j, \dot{f}_1^j) \sqsubseteq_n X_{N \cap U}\big).$$

Since $\mathcal{P}$ preserves p-points we can assume that for every $j \leq k$ $i = 0, 1$, p_0 decides whether $\dot{f}_i^j$ is constant or finite-to-one mod U. For each n try to choose $q_n \geq p_n$ such that

1. q_n is N-generic,
2. there exists $X_n \in N \cap U$ such that $q_n \Vdash \forall j \leq k\; (\dot{f}_0^j, \dot{f}_1^j) \sqsubseteq_0 X_n$ & $\dot{f}_i^j[X_n] \cap f_{1-i}^j[n] = \emptyset$.

Given n we proceed as follows: First find $q' \geq p_n$, $X \in N \cap U$, $l_i^j \in \omega$ for $j \leq k$, $i = 0, 1$ such that

1. $q' \Vdash \forall j \leq k\; (\dot{f}_0^j, \dot{f}_1^j) \sqsubseteq_0 X$,
2. for every $j \leq k$, $i = 0, 1$ either
   (a) $q' \Vdash \dot{f}_i^j$ is finite-to-one on X or
   (b) $q' \Vdash \dot{f}_i^j$ is constant on X with value l_i^j.

Next choose $q'' \geq q'$ and m so that either

(1) $q'' \Vdash \forall i = 0, 1\ \forall j \leq k\ \min(\dot{f}_i^j[X \setminus m]) > \max(f_i^j(k) : k \leq n)$ (finite-to-one case) or

(2) $l_i^j \in f_i^j[n]$ (constant case).

Finally, let $q_n \geq q''$ be N-generic and let $X_n = X \setminus m$. Note that this construction will succeed for all sufficiently large n, that is when $l \in f_i^j[n]$. Since U is a p-point there is infinitely many n such that $X_{N \cap U} \setminus n \subseteq X_n$. Choose $n^\star$ to be such an n and let $q = q_{n^\star}$.

Fix $j \leq k$ and let $n_j = \min(n : (f_0^j, f_1^j) \sqsubseteq_n X_{N \cap U})$. We have to show that $q \Vdash (\dot{f}_0^j, \dot{f}_1^j) \sqsubseteq_{n_j} X_{N \cap U}$. Let $x, y \in X_{N \cap U} \setminus n_j$.

Case 1 $x, y \leq n$. In this case $q \Vdash \dot{f}_i^j \restriction n = f_i^j \restriction n$ and we know that $f_i^j(x) \neq f_{1-i}^j(y)$ for $i = 0, 1$.

Case 2 $x, y > n$. In this case $x, y \in X_n$ and since $q \Vdash (\dot{f}_0^j, \dot{f}_1^j) \sqsubseteq_0 X_n$, it follows that $q \Vdash \dot{f}_{1-i}^j(x) \neq \dot{f}_i^j(y)$

Case 3 $x \leq n < y$. In this case we have two possibilities: either one of the functions $\dot{f}_0^j, \dot{f}_1^j$ is forced to be constant on X_n or both are forced to be finite-to-one on X_n. In the first case $q \Vdash \exists k \leq n\ \dot{f}_i^j(y) = f_i^j(k)$. Thus $q \Vdash \dot{f}_{1-i}^j(x) \neq \dot{f}_i^j(y)$ since $f_i^j(k) \neq f_{1-i}^j(x)$. If both functions are finite-to-one then $q \Vdash \forall i = 0, 1\ f_i^j(x) \leq \max(f_i^j(k) : k \leq n) < \dot{f}_{1-i}^j(y)$. ⊣

Now Theorem 13 follows readily from Lemma 16 and the results of this section.

Acknowledgement. We would like to thank Mauro Di Nasso and Marco Forti for introducing us to the problem, and helpful comments concerning the draft versions of this paper. We would also like to thank Alan Dow and Andrzej Roslanowski for careful reading the early versions of the paper.

REFERENCES

[1] Tomek Bartoszyński and Haim Judah, ***Set Theory: On the Structure of the Real Line***, A K Peters, Wellesley, MA, 1995.

[2] Andreas Blass and Saharon Shelah, *Near coherence of filters. III. A simplified consistency proof*, ***Notre Dame Journal of Formal Logic***, vol. 30 (1989), no. 4, pp. 530–538.

[3] Mauro Di Nasso and Marco Forti, *Topological and nonstandard extensions*, ***Monatshefte für Mathematik***, vol. 144 (2005), no. 2, pp. 89–112.

[4] ———, *Ultrafilter semirings and nonstandard submodels of the Stone-Čech compactification of the natural numbers*, ***Logic and Its Applications*** (Yi Zhang, editor), Contemporary Mathematics, vol. 380, American Mathematical Society, Providence, RI, 2005, pp. 45–51.

[5] ———, *Hausdorff ultrafilters*, ***Proceedings of the American Mathematical Society***, vol. 134 (2006), no. 6, pp. 1809–1818.

[6] Marco Forti, Mauro Di Nasso, and Vieri Benci, *Hausdorff nonstandard extensions*, ***Boletim da Sociedade Paranaense de Matemática. 3rd Série***, vol. 20 (2002), no. 1-2, pp. 9–20.

[7] Thomas Jech, ***Set Theory***, Pure and Applied Mathematics, Academic Press, New York, 1978.

[8] Arnold W. Miller, *Rational perfect set forcing*, ***Axiomatic Set Theory (Boulder, Colo., 1983)***, Contemporary Mathematics, vol. 31, American Mathematical Society, Providence, RI, 1984, pp. 143–159.

[9] Saharon Shelah, ***Proper and Improper Forcing***, second ed., Perspectives in Mathematical Logic, Springer-Verlag, Berlin, 1998.

NATIONAL SCIENCE FOUNDATION
4201 WILSON BLVD.
ARLINGTON, VA 22230, USA
E-mail: tomek.bartoszynski@gmail.com

DEPARTMENT OF MATHEMATICS
HEBREW UNIVERSITY
JERUSALEM, ISRAEL
E-mail: shelah@math.huji.ac.il
URL: `http://math.rutgers.edu/~shelah/`

ZERO-GROUPS AND MAXIMAL TORI

ALESSANDRO BERARDUCCI

Abstract. We give a presentation of various results on zero-groups in o-minimal structures together with some new observations. In particular we prove that if G is a definably connected definably compact group in an o-minimal expansion of a real closed field, then for any maximal definably connected abelian subgroup T of G, G is the union of the conjugates of T. This can be seen as a generalization of the classical theorem that a compact connected Lie group is the union of the conjugates of any of its maximal tori.

§1. Introduction. We consider groups definable in an o-minimal expansion $\mathbf{M} = (M, <, +, \cdot, \dots)$ of a real closed field $(M, <, +, \cdot)$. Classical examples of such groups are the (real)algebraic subgroups of the general linear group $GL(n, M)$. Identifying the algebraically closed field $C = M[\sqrt{-1}]$ with M^2 in the standard way, we also obtain all the algebraic subgroups of $GL(n, C)$. Less classical examples of definable groups can be found in [18] or in [14]. By [16] each definable group can be equipped with a unique group topology which makes it into a definable manifold (see Definition 3.1). A notion of definable compactness for definable manifolds (and more generally for definable spaces) can be introduced as in [14]. (In the semialgebraic case the definably compact spaces are the complete spaces of [6].) One has also a notion of definable connectedness: a definable group is definably connected if it has no proper definable subgroups of finite index [16]. Unless $M = \mathbb{R}$, definable compactness (or connectedness) does not imply compactness (or connectedness). In this note we prove:

THEOREM 6.12. *If G is a definably connected definably compact group in an o-minimal expansion $\mathbf{M}$ of a real closed field, then for any maximal definable abelian subgroup T of G, G is the union of the conjugates of T.*[1]

Partially supported by Progetto MIUR, Cofin 2004, Metodi di Logica in Algebra, Analisi e Geometria. I thank Jean-Pierre Ressayre and the Équipe de Logique of the University of Paris VII for their invitation and kind hospitality in the period March 5 - April 5, 2005.

[1]We learned from a referee that a recent paper of M. Edmundo [8] contains another proof of this result which uses the classification of definable semisimple groups.

Logic Colloquium '04
Edited by A. Andretta, K. Kearnes, and D. Zambella
Lecture Notes in Logic, 29

This can be seen as a generalization of the classical theorem that a compact connected Lie group is the union of the conjugates of any of its maximal tori. Indeed the classical theorem can be deduced from our theorem taking $M = \mathbb{R}$, after some definability considerations (Corollary 6.13).

We show more generally that if G is a definably connnected definably compact group and $H < G$ is a definable subgroup of G such that the (o-minimal) Euler characteristic of G/H is different from zero, then G is the union of the conjugates of H (Theorem 6.5).

Under the hypothesis of Theorem 6.12 it is not difficult to prove that the "Weyl group" $W(G) := N_G(T)/T$ is finite (Theorem 6.14).

From Theorem 6.12 and earlier work on definable abelian groups one can obtain a result of M. Otero, also proved by M. Edmundo in [8] by different methods, stating that every definably compact definably connected group G is divisible (Corollary 6.15).

The proof of Theorem 6.12 makes use of an o-minimal version of Lefschetz fixed point theorem (see [7] or [2, Theorem 3.3]) and of the notion of 0-group introduced by Strzebonski in [18]. Strebonski generalized many classical results on p-groups to the case of 0-groups. It turns out that, in the definably compact case, the 0-groups are exactly the definably connected abelian groups (Corollary 5.7).

Granted the appropriate background the proof of Theorem 6.12 is rather short, but we have taken this opportunity to give an exposition, with bibliographical references, of the relevant notions and results, and to make some side observations.

§2. Euler characteristic. We assume some familiarity with the basic notions of o-minimality (see [19]). Fix in the sequel an o-minimal structure $\boldsymbol{M} = (M, <, +, \cdot, \dots)$ expanding a real closed field (the dots represent possible additional structure besides the field structure). Although many of the results we use remain true also for more general o-minimal structures, working over a field simplifies some proofs and moreover the known proofs of the fixed point theorem (Theorem 6.4) do make use of the field structure. A subset $X \subset M^k$ is definable if it is (first order) definable in $\boldsymbol{M}$ possibly with parameters. For instance, if $\boldsymbol{M}$ is a real closed field $(M, <, +, \cdot)$ without additional structure, the definable sets are exactly the semialgebraic sets. We give to M the topology generated by the open intervals and to M^n the product topology. To each definable set X one can attach two invariants: its dimension $\dim(X) \in \mathbb{N}$ and its o-minimal Euler characteristic $E(X) \in \mathbb{Z}$.

DEFINITION 2.1 (see [19]). The *dimension* of X is $\geq n$ iff X contains a subset definably homeomorphic to an open subset of M^n. The (o-minimal) *Euler characteristic* $E(X)$ is defined as the number of even dimensional cells in X

minus the number of the odd dimensional cells in X, relative to any given cell decomposition.

The name "Euler characteristic" may be slightly misleading, since $E(X)$ does not always coincide with the classical Euler characteristic $\chi(X)$, even when the underlying o-minimal structure $\boldsymbol{M}$ is based on the real numbers. For instance an open interval $I \subset \mathbb{R}$ has $E(I) = -1$ and $\chi(I) = 1$. To understand why this is the case one needs to remind that o-minimal cells do not include the boundary, so an open interval is a odd dimensional cell. This discrepancy is reflected in the fact that, while χ is invariant under homotopies, E satisfies instead the following:

PROPOSITION 2.2 (see [19, Chapter 4 (1.3), (2.4)]). $\dim(X)$ *and* $E(X)$ *are invariant under definably bijections, not necessarily continuous.*

Following [18] we can now define $\dim(G/H)$ and $E(G/H)$ where G is a definable group and $H < G$ is a definable subgroup.

DEFINITION 2.3. A *definable group* is a definable set $G \subseteq M^k$ together with a definable group operation. We do not require the group operation to be continuous in the topology induced from M^k. Let G be a definable group, and let H be a definable subgroup, not necessarily normal. Let G/H be the set of left cosets of H. By "definable choice" (see [19, Chapter 6 (1.2) p. 94]) there is a definable function $\imath$ with domain G such that $\imath(g) = \imath(h)$ iff $gH = hH$ (one can also require $\imath(g) \in gH$). So whenever convenient we can identify G/H with the definable set $\imath(G)$ (identifying $gH \in G/H$ with $\imath(g)$). Different choices of $\imath$ give rise to definable bijections, so as a definable set G/H is only defined up to definable bijections. However, since the o-minimal Euler characteristic and the dimension are invariant under definable bijections, $E(G/H)$ *and* $\dim(G/H)$ *are well defined.*

It is possible to define, besides $E(X)$, another invariant $E'(X)$ which more closely resembles the classical Euler characteristic and is invariant under definable homotopies. To this aim it suffices to replace, in the classical definition $\chi(X) = \Sigma_i(-1)^i \operatorname{rank} H_i(X)$, the classical homology group H_i with the o-minimal homology groups of Woerheide [20] (which are naturally isomorphic the classical ones when the o-minimal structure is based on the reals, see [1, Proposition 3.2]). When X is a closed and bounded subset of M^k one can easily prove using the triangulation theorem that $E(X) = E'(X)$ (see [2, p. 788]), so in particular if $M = \mathbb{R}$ and $X \subset \mathbb{R}^k$ is compact, then $E(X) = E'(X) = \chi(X)$. Thus $E(X)$ is a tool which allows one to use combinatorial arguments typical of finite groups theory (thanks to its invariance under definable bijections), while at the same time permitting to draw conclusions of topological nature (thanks to the fact that it coincides with $E'(X)$ when X is closed and bounded). We will implicitly use these facts in the sequel. In fact the results relying on the fixed point theorem depend on the use of E'.

§3. Definable spaces. The notion of definable space is discussed in [19]. We give here an apparently weaker definition that is equivalent up to isomorphisms.

DEFINITION 3.1. Let X be a definable set and let τ be a topology on X. We say that (X, τ) is a *definable space* if there is a finite cover of X by definable sets $U_1, \dots, U_k$, open in τ, and a natural number n, such that for each U_i there is a definable function $\varphi_i : U_i \to V_i$ which is a homeomorphism between U_i (with the τ topology) and a subset V_i of M^n (where M has the topology generated by the open intervals, and M^n has the product topology). If moreover the V_i are open in M^n, we say that (X, τ) is a *definable manifold* of dimension n. The collection of the "*local charts*" (U_i, φ_i, n) is called an *atlas* of (X, τ). So each definable space or manifold has a finite atlas.

The easiest example of definable space is a *definable subspace of* M^n, namely a definable set $X \subset M^n$ with the topology inherited by the ambient space M^n. A definable space is *affine* if it can be definably embedded in M^n for some n (namely if it is definably homeomorphic to a definable subset of M^n considered as a definable subspace of M^n).

We will later need:

THEOREM 3.2 ([17], see also [19, Chapter 10 (1.8)]). *A necessary and sufficient condition for a definable space to be affine is that it is regular, namely its points are closed and for every point p and closed set C there are open neighbourhoods of p and C which are disjoint.*

Since each open set is a union of definable open sets, it is easy to see that a definable space is regular if and only if it is *definably regular*, namely its points are closed and for every point p and definable closed set C there are definable open neighbourhoods of p and C which are disjoint.

LEMMA 3.3. *Let (X, τ) be a definable manifold of dimension n and let $Y \subset X$ be a subset of X of dimension m, equipped with the subspace topology. Then Y is a definable (sub)manifold of X if and only if every point $p \in Y$ has a definable open neighbourhood $O \subset Y$ definably homeomorphic to an open subset of M^m.*

So although definable manifolds are required to have a finite atlas, for submanifolds it is not necessary to require this finiteness condition as it is always automatically ensured.

PROOF. The lemma was proved in [1, Proposition 4.2] in the case when $(X, \tau) = M^n$. The general case can be reduced to this special case using the fact that X is covered by finitely many open sets definably homeomorphic to open subsets of M^n. ⊣

Our main source of definable spaces, besides the subspaces of M^n, are the definable groups. Recall that for a definable group $G \subseteq M^k$ we do not require the group operation to be continuous in the topology induced from the ambient space M^k. The good topology on definable groups is not the topology of the ambient space but the one given by the following:

THEOREM 3.4. [16, Proposition 2.5] *If G is a definable group, then there is a (unique) topology τ on G, called the definable manifold topology of G, such that:*

1. *(G, τ) is a topological group (i.e. multiplication and inversion are continuous);*
2. *(G, τ) is a definable manifold of dimension n, where n is the (o-minimal) dimension of G.*

The uniqueness of the definable group topology follows from:

LEMMA 3.5. [13, Lemma 1.11] *If $f : H \to G$ is a definable group homomorphism between definable groups equipped with definable manifold topologies, then f is continuous.*

For definable spaces we have the following notion of definable compactness:

DEFINITION 3.6. [14] An Hausdorff definable space (X, τ) is *definably compact* iff for every definable function $f : I \to X$, where $I = (a, b) \subset M$ is an open interval, $\lim_{t \to b^-} f(t)$ exists in (X, τ).

Since by o-minimality every definable function $f : I \to (X, \tau)$ is piecewise continuous, without loss of generality we can assume f continuous.

PROPOSITION 3.7. [14, Theorem 2.1] *Let $X \subset M^n$ be a definable set with the topology induced by the ambient space M^n. Then X is definably compact iff it is closed and bounded.*

The closed and bounded subsets of M^n need not be compact if $M \neq \mathbb{R}$, but they behave in many respects like compact sets within the definable category. For instance the image of a closed and bounded set under a continuous definable function is closed and bounded (see [19, Chapter 6 (1.9) p. 95] for o-minimal structures expanding an ordered abelian group, and [14] for arbitrary o-minimal structures).

LEMMA 3.8. *Let $f : (X, \tau) \to (Y, \mu)$ be a definable continuous surjective function between definable spaces and assume that (X, τ) is definably compact. Then (Y, μ) is definably compact and f is a closed map.*

PROOF. Let $\gamma : I \to Y$ be a definable function, $I = (a, b)$. By definable choice γ can be lifted to a definable function $\sigma : I \to X$ with $f \circ \sigma = \gamma$. Since σ has a limit in X (and γ and σ are piecewise continuous by o-minimality), $f \circ \sigma$ has a limit in Y. So Y is definably compact. Since a closed subset of a definably compact set is definably compact, the same argument shows that f is a closed map. ⊣

THEOREM 3.9. [16, Corollary 2.8] *Let G be a definable group and let $H < G$ be a definable subgroup. Then H is closed in the definable group topology of G.*

Moreover we have:

THEOREM 3.10. [15, Lemma 2.6] *Let G be a definable group and $H < G$ be a definable subgroup. Then the definable group topology on H coincides with the topology as a subspace of G*

The proof in [15] is in term of generic elements. The next lemma yields a proof based on Lemma 3.3.

LEMMA 3.11. *Let $f: H \to G$ be an injective definable morphism of definable groups. Then f is a homeomorphism from H to $f(H)$, where H and G have the definable group topology and $f(H) \subseteq G$ has the subspace topology.*

PROOF. $f: H \to G$ is continuous by Lemma 3.5. Let $K = f(H) \subseteq G$ and let $m = \dim H = \dim K$.

By the cell decomposition theorem K contains an open subset $V \subset K$ (in the topology inherited from G) definably homeomorphic to an open subset of M^m (take a cell of dimension m in the intersection of K with a local chart of G).

Now K is a subgroup of G and since it has the subspace topology it is also a topological subgroup. This implies that the translates of V in K are definably homeomorphic to each other, and since they cover K, we get that every point of K has a neighbourhood definably homemomorphic to an open subset of M^m. The number of such translates need not be finite, but nevertheless by Lemma 3.3 we obtain that K is a definable submanifold of G. We can then apply Lemma 3.5 to conclude that $f^{-1}: K \to H$ is continuous, thus finishing the proof. ⊣

If H is definably compact Lemma 3.11 has a shorter proof. Indeed in this case the continuous map $f: H \to G$ is a closed map (by Lemma 3.8), hence an homeomorphism onto its image.

We will later need:

DEFINITION 3.12. A definable space (X, τ) is *definably connected* if and only if it has no definable proper non-empty clopen subset.

REMARK 3.13. Each definable space (X, τ) is a finite union of maximal definably connected subsets, called its definably connected components.

PROOF. Immediate from the cell decomposition theorem when X is a subspace of M^k. The general case follows working in the local charts. ⊣

REMARK 3.14. A definable space (X, τ) is definably connected if and only if it is definably path connected, i.e. any two points can be joined by a definable continuous path. (See [19, Chapter 6, Proposition 3.2, p. 100] for the case of subspaces of M^k.)

§4. **Homogeneous spaces.** Besides definable groups, another source of definable manifolds are the definable homogeneous spaces, namely the definable sets on which a definable group acts transitively by a definable action.

THEOREM 4.1. [13, Theorem 2.11] *Let G be a definable group, and let $\alpha\colon G \times V \to V$ be a definable transitive action on the definable set V.*[2] *There is a topology τ on V such that:*

1. *(V, τ) is a definable manifold;*
2. *the action $\alpha\colon G \times V \to V$ is continuous (where G has the definable manifold topology).*

COROLLARY 4.2. [13, Corollary 2.14] *Let G be a definable group, and let $H < G$ be a definable subgroup, not necessarily normal. There is a topology τ on G/H such that:*

1. *$(G/H, \tau)$ is a definable manifold;*
2. *the natural action $L\colon G \times G/H \to G/H$ given by left multiplication $L(g_1, g_2H) = g_1g_2H$ is continuous (where G has the definable manifold topology).*

THEOREM 4.3. *Let G be a definable group and let $H < G$ be a definable subgroup. The following topologies on G/H coincide:*

1. *The topology τ on G/H given by Theorem 4.2;*
2. *The quotient topology ν on G/H (where G has the definable manifold topology);*
3. *If H is normal in G, the group G/H has its own definable manifold topology given by Theorem 3.4, which also coincides with the quotient topology.*

Thanks to this result we can speak of the *definable manifold topology* on G/H without ambiguity.

PROOF. The projection $\pi\colon G \to (G/H, \tau)$ is continuous because $\pi(g) = gH = L(g, 1H)$ and L is continuous. Since the kernel is H, passing to the quotient we have an induced continuous map $id\colon (G/H, \nu) \to (G/H, \tau)$ where ν is the quotient topology and id is the identity. To prove that $\nu = \tau$ it remains to show that $id\colon (G/H, \nu) \to (G/H, \tau)$ is an open map (hence an homeomorphism). By definition of the quotient topology this is equivalent to say that $\pi\colon G \to (G/H, \tau)$ is an open map, and this is ensured by Theorem 4.4 below. Finally if $H \lhd G$, we know that G/H has a unique group topology which is at the same time a definable manifold topology. Since the quotient topology ν on G/H is a group topology, and we have just shown that it coincides with the definable manifold topology τ, the proof is finished. $\dashv$

[2] We recall that the action α is transitive if for every $v, v' \in V$ there is $g \in G$ with $\alpha(g, v) = v'$.

It remains to prove:

THEOREM 4.4. *Let* $\alpha: G \times V \to V$ *be a definable transitive action and let* τ *be the definable manifold topology on* V *given by Theorem* 4.1. *Let* $v \in V$. *The continuous map* $\gamma: G \to (V, \tau)$ *defined by* $\gamma(g) = \alpha(g, v)$, *is an open map.*

PROOF. We say that γ is open at $p \in G$ if for every open neighborhood O of p, $\gamma(O)$ contains $\gamma(p)$ in its interior. Clearly if γ is open at every point then it is an open map. Using the transitivity and the continuity of the action of G (on both G and V), it is easy to see that if γ is open at some point then it is open at every point. So it suffices to prove that γ is open at some point. We will work on local charts and we will use the following easy consequence of the trivialization theorem (see [19, Chapter 9 (1.2) p. 142] for the statement of the trivialization theorem):

CLAIM 4.5. Let $X, Y \subseteq M^k$ be definable sets with the topology induced by M^k and let $f: X \to Y$ be a continuous onto map. Then f is open at some point of X.

To prove the claim recall that $f: X \to Y$ is definably trivial if there is a definable homeomorphism $\sigma: X \to Y \times F$ such that $f = \pi_1 \circ \sigma$, where $\pi_1: Y \times F \to Y$ is the projection. Clearly if f is definably trivial it is an open map, since so are the projections. In the general case, by the trivialization theorem and the surjectivity of f we can find an open set O of Y such that the restriction of f to $f^{-1}(O)$ is a definably trivial map onto O, and the claim follows.

To finish the proof consider a cell decomposition of G compatible with the open sets of a finite atlas of G and with the γ-preimages of the open sets of a finite atlas of V. At least one cell of this decomposition of G is mapped by γ into a subset of V containing a non-empty open set $P \subset V$. Let $X = \gamma^{-1}(P) \subseteq G$. The restriction of γ to X is a surjective continuous map $\gamma_{|X}: X \to P$. Moreover by our costruction X is contained in a single chart of G, and P is contained in a single chart of V. So using the claim we can conclude that $\gamma_{|X}$ is open at some point p of X, so a fortiori γ is open at p. ⊣

REMARK 4.6. If G is definably compact the proof of Theorem 4.3 can be simplified. In fact by Lemma 3.8 and the continuity of π, if G is definably compact, also $(G/H, \tau)$ is definably compact and the continuous injective map $id: (G/H, v) \to (G/H, \tau)$ is a closed map, hence an homeomorphism.

§5. Zero-groups.

DEFINITION 5.1. G is a 0-*group* if it is a definable group and for every proper definable subgroup $H < G$ we have $E(G/H) = 0$ (we allow H to be the trivial group).

Note that the trivial group is a 0-group since it has no proper subgroups.

THEOREM 5.2. [16, Proposition 2.12] *Let G be a definable group. Put on G the definable manifold topology. Let $G^0 \subset G$ be the definably connected component of the identity of G. Then G^0 is a normal subgroup of G and it is the smallest definable subgroup of G of finite index. Thus G is definably connected if and only if it has no proper definable subgroups of finite index.*

The definition of definable connectedness given above (Definition 3.12) is thus consistent with the following:

DEFINITION 5.3. A definable group G is *definably connected* iff G has no definable proper subgroup of finite index.

REMARK 5.4. If G is a 0-group, then G is definably connected.

PROOF. Let $G^0 < G$ be the definably connected component of the identity. Then $[G : G^0]$ is finite and therefore $E(G/G^0) = [G : G^0] \neq 0$. If $G^0 \neq G$ this would contradict the definition of 0-group. ⊣

THEOREM 5.5. [18, Corollary 5.7] *If G is a 0-groups, then G is abelian.*

THEOREM 5.6 ([7] or [2, Theorem 3.3]). *If G is a definably compact infinite group, $E(G) = 0$.*

COROLLARY 5.7. *Let G be a definably compact group. Then G is a 0-group if and only if it is abelian and definably connected.*

PROOF. Suppose G is definably compact, abelian and definably connected. We must prove that G is a 0-group. Let $H < G$ be a proper subgroup (if H does not exist, G is the trivial group, which is a 0-group). Since G is abelian $H \lhd G$. The group G/H is definably compact since so is G, and it is infinite since G is definably connected. So $E(G/H) = 0$ by Theorem 5.6. ⊣

Strzebonki gave an example of a definable (even real semialgebraic) 0-group which is not definably compact [18, Example 5.3]. The 0-groups which are definably compact are exactly the Strzebonski tori defined below:

DEFINITION 5.8. G is a Strzebonski torus if and only if G and all definably connected subgroups of G are 0-groups.

Since 0-groups are definably connected, this is equivalent to the definition given by Strzebonski in [18]: G is a (Strzebonski) torus if G is a 0-group and every $H < G$ contains a 0-group $K < H$ with $[H : K]$ finite.

PROPOSITION 5.9. *Let G be a definable group. Then G is a Strzebonski torus iff G is definably compact abelian and definably connected.*

PROOF. Assume G is a Strzebonski torus. Then in particular is a 0-group, so it is abelian and definably connected. It remains to show that G is definably compact. We reason by contradiction using a result of Peterzil and Steinhorn [14]: if a definably group G is not definably compact, then it contains a definable one-dimensional torsion free subgroup $H < G$. We can assume that H is definably connected (as otherwise replace it with its component at the identity). Since groups with $E = 0$ have elements of every prime order (by [18]), it follows that $E(H) \neq 0$, so G is not a Strzebonski torus.

For the opposite direction suppose G is definably compact abelian and definably connected. We have already proved that G is a 0-group. To show that it is a torus we must show that if $H < G$ is definably connected, then H is a 0-group. This is clear since H satisfies the same assumptions used to show that G is a 0-group. ⊣

§6. 0-Sylow subgroups and maximal tori.

THEOREM 6.1. *Let G be a definably compact group and let $H < G$ be a definable subgroup. If $E(G/H) \neq 0$ and $f : G/H \to G/H$ is a definable continuous map definably homotopic to the identity (with respect to the definable manifold topology of G/H), then f has a fixed point.*

PROOF. Let us first observe:

CLAIM 6.2. G/H is a regular topological space (we do not need definable compactness here).

Indeed each definable manifold, although it does not need to be Hausdorff, it is certainly T_1 (i.e. its points are closed). So G with the definable manifold topology is T_1. Now we use the fact that if G is a T_1 topological group and $H < G$ is a closed subgroup, then G/H is regular (see [11, Chapter 1, Theorem 1.6]).

By Theorem 3.2 we can thus identify G/H with a definable submanifold of M^k. For such manifolds a definable version of the singular homology groups has been developed by Woerheide ([20]), and a notion of orientability can then be defined as in [2] in terms of the definable homology groups. A routine argument already used in [2] (for G instead of G/H) shows:

CLAIM 6.3. G/H is definably orientable.

The idea is that (as for classical homogeneous spaces) an orientation on G/H is obtained by choosing a local orientation at a point and extending it consistently to the whole space G/H using the transitivity of the action of G on G/H.

To conclude the proof of Theorem 6.1 it now suffices to invoke Theorem 6.4 below. ⊣

THEOREM 6.4 ([7] or [2, Theorem 3.3]). *Let $X \subseteq M^k$ be a definable set. Suppose that X, with the subspace topology from M^k, is a definably compact definably oriented definable manifold with $E(X) \neq 0$. Let $f : X \to X$ be a definable continuous map definably homotopic to the identity. Then f has a fixed point.*

We can now prove:

THEOREM 6.5. *Let G be a definably connected definably compact group and let $H < G$ be a definable subgroup with $E(G/H) \neq 0$. Then $G = \bigcup_{g \in G} gHg^{-1}$.*

PROOF. Let $g \in G$. We want to prove that $g \in \bigcup_{x \in G} xHx^{-1}$. Consider the map $L_g : G/H \to G/H$ defined by $L_g(xH) = \alpha(g, xH) = gxH$. We claim that L_g is definably homotopic to the identity. Granted the claim, by Theorem 6.1 L_g has a fixed point $xH \in G/H$. So $gxH = xH$ and therefore $g \in xHx^{-1}$. To prove the claim recall, by 3.14, that G is definably path connected. So there is a definable continuous function $t \mapsto g_t$, $t \in [0,1]^M$, with $g_0 = 1, g_1 = g$. Hence $t \mapsto L_{g_t}$ is a definable homotopy between $id = L_{g_0}$ and $L_g = L_{g_1}$. ⊣

To apply the above results to 0-subgroups we recall some results of Strzebonski.

DEFINITION 6.6. Let G be a definable group. A 0-group $H < G$ is called 0-Sylow if it is a maximal 0-subgroup of G.

REMARK 6.7. [18, Theorem 2.14] If $K < H < G$ are definable groups (no normality assumptions), there is a definable bijection from $G/H \times H/K$ to G/K. So $E(G/H)E(H/K) = E(G/K)$ and $\dim(G/H) + \dim(H/K) = \dim(G/K)$.

PROPOSITION 6.8 (see [18, Remark 2.18]). *If H, G are 0-groups and H is a proper subgroup of G, then* $\dim(H) < \dim(G)$.

PROOF. By Remark 6.7 $\dim(G) = \dim(G/H) + \dim(H)$. Since G is a 0-group, $E(G/H) = 0$. So G/H is infinite (because for a finite set X, $E(X) = \mathrm{Card}(X)$). But then $\dim(G/H) > 0$ (since infinite definable sets have positive dimension) and the result follows. ⊣

COROLLARY 6.9. [18, Remark 2.18] *Every 0-subgroup is contained in a 0-Sylow.*

THEOREM 6.10. [18] *Let G be a definable group and let $H < G$ be a definable subgroup which is a 0-group.*

1. *If $E(G/H) = 0$, then there is a 0-subgroup K of G with $H < K < N_G(H)$ and $K \neq H$.*
2. *If $E(G/H) \neq 0$, then H is 0-Sylow.*
3. *Thus $E(G/H) \neq 0$ if and only if H is 0-Sylow.*

PROOF. Part 1. is ([18, Theorem 2.14]). Part 2. follows immediately from the definitions and Remark 6.7. ⊣

THEOREM 6.11. *Let G be definably compact and definably connected. Let $T < G$ be a 0-Sylow of G. Then $G = \bigcup_{x \in G} xTx^{-1}$. (Moreover by* [18, *Theorem* 2.21] *each two 0-Sylow subgroups are conjugates.)*

PROOF. By Theorem 6.5. ⊣

Since in the definably compact case the 0-groups are the definably connected abelian groups the above theorem has the following equivalent formulation:

THEOREM 6.12. *If G is a definably connected definably compact group, then for any maximal definably connected abelian subgroup T of G, G is the union of the conjugates of T.*

If the o-minimal structure is an expansion of $\mathbb{R}$ we obtain the following classical result:

COROLLARY 6.13. *If G is a compact connected Lie group, then for any maximal abelian connected closed subgroup of $H < G$, G is the union of the conjugates of H.*

PROOF. The only thing to observe is that we do not need any definability assumptions. The definability comes for free since any compact Lie group G is isomorphic to a compact subgroup K of $GL(n, \mathbb{R})$ for some n (see [4, Chapter 3, Theorem 4.1, p. 136]) and any such K is a (real)algebraic subgroup of $GL(n, \mathbb{R})$ [5, Proposition 2, p. 230], hence it is definable in the o-minimal structure $(\mathbb{R}, <, +, \cdot)$. ⊣

Let $N_G(T)$ be the normalizer of T in G. In the hypothesis of Theorem 6.12 we have:

THEOREM 6.14. *The "Weyl group" $W(G) := N_G(T)/T$ is finite.*

PROOF. Suppose $W = W(G)$ is infinite. Then so is its definably connected component $W^0 < W$. Since G is definably compact, so is W^0. Hence $E(W^0) = 0$. Now $E(W) = E(W/W^0)E(W^0) = 0$. By definition this means $E(N_G(T)/T) = 0$. Hence, by Theorem 6.10, T is not a zero-Sylow, a contradiction. ⊣

A compact connected abelian Lie group is isomorphic to a torus, namely a product of finitely many copies of $\mathbb{R}/\mathbb{Z}$. This suggests that a definably compact definably connected abelian definable group G of (o-minimal) dimension n should resemble an n-dimensional torus $(\mathbb{R}/\mathbb{Z})^n$. This analogy is partially confirmed in [9], where it is shown that such a group G has the same torsion of an n-dimensional torus. Further information on G comes by considering quotients. Assuming the o-minimal structure sufficiently saturated, in [3] it is shown that G has a smallest type-definable subgroup $G^{00} < G$ of bounded index (which is moreover normal and divisible). By [10, Theorem 8.1] G^{00} is torsion free and G/G^{00} isomorphic to the torus $(\mathbb{R}/\mathbb{Z})^n$.

However the analogy with tori has its limitations: for instance there are definably compact definably connected abelian definable group G of dimension $n > 1$ without any proper infinite definable subgroup (see [14]). Hence, given a definably connected group G of dimension > 1, we have no control on how big a minimal 0-subgroup H is: it is not necessarily the case that $\dim(H) = 1$ (H could be G itself).

COROLLARY 6.15. [12, 8] *If G is a definably connected definably compact group, then G is divisible.*

PROOF. By Corollary 6.12 we can reduce to the case where G is abelian. The divisibility of definably connected abelian groups follows from the results of [18], as observed in [9]. In fact for $k \in \mathbb{N}$ define $p_k: H \to H$ by $p_k(x) = x^k$. Since H is abelian p_k is a homomorphism. By [18] the torsion subgroups $\operatorname{Ker} p_k$ is finite, and hence $\dim H = \dim \operatorname{Im} p_k$. Since H is definably connected $H = \operatorname{Im} p_k$, hence H is divisible. ⊣

REFERENCES

[1] A. BERARDUCCI and M. OTERO, *o-minimal fundamental group, homology and manifolds*, ***Journal of the London Mathematical Society. Second Series***, vol. 65 (2002), no. 2, pp. 257–270.

[2] ———, *Transfer methods for o-minimal topology*, ***The Journal of Symbolic Logic***, vol. 68 (2003), no. 3, pp. 785–794.

[3] A. BERARDUCCI, M. OTERO, Y. PETERZIL, and A. PILLAY, *A descending chain condition for groups definable in o-minimal structures*, ***Annals of Pure and Applied Logic***, vol. 134 (2005), no. 2-3, pp. 303–313.

[4] T. BRÖCKER and T. TOM DIECK, ***Representations of Compact Lie groups***, vol. 98, Springer-Verlag, New York, 1985.

[5] C. CHEVALLEY, ***Théorie des groupes de Lie. Tome III. Théorèmes généraux sur les algèbres de Lie***, Hermann & Cie, Paris, 1955.

[6] H. DELFS and M. KNEBUSCH, *Semialgebraic topology over a real closed field. II. Basic theory of semialgebraic spaces*, ***Mathematische Zeitschrift***, vol. 178 (1981), no. 2, pp. 175–213.

[7] M. EDMUNDO, *o-minimal cohomology and definably compact definable groups*, Manuscript Dec. 16, 2001, Revised May 6, 2003, 85 pp.

[8] ———, *A remark on divisibility of definable groups*, ***Mathematical Logic Quarterly***, vol. 51 (2005), no. 6, pp. 639–641.

[9] M. EDMUNDO and M. OTERO, *Definably compact abelian groups*, ***Journal of Mathematical Logic***, vol. 4 (2004), no. 2, pp. 163–180.

[10] E. HRUSHOWSKI, Y. PETERZIL, and A. PILLAY, *Groups, measures, and the NIP*, July 15, 2005, pp. 44.

[11] M. MIMURA and H. TODA, ***Topology of Lie groups. I, II***, vol. 91, American Mathematical Society, Providence, RI, 1991, Translated from the 1978 Japanese edition by the authors.

[12] M. OTERO, *On divisibility in definable groups*, RAAG preprint server n. 147, 2004.

[13] Y. PETERZIL, A. PILLAY, and S. STARCHENKO, *Definably simple groups in o-minimal structures*, ***Transactions of the American Mathematical Society***, vol. 352 (2000), no. 10, pp. 4397–4419.

[14] Y. PETERZIL and C. STEINHORN, *Definable compactness and definable subgroups of o-minimal groups*, ***Journal of the London Mathematical Society. Second Series***, vol. 59 (1999), no. 3, pp. 769–786.

[15] ———, *Definable homomorphisms of abelian groups in o-minimal structures*, ***Annals of Pure and Applied Logic***, vol. 101 (2000), no. 1, pp. 1–27.

[16] A. PILLAY, *On groups and fields definable in o-minimal structures*, ***Journal of Pure and Applied Algebra***, vol. 53 (1988), no. 3, pp. 239–255.

[17] R. ROBSON, *Embedding semi-algebraic spaces*, ***Mathematische Zeitschrift***, vol. 183 (1983), no. 3, pp. 365–370.

[18] A. STRZEBONSKI, *Euler characteristic in semialgebraic and other o-minimal groups*, ***Journal of Pure and Applied Algebra***, vol. 96 (1994), no. 2, pp. 173–201.

[19] L. van den Dries, ***Tame Topology and o-Minimal Structures***, London Mathematical Society Lecture Note Series, vol. 248, Cambridge University Press, Cambridge, 1998.

[20] A. Woerheide, ***o-minimal homology***, Ph.D. thesis, University of Illinois, UrbanaŰChampaign, 1996.

UNIVERSITÀ DI PISA
LARGO BRUNO PONTECORVO 5
56127 PISA
URL: www.dm.unipi.it/~berardu

CONSTRUCTIVE SET THEORY WITH OPERATIONS

ANDREA CANTINI AND LAURA CROSILLA

Abstract. We present an extension of constructive Zermelo–Fraenkel set theory [2]. Constructive sets are endowed with an applicative structure, which allows us to express several set theoretic constructs uniformly and explicitly. From the proof theoretic point of view, the addition is shown to be conservative. In particular, we single out a theory of constructive sets with operations which has the same strength as Peano arithmetic.

§1. Introduction. The general context of this article is an investigation of foundational frameworks for constructive mathematics in the style of Bishop. In the early seventies, prompted by Bishop's monograph *Foundations of Constructive Analysis* [9], various formal systems emerged to provide a clear and precise basis for the newborn mathematical style. Completely new systems were introduced like Feferman's explicit mathematics [16] and Martin–Löf's type theory [23, 24]. In addition, various systems in the form of the Zermelo–Fraenkel axiomatization but based on intuitionistic logic were put forward for example by Friedman [20, 21], Myhill [27] and Aczel [2, 4, 5]. The latter author was motivated by Myhill's *Constructive set theory* [28] in which a system of sets and functions was introduced adhering to a (generalized) predicative approach. Constructive set theory thus avoided the use of unbounded forms of separation and of the full axiom of powerset. Aczel brought these ideas in a system expressed in the language of Zermelo–Fraenkel set theory and also provided a natural interpretation for it in a version of Martin–Löf's type theory with a universe of small types and an inductive type built over it.

In this paper we propose an extension of Aczel's theory which could be seen as an attempt to provide a bridge between constructive set theory and Feferman's explicit mathematics.[1] The constructive set theory we introduce embodies in fact some of the characteristic features of explicit mathematics,

The research is part of a project supported by MIUR and the University of Florence. The second author is supported by a research fellowship (2003-2005) within the same project. The paper was presented by the first author at the Proof Theory Session, Logic Colloquium 2005, Turin. We thank the anonymous referee for useful comments.

[1]We would like to stress the 'experimental' character of this extension.

Logic Colloquium '04
Edited by A. Andretta, K. Kearnes, and D. Zambella
Lecture Notes in Logic, 29

like the presence of non-extensional elements and the uniformity of the set constructors. At the same time it can be seen as capturing an aspect of Myhill's original presentation, in allowing for a primitive notion of function. In particular, we define a conservative extension, which we name ***COST***, of Aczel's system ***CZF***, which is enriched with urelements as well as operations. The urelements represent elements of a combinatory algebra extended with natural numbers. The notion of operation is quite general and it is shown to be essentially non-extensional. By combining these features, ***COST*** allows to apply sets to sets and to state some basic axioms in a uniform way. There are operations yielding a restricted form of Δ_0-separation and there is a certain interplay between sets and operations (which must be limited, in order to avoid paradoxes). It turns out that all operations needed for set–theoretic analogues of the type–theoretic interpretation of ***CZF*** can be defined via the applicative structure of ***COST*** (see [31]).

The motivation for constructive set theory with operations is many-fold; we here list some points.

(1) Constructively, it is natural to allow for a primitive non-extensional notion of operation, independent from the extensional notion of set. The idea is that one can perform mathematics as usual, based on the extensional notion of set, and simultaneously stress the computational aspects, by allowing also for non-extensional operations.
(2) Natural numbers and recursive functions are considered as first class objects, and therefore are postulated as primitive.
(3) Uniformity is achieved for the usual set theoretic constructs. This seems to be in accordance with fundamental approaches to constructive foundations as represented by constructive type theory and explicit mathematics.
(4) As we shall see, the separation between natural numbers and sets has immediate benefits from a proof–theoretic point of view: we can in fact single out a subsystem of ***COST***, named $\boldsymbol{COST}_b$, which contains *full* $\in$*–induction*, but only *set induction on natural numbers* (regarded as urelements). It can then be shown that $\boldsymbol{COST}_b$ is conservative over Peano arithmetic. We thus obtain a strictly predicative system. Interestingly, from a technical point of view, this weak system still has the expressive power to represent a general theory of inductive definitions as in [6], without limitations, thanks to the fact that the $\in$-induction principle is unrestricted.

As to item (3) above, we observe that ***COST*** is not entirely satisfactory from the point of view of explicit mathematics: for there remain strong collection and subset collection as non-uniform essential principles. We note, though, that *some results in this paper only require the subsystems of* $\boldsymbol{COST}_b$ *(or of* $\boldsymbol{COST}^-$*) which are obtained by removing the collection principles, thus making* *use only of the uniform operations of direct image and exponentiation (see for example section 3).*

$\boldsymbol{COST}$ and $\boldsymbol{COST}_b$ take inspiration from the operational set theory $\boldsymbol{OST}$ presented by Feferman [18], whence our acronym. An intuitionistic impredicative set theory with operations has been investigated by Beeson in [8].

As to the contents of this paper, §2 describes the language and the axioms of $\boldsymbol{COST}$ and $\boldsymbol{COST}_b$, while §§3–4 collect elementary facts linking the set theoretic and the applicative structures. In particular, it is shown that extensionality and totality for operations are inconsistent with $\boldsymbol{COST}_b$, as consequence of the fact that we can prove the existence of an operation testing equality over the universe. As an example of the capability of the theory to represent inductive definitions, we hint at the construction in $\boldsymbol{COST}_b$ of a form of self–validating semantics introduced by Rathjen in [31] and called the "formula–as–classes interpretation". In §5 we investigate the relationship between set–theoretic functions and operations, by recalling Beeson's axiom $\boldsymbol{FO}$ which states that every function is an operation. We also briefly discuss the status of various forms of choice principles for operations. The subsequent section shows that $\boldsymbol{COST}$ is conservative over $\boldsymbol{CZF}^{op}$, a version of $\boldsymbol{CZF}$ which is built over a set of urelements, endowed with the structure of total extensional combinatory algebra[2]. The proof makes essential use of an inductive definition, in $\boldsymbol{CZF}^{op}$, of a suitable applicative relation. This relation endows the whole universe with a structure corresponding to a partial combinatory algebra, extended by suitable logical operations and set theoretic constructors. Crucially, the interpretation is independent from the strength of the principle of induction on natural numbers so that also $\boldsymbol{COST}_b$ is interpretable into a corresponding subsystems of $\boldsymbol{CZF}^{op}$ named $\boldsymbol{CZF}^{op}_b$.

§§7.1–8 yield the final reduction. The main result is formally a realizability interpretation theorem reducing $\boldsymbol{CZF}^{op}_b$ ($\boldsymbol{CZF}^{op}$) to a classical axiomatic theory of abstract self-referential truth, $\boldsymbol{T_c}$ ($\boldsymbol{T_f}$), which is conservative over $\boldsymbol{PA}$ (over $\boldsymbol{ID_1}$, respectively). The result is inspired by Aczel's type theoretic interpretation [2, 4]; however, viewing the construction from a model–theoretic perspective we directly simulate a suitable set–theoretic universe, $V_{\mathcal{U}}$, in a *classical Frege structure* (cf. [19, 11]). As a consequence, the type–theoretic machinery is avoided, the notion of *small type* is replaced by the notion of *total classification* (in short, *class*) in the sense of $\boldsymbol{T_c}$, and the type theoretic closure conditions are replaced by corresponding closure conditions on classes. It turns out that sets in $V_{\mathcal{U}}$ are hereditarily generated as families of sets indexed by classes (where families are explicitly given via operations of the underlying

[2]That the structure of the urelements we have chosen to base $\boldsymbol{COST}$ and $\boldsymbol{CZF}^{op}$ upon is both extensional and total, is only due to the fact that the proof theoretic result in §6 is thus stronger. One could instead opt for a theory in which the urelements form a partial non-extensional combinatory algebra.

combinatory algebra with values in $V_{\mathcal{U}}$). This has the effect of somewhat simplifying the technical details[3].

§2. The system *COST*.

2.1. The language of *COST*. The language of ***COST*** is the following applicative extension, $\mathcal{L}^R$, of the usual first order language of Zermelo-Fraenkel set theory, $\mathcal{L}$.

The language includes:

- the combinators K and S;
- constants $0, SUC, PR, D$ (for zero, successor, predecessor and definition by cases on natural numbers);
- a ternary predicate symbol, App, for application, and the following unary predicates $\mathcal{S}$ (sets), $\mathcal{N}$ (natural numbers) and $\mathcal{U}$ (urelements representing operations). $App(x, y, z)$ is read as x applied to y yields z.

The language includes in addition the following constants:

- el for the ground operation representing membership;
- pair, un, im, exp, sep, for set operations;
- $\varnothing$, Nat, Uap and Ur, set constants.

As part of the usual set theoretic language we also have the predicate symbols $\in$ and $=$. The logical symbols are all the intuitionistic operators: $\bot, \wedge, \vee, \rightarrow, \exists, \forall$.

For convenience we also use the bounded quantifiers $\exists x \in y$ and $\forall x \in y$, as abbreviations for $\exists x(x \in y \wedge \dots)$ and $\forall x(x \in y \rightarrow \dots)$.

As customary, we define $\varphi \leftrightarrow \psi$ by $(\varphi \rightarrow \psi) \wedge (\psi \rightarrow \varphi)$ and $\neg\varphi$ by $\varphi \rightarrow \bot$.

Terms and formulas. Terms and formulas are inductively defined as usual.

To increase perspicuity, we consider a definitional extension of $\mathcal{L}^R$ with application terms, defined inductively as follows.

(i) Each variable and constant is an application term.
(ii) If t, s are application terms then ts is an application term.

Application terms will be used in conjunction with the following abbreviations.

(i) $t \simeq x$ for $t = x$ when t is a variable or constant.

[3]Of course, we do not claim that the conceptual virtue underlying Aczel'interpretation – that of giving a meaning to set theoretic statements in terms of a systematic constructive semantics – is shared by the present approach. Our motivation is rather the technical attempt to explore the relation between the constructive notion of set as presented in Zermelo–Fraenkel set theory and as arising in other contexts, like the framework of explicit mathematics. However, the use of classical systems as a means for computing the proof theoretic strength of suitable subsystems of constructive set theory is well-established (see e.g. [14], where formal fragments of Martin-Löf's type theory needed for realizability are developed in classical theories of known strength, like Peano arithmetic with reflection or suitable fixed point theories).

(ii) $ts \simeq x$ for $\exists y \exists z\, (t \simeq y \wedge s \simeq z \wedge App(y, z, x))$.
(iii) $t \downarrow$ for $\exists x\, (t \simeq x)$.
(iv) $t \simeq s$ for $\forall x\, (t \simeq x \leftrightarrow s \simeq x)$.
(v) $\varphi(t, \dots)$ for $\exists x\, (t \simeq x \wedge \varphi(x, \dots))$.
(vi) $t_1 t_2 \dots t_n$ for $(\dots(t_1 t_2)\dots)t_n$.

We recall that a formula of $\mathcal{L}^R$ is *bounded*, or Δ_0, if and only if all quantifiers occurring in it, if any, are bounded. We say that a formula of $\mathcal{L}^{op}$ is *App-bounded*, or Δ_0^{App}, if and only if it is bounded and in addition it contains application App restricted to urelements. Formally Δ_0^{App} is inductively generated by means of sentential operators and bounded quantifiers from formulas of the form $t_1 = t_2, t_1 \in t_2, \mathcal{S}(t), \mathcal{N}(t), \mathcal{U}(t)$ and $App(t_1, t_2, t_3) \wedge \mathcal{U}(t_1) \wedge \mathcal{U}(t_2) \wedge \mathcal{U}(t_3)$.

Relations and set theoretic functions. The notion of relation between two sets is defined in the usual way in set theory: a relation R between sets a and b is a set of ordered pairs the first component of which is an element of a and the second an element of b. Ordered pairs are defined in the usual way. We say that R is *total* if $\forall x \in a\, \exists y \in b\, R(x, y)$ (where $R(x, y)$ is an abbreviation for $\langle x, y\rangle \in R$).

The domain of a relation R, is defined as the collection $Dom(R) := \{x : \exists y\, R(x, y)\}$. The range of R is $Ran(R) := \{y : \exists x\, R(x, y)\}$. As we shall see later, remark 4.1, *Dom* and *Ran* correspond to *internal operations* in ***COST***, mapping sets into sets.

Together with the notion of operation, we also have a notion of set theoretic function which we can express by a formula, $Fun(F)$, stating that F is a set encoding a total binary relation which satisfies the uniqueness condition: $\forall x \forall y \forall z\, (\langle x, y\rangle \in F \wedge \langle x, z\rangle \in F \to y = z)$.

The relation between the notions of set theoretic function and operation will be discussed in section 5.

REMARK 2.1. Officially we only have one sort of variables and three predicates representing sets, urelements and natural numbers. For convenience, though, we use distinct variables to distinguish among them. In particular, we use the variables $u, v, x, y, z, \dots$ as generic variables. We write $a, b, \dots$ for sets, i.e. for those x for which $\mathcal{S}(x)$ holds, but denote by $F, G, \dots$ those sets which are functions. By $f, g, \dots$ we denote *urelements as well as sets*, insofar they are used as operations (i.e. they formally occur as operators in application terms or as first coordinates in App-contexts). The variables $p, q, \dots$ are used to indicate elements of the applicative structure, i.e. those x for which $\mathcal{U}(x)$ holds. Finally, we write $k, m, n, \dots$ for those x for which $\mathcal{N}(x)$ holds.

The convention on the variables will only be applied when it is believed no misunderstanding can arise.

Truth values. We may represent false and truth by the empty set and the singleton empty set, respectively, that is we let $\bot := \varnothing$ and $\top := \{\varnothing\}$, also denoted by $\boldsymbol{1}$.

We let Ω to be the *class* $\mathcal{P}\boldsymbol{1}$, the powerset of $\boldsymbol{1}$. Then $x \in \Omega$ is an abbreviation for $\bot \subseteq x \subseteq \top$. Ω intuitively represents the class of truth values (or of propositions). Note that Ω cannot be taken to be a set, or otherwise, by means of exponentiation we would obtain full powerset (see Aczel [2, Proposition 2.3]).

Further conventions: for a and b sets or classes, write $f : a \to b$ for $\forall x \in a\,(fx \in b)$ and $f : \boldsymbol{V} \to b$ for $\forall x\,(fx \in b)$. By $f : a^2 \to b$ and $f : \boldsymbol{V}^2 \to b$ we indicate $\forall x \in a\, \forall y \in a\,(fxy \in b)$ and $\forall x\, \forall y\,(fxy \in b)$, respectively. This can be clearly extended to arbitrary exponents $n > 2$. Finally, $f : a \to \boldsymbol{V}$ means that f is everywhere defined on the set a.

DEFINITION 2.2 (The theory ***COST***). ***COST*** is the $\mathcal{L}^R$ theory whose principles are all the axioms and rules of first order intuitionistic logic with equality, plus the following principles.

(1) *Ontological axioms and extensionality*
  (a) $\neg(\mathcal{U}(x) \wedge \mathcal{S}(x))$
  (b) $\mathcal{U}(x) \vee \mathcal{S}(x)$[4]
  (c) $\mathcal{N}(x) \to \mathcal{U}(x)$
  (d) $x \in y \to \mathcal{S}(y)$
  (e) $\forall x\,(x \in a \leftrightarrow x \in b) \to a = b$
(2) *General applicative axioms and* $\mathcal{N}$*-closure*
  (a) $App(x, y, z) \wedge App(x, y, w) \to z = w$
  (b) $Kxy = x \wedge Sxy\downarrow \wedge Sxyz \simeq xz(yz)$
  (c) $\mathcal{N}(0) \wedge \forall x\,(\mathcal{N}(x) \to \mathcal{N}(SUCx) \wedge SUC\,x \neq 0)$
  (d) $PR\,0 = 0 \wedge \forall n\,(\mathcal{N}(PR\,n) \wedge PR\,(SUC\,n) = n)$
  (e) $Dxynn = x \wedge (n \neq m \to Dxynm = y)$
  (f) $\exists r\,App(p, q, r)$
  (g) $\forall r\,(pr \simeq qr) \to p = q$
  (h) $\mathcal{U}(K) \wedge \mathcal{U}(S) \wedge \mathcal{U}(SUC) \wedge \mathcal{U}(PR) \wedge \mathcal{U}(D)$
(3) *Membership operation*
  (a) $\mathsf{el} : \boldsymbol{V}^2 \to \Omega$ and $\mathsf{el}\,xy \simeq \top \leftrightarrow x \in y$
(4) *Set constructors*
  (a) $\mathcal{S}(\varnothing) \wedge \forall x\,(x \notin \varnothing)$
  (b) $\mathcal{S}(Ur) \wedge \forall x\,(x \in Ur \leftrightarrow \mathcal{U}(x))$
  (c) $\mathcal{S}(Nat) \wedge \forall x\,(x \in Nat \leftrightarrow \mathcal{N}(x))$
  (d) $\mathcal{S}(Uap) \wedge \forall x\,(x \in Uap \leftrightarrow \exists p\, \exists q\, \exists r\,(App(p, q, r) \wedge x = \langle p, q, r\rangle))$
  (e) $\mathcal{S}(\mathsf{pair}\,xy) \wedge \forall z\,(z \in \mathsf{pair}\,xy \leftrightarrow z = x \vee z = y)$

[4]This can be constructively justified as follows: each element explicitly carries the information determining its ontological category (that is, it has a canonical form). This is indeed the case in the model we build within the system $\boldsymbol{CZF}^{op}$.

(f) $\mathcal{S}(\mathsf{un}\, a) \wedge \forall z\, (z \in \mathsf{un}\, a \;\leftrightarrow\; \exists y \in a(z \in y))$
(g) $(f : a \to \Omega) \to \mathcal{S}(\mathsf{sep}\, fa) \wedge \forall x\, (x \in \mathsf{sep}\, fa \;\leftrightarrow\; x \in a \wedge fx \simeq \top)$
(h) $(f : a \to V) \to \mathcal{S}(\mathsf{im}\, fa) \wedge \forall x\, (x \in \mathsf{im}\, fa \leftrightarrow \exists y \in a(x \simeq fy))$
(i) $\mathcal{S}(\mathsf{exp}\, ab) \wedge \forall x(x \in \mathsf{exp}\, ab \;\leftrightarrow\; (Fun(x) \wedge Dom(x) = a \wedge Ran(x) \subseteq b))$

(5) *Induction*
(i) *Nat*–induction is the axiom schema:

$$\varphi(0) \wedge \forall x\, (\mathcal{N}(x) \to (\varphi(x) \to \varphi(SUC\, x))) \to \forall x\, (\mathcal{N}(x) \to \varphi(x));$$

(ii) $\in$–induction is the axiom schema:

$$\forall x(\mathcal{S}(x) \to (\forall y \in x\, \varphi(y) \to \varphi(x))) \to \forall x(\mathcal{S}(x) \to \varphi(x))^5;$$

in (i)-(ii) φ ranges over arbitrary formulas.

(6) *Collection Principles*
(a) Subset Collection: for every formula $\varphi(x, y, u)$,

$$\exists c\, \forall u\, \big(\forall x \in a\, \exists y \in b\; \varphi(x, y, u) \;\to\; \exists d \in c\; (\forall x \in a\, \exists y \in d\; \varphi(x, y, u) \;\wedge\; \forall y \in d\, \exists x \in a\; \varphi(x, y, u))\big);$$

(b) Strong Collection scheme: for every formula $\varphi(x, y)$,

$$\forall x \in a\, \exists y\; \varphi(x, y) \to \exists b\; (\forall x \in a\, \exists y \in b\; \varphi(x, y) \;\wedge\; \forall y \in b\, \exists x \in a\; \varphi(x, y)).$$

REMARK 2.3. The last three axioms of (2) state that the urelements form a total extensional combinatory algebra, and that the representatives of some operations can be chosen to be objects. We have chosen to include these axioms because they are validated in the model of section 6, so that we obtain there a more general result.

REMARK 2.4. Recall that *Replacement* is the principle

$$\forall x \in a\, \exists! y\; \varphi(x, y) \to \exists b\; \forall y\; (y \in b \quad\leftrightarrow\quad \exists x \in a\; \varphi(x, y)).$$

Observe that im only allows to recast a bounded form of replacement, due to the restriction on separation (see Proposition 3.3). This might be enough for many purposes. In particular if we remove the principle of full induction on sets, we obtain a theory analogous to Friedman's ***B*** (without the principle of dependent choice) [21]. If we wish to have more freedom and allow also replacement for arbitrary formulas, then we need to postulate it separately. Here, in the style of ***CZF*** we have allowed for a strengthened form of collection.[6] We have also introduced a strengthening of exponentiation, subset collection. Of course, in the presence of these principles, the non-uniform versions of direct image and exponentiation become redundant.

---

[5]Observe that, given the ontological axioms, this schema implies (and is actually equivalent to) $\forall x(\forall y \in x\, \varphi(y) \to \varphi(x)) \to \forall x \varphi(x)$.

[6]The delicate relation between collection and replacement in the context of ***CZF*** has been investigated in the recent work by Rathjen [30].

REMARK 2.5 (On operations as a primitive sort). In the present theory operations *do not form a separate ontological category, but they are represented by elements of the set theoretic universe* (either atoms or sets). This is somewhat unsatisfactory from a conceptual point of view. It would be more appropriate to introduce a suitable primitive predicate $Op(x)$ and require (i) that every object in the universe is either an atom or a set or else an operation; (ii) that the three sorts are disjoint; (iii) that whenever $xy \simeq z$, then x is an operation. It is not unreasonable from a constructive perspective to assume that every operation is represented (or named) by a set or an atom. Once this is accepted, the applicative structure can be mirrored at the level of sets and atoms and this idea exactly underlies the present choice.

2.2. Bounded induction for natural numbers. We now place some restrictions on the complexity of the formula in the principle of induction on the natural numbers. As we shall see, this has the effect of reducing considerably the proof theoretic strength of the theory considered. In the context of $\boldsymbol{COST}$ we may formulate a restricted principle, called *restricted Nat–induction*:[7]

$$f : Nat \to \Omega \wedge f0 \simeq \top \wedge \forall x \in Nat\, (fx \simeq \top \to f(SUCx) \simeq \top) \\ \to \forall x \in Nat\, (fx \simeq \top).$$

DEFINITION 2.6. $\boldsymbol{COST}_b$ is the system obtained from $\boldsymbol{COST}$ by removing full induction on the natural numbers and replacing it by restricted Nat–induction. $\boldsymbol{COST}^-$ is $\boldsymbol{COST}$ without induction axioms (on sets and numbers).

§3. Elementary facts.

First of all, as a consequence of the axioms for combinators, the universe of sets is closed under abstraction and recursion for operations [33]:

LEMMA 3.1 ($\boldsymbol{COST}^-$). (i) *For each term t, there exists a term $\lambda x.t$ with free variables those of t other than x and such that*

$$\lambda x.t \downarrow \wedge (\lambda x.t)y \simeq t[x := y].$$

(ii) (*Second recursion theorem*) *There exists a term* **rec** *with*

$$\boldsymbol{rec} f \downarrow \wedge (\boldsymbol{rec} f = e \to ex \simeq fex).$$

We now show that the logical operations generating bounded formulas are mirrored by internal operations.

LEMMA 3.2 ($\boldsymbol{COST}^-$). *There are application terms* eq, and, all, exists, imp, or, ur, nat, set, uap *such that*

[7]Clearly, this principle is an operational version of the usual induction axiom on the natural numbers:

$$0 \in a \wedge \forall x \in Nat\, (x \in a \to SUCx \in a) \to \forall x \in Nat\, (x \in a).$$

(i) $\mathsf{eq} : V^2 \to \Omega$ *and* $\mathsf{eq}xy \simeq \top \leftrightarrow x = y$;
(ii) $x \in \Omega \wedge y \in \Omega \to \mathsf{and}xy \in \Omega \wedge (\mathsf{and}xy \simeq \top \leftrightarrow (x \simeq \top \wedge y \simeq \top))$;
(iii) $(f : a \to \Omega) \to (\mathsf{all}fa \in \Omega) \wedge (\mathsf{all}fa \simeq \top \leftrightarrow \forall x \in a\,(fx \simeq \top))$;
(iv) $(f : a \to \Omega) \to \mathsf{exists}fa \in \Omega \wedge (\mathsf{exists}fa \simeq \top \leftrightarrow \exists x \in a\,(fx \simeq \top))$;
(v) $x \in \Omega \wedge (x = 1 \to y \in \Omega) \to \mathsf{imp}xy \in \Omega \wedge (\mathsf{imp}xy \simeq \top \leftrightarrow (x \simeq \top \to y \simeq \top))$;
(vi) $x \in \Omega \wedge y \in \Omega \to \mathsf{or}xy \in \Omega \wedge (\mathsf{or}xy \simeq \top \leftrightarrow (x \simeq \top \vee y \simeq \top))$;
(vii) $\mathsf{ur} : V \to \Omega$ *and* $\mathsf{ur}x \simeq \top \leftrightarrow \mathcal{U}(x)$;
(viii) $\mathsf{nat} : V \to \Omega$ *and* $\mathsf{nat}x \simeq \top \leftrightarrow \mathcal{N}(x)$;
(ix) $\mathsf{set} : V \to \Omega$ *and* $\mathsf{set}x \simeq \top \leftrightarrow \mathcal{S}(x)$.
(x) $\mathsf{uap} : V^3 \to \Omega$ *and* $\mathsf{uap}xyz \simeq \top \leftrightarrow App(x,y,z) \wedge \mathcal{U}(x) \wedge \mathcal{U}(y) \wedge \mathcal{U}(z)$.

PROOF. (i): choose $\mathsf{eq}\,ab := \mathsf{el}\ a(\mathsf{pair}\,bb)$.

(ii): Let $x \cap y = \mathsf{sep}\,(\lambda u.\mathsf{el}\,uy)x$ and choose $\mathsf{and}xy := x \cap y$.

(iii): $ha = \{y \in a : y = \top\}$ is a set by uniform separation. Observe that $a \subseteq b$ is equivalent to $a \cap b = a$ and hence there is an application term sub such that $\mathsf{sub}\,ab \simeq \top$ iff $a \subseteq b$. So $\bigwedge a = \{x \in \top : a \subseteq ha\}$ is a set again by making use of sep and we can choose $\mathsf{all}\,fa = \bigwedge(\mathsf{im}\,fa)$.

(iv): let $\mathsf{exists}fa = \mathsf{un}\,(\mathsf{im}\,fa)$.

(v): observe that $c := c(x,y,u) = \{w \in \{y\} : u \in x\}$ is a set by pair-, sep- and el-axioms. Hence $\mathsf{imp}xy = \{u \in \top : \forall v \in c\,(u \in v)\}$ is a set (by the axioms involving $\emptyset$, sep, pair, el and (iii) above). It is easily checked that $\mathsf{imp}xy \simeq \top \leftrightarrow (x \simeq \top \to y \simeq \top)$.

(vi): apply exists, and, imp.

(vii)–(x): apply the ontological axioms, the axioms for el and the existence of Ur, Nat, Uap. In the case (ix) also apply implication imp (v). ⊣

PROPOSITION 3.3 (***COST***$^-$). (i) *For each Δ_0^{App} formula φ, there is an application term f_φ such that, if $\boldsymbol{FV}(\varphi) = \{x_1, \dots, x_k\}$, then $f_\varphi : V^k \to \Omega$ and*

$$f_\varphi\, x_1 \dots x_k \simeq \top \leftrightarrow \varphi(x_1, \dots, x_k).$$

(ii) *To each Δ_0^{App} formula $\varphi(x, y_1 \dots y_k)$, we can associate an application term c_φ such that*

$$c_\varphi a y_1 \dots y_k \downarrow \wedge \forall u(u \in c_\varphi a y_1 \dots y_k \leftrightarrow u \in a \wedge \varphi(u, y_1, \dots, y_k)).$$

PROOF. (i) According to the inductive definition of Δ_0^{App}, if φ is one of the formulas $\mathcal{S}(x), \mathcal{U}(x), \mathcal{N}(x), x = y, x \in y, App(x,y,z) \wedge \mathcal{U}(x) \wedge \mathcal{U}(y) \wedge \mathcal{U}(z)$, we apply lemma 3.2 and the appropriate axioms. For the compound formulas, a simple induction applies, making use of the internal representation of the logical operations.

(ii) follows from (i) and explicit separation. ⊣

COROLLARY 3.4 ($\boldsymbol{COST}_b$). (i) *Restricted Nat–induction implies induction on the natural numbers for Δ_0^{App}–formulas;*
(ii) $\forall n\,(n = 0 \vee SUC(PRn) = n)$;
(iii) $\forall n\,\forall m\,(SUC\,n = SUC\,m \rightarrow n = m)$.
(iv) *Heyting Arithmetic* $\boldsymbol{HA}$ *is interpretable in* $\boldsymbol{COST}_b$.

PROOF. (i) Apply restricted *Nat*–induction and the previous proposition.
(ii)-(iii) By restricted *Nat*–induction and $\mathcal{N}$–closure axioms.
(iv) Choose *Nat* as domain of the interpretation and apply (i)-(iii) and lemma 3.1 for interpreting plus and times. ⊣

REMARK 3.5 (Bounded separation with application terms). We are allowed to use application terms as genuine terms insofar as they are defined. In the special case of separation, if t, s are application terms such that $t\downarrow$, t is a set, $s\downarrow$ and $s : t \rightarrow \Omega$, then there exists an application term $r := \mathrm{sep}\,ts$ such that

$$\mathcal{S}(r) \wedge \forall u(u \in r \leftrightarrow u \in t \wedge su \simeq \top).$$

Instead of r, we write $\{u \in t : su \simeq \top\}$. Similarly, if φ is Δ_0^{App} with free variables x, y, and t, s are application terms such that $t\downarrow$, t is a set, $s\downarrow$, then there exists an application term $r_\varphi := c_\varphi ts$ such that

$$\mathcal{S}(r_\varphi) \wedge \forall u(u \in r_\varphi \leftrightarrow u \in t \wedge \varphi(u, s)).$$

Instead of r_φ, we again stick to the more familiar and perspicuous notation

$$\{u \in t : \varphi(u, s)\}.$$

As a consequence of proposition 3.3, we can also express an operator representing definition by cases on the universe for App–bounded predicates.

LEMMA 3.6 ($\boldsymbol{COST}^-$). *Let $\varphi(x, y)$ be Δ_0^{App} (with the free variables shown). Then there exists an operation D_φ such that $D_\varphi abu\downarrow$ and*

$$D_\varphi abuv = \begin{cases} a, & \text{if } \varphi(u, v); \\ b, & \text{else} \end{cases} \tag{3.1}$$

Hence if $\varphi(u, v) \vee \neg\varphi(u, v)$, D_φ is total.

PROOF. By uniform bounded separation and uniform union, there exists a total operation D_φ such that

$$D_\varphi = \lambda a \lambda b \lambda u \lambda v.\{x \in a : \varphi(u, v)\} \cup \{x \in b : \neg\varphi(u, v)\}.$$

The definition of lambda abstraction implies $D_\varphi abu\downarrow$. Clearly D_φ satisfies (3.1). ⊣

As a special case, we obtain a global test for equality, definition by cases on $\mathcal{N}$ and a test for being an urelement or a set.

LEMMA 3.7. (1) ***COST***$^-$ *proves that there exists an operation* EQ *such that* $\mathsf{EQ}u\downarrow$ *and*

$$\mathsf{EQ}uv = \begin{cases} \top, & \text{if } u = v; \\ \bot, & \text{else} \end{cases} \tag{3.2}$$

(2) ***COST***$^-$ *proves that there exists an operation* D_N^* *such that* $D_N^* abu\downarrow$ *and if* $u \in Nat$, $v \in Nat$

$$D_N^* abuv = \begin{cases} a, & \text{if } u = v, \\ b, & \text{else} \end{cases}$$

(3) ***COST***$^-$ *proves that there exists a total function* ***Test***, *such that* $\textbf{\textit{Test}}\, xyz\downarrow$ *and*

$$\textbf{\textit{Test}}\, xyz = \begin{cases} x, & \text{if } \mathcal{U}(z); \\ y, & \text{if } \mathcal{S}(z) \end{cases} \tag{3.3}$$

REMARK 3.8. We cannot use D_N^* in place of the operator D of axiom (e) and (h) of definition 2.2, because we cannot see how to prove $D_N^* \in Ur$, i.e. the applicative axiom (h) of the same subsection.

In the following we shall make use of the usual notation $\bigcup$ for the uniform operation of union, un, and write $\cup$ for the obvious definition of a uniform version of binary union.

REMARK 3.9 (*On recursive definitions*). Standard recursive set theoretic definitions can be rendered quite smoothly and uniformly by means of the applicative structure. Let us consider two examples. By uniform union, image, lemma 3.7 and the second recursion theorem, there exists an operation ***TC*** of *transitive closure* such that

$$\textbf{\textit{TC}}\, x \simeq \begin{cases} \emptyset, & \text{if } \mathcal{U}(x); \\ x \cup \bigcup\{\textbf{\textit{TC}}\, y : y \in x\}, & \text{if } \mathcal{S}(x) \end{cases}$$

Of course $\in$–induction implies

$$\forall x\, (\mathcal{S}(x) \rightarrow \textbf{\textit{TC}}\, x\downarrow \wedge \mathcal{S}(\textbf{\textit{TC}}\, x)).$$

A similar argument allows us to define the *support operation*, which singles out the urelements which contribute to the build up of a set. In fact, it is possible to find an operation ***sp*** such that

$$\textbf{\textit{sp}}\, x = \begin{cases} \{x\}, & \text{if } \mathcal{U}(x); \\ \bigcup_{y \in x} \textbf{\textit{sp}}\, y, & \text{if } \mathcal{S}(x) \end{cases}$$

REMARK 3.10. In $\textbf{\textit{COST}}_b$ we have two notions of natural number: on the one hand, *the set Nat* of natural numbers (as urelements) on which we are allowed to induct only for bounded conditions; on the other hand, the *class*

ω of finite ordinals, which is inductively defined and to which we can lift full $\in$–induction.[8]

3.1. Refuting extensionality and totality of operations. The combination of operations and sets needs to be accomplished with care. The following arguments show that totality and extensionality can not be assumed in general (even if they have been required to hold for the urelements). We also show that separation can not be extended to arbitrary Δ_0 formulas.

First of all, let us recall a (folklore) fact about *partial combinatory algebras* (pcas for short). By a pca we understand a non-empty set endowed with a partial binary function (i.e. application) and two special elements K and S satisfying the standard axioms for combinators (see definition 2.2). A pca is *extensional* if it satisfies

$$\forall x\,(fx \simeq gx) \rightarrow f = g. \tag{3.5}$$

LEMMA 3.11. *Extensional* pcas *satisfy the fixed point property for total operations*: *if g is a total operation, then for some e, $ge = e$.*

PROOF. By lemma 3.1, there exists e such that $e \downarrow$ and $\forall x(ex \simeq gex)$. By totality of g, also $ge \downarrow$. Hence by (3.5), $e = ge$. ⊣

PROPOSITION 3.12. ***COST***$^-$ *refutes extensionality for operations* (3.5).

PROOF. By (3.3), there exists a *total* operation f satisfying

$$fx = \begin{cases} \emptyset, & \text{if } \mathcal{U}(x); \\ 0, & \text{if } \mathcal{S}(x). \end{cases}$$

Clearly f has no fixed point by the axioms of definition 2.2 (1)(a)-(c), (2)(c) and (4)(a). Hence extensionality for operations must fail. ⊣

PROPOSITION 3.13. ***COST***$^-$ *refutes totality of application for operations, i.e. it proves $\neg\forall x\,\forall y\,\exists z\,App(x, y, z)$.*

---

[8] What is the relation between *Nat* and ω? Of course, we can define (recursion theorem, definition by cases on *Nat*) an operation ι such that

$$\iota x = \begin{cases} \emptyset, & \text{if } x = 0; \\ \iota y \cup \{\iota y\}, & \text{if } x = SUCy. \end{cases} \tag{3.4}$$

Observe however that, even if we can prove that ι is surjective, we cannot prove that ι is *total* and *injective* on *Nat*: we ought to use an instance of *Nat*–induction which is only allowed in ***COST***, but not in ***COST***$_b$; on the other hand in ***COST*** *Nat* is isomorphic to ω via ι and this also ensures that ω is a set. Conversely, if we try to embed ω into *Nat* within ***COST***$_b$, we cannot expect to find an operation, since we need a test on ω, and ω is not definable by bounded formulas unless we have a suitable axiom (of infinity). Instead, we can proceed by usual set recursion and find a class function embedding ω into *Nat*. Of course, that the map is *surjective* only follows by a form of *Nat*–induction, which is not available in ***COST***$_b$ (here *Nat* plays the role of a non-standard copy of the natural numbers, ω being standard).

PROOF. Using (3.2), we can define g such that

$$gx = \begin{cases} \top, & \text{if } x = \bot; \\ \bot, & \text{else.} \end{cases}$$

But, if application is total there is e such that $ge = e$. Then

$$\begin{aligned} e = \bot &\Rightarrow ge = e = \top, \\ &\Rightarrow e \neq \bot; \text{ but} \\ e \neq \bot &\Rightarrow ge = e = \bot. \end{aligned}$$

Hence $e \neq \bot \ \Leftrightarrow e = \bot$: contradiction. ⊣

REMARK 3.14. Propositions 3.12-3.13 correspond to a structural property of pcas: *every pca with discrete universe is non-extensional and strictly partial.* (That the universe is discrete means that the algebra contains an operation testing equality for elements, as in lemma 3.7).

PROPOSITION 3.15. ***COST***$^-$ *with uniform separation for conditions containing* $\simeq$ *proves* $\bot$.

PROOF. By uniform separation including $\simeq$-conditions, there would exist a total operation g such that

$$gfz = \{x \in \top \mid fz \simeq x\}. \tag{3.6}$$

By lemma 3.1 (second recursion theorem), there exists some e such that $gez \simeq ez$. Since g is total, e is total and hence $ee\downarrow$ and is a set satisfying $ee = \{x \in \top : ee = x\}$. Note that ee must be empty and hence by extensionality, $ee = \emptyset$. Then $ee \in ee$: contradiction!. ⊣

REMARK 3.16. Observe that $\in$*-induction is not used* in the proof above, but only the axiom on the empty set. The argument above was originally used by Gordeev in order to refute class extensionality in Feferman's systems of Explicit Mathematics (see [7]); for the role of uniformity and refinements, see [12].

§4. How to use *COST*.

Although ***COST*** and its subsystems have here been introduced for different reasons, we have already observed that we can take advantage of these theories for expressing in a natural way metatheoretical constructions, e.g. we can recast Rathjen's *formulas-as-classes interpretation* [31].

Already Beeson observed the link between his version of intuitionistic set theory with rules and a variant of set recursion (see Beeson [8] and Normann [29]). In [31] Rathjen introduces a form of extended set recursion (inspired by [26]), named $E_{\mathcal{P}}$–computability. According to this form of set recursion, exponentiation is taken as one of the basic operations which are used to define set computability. Therefore, for a and b sets, ${}^a b$ is computable. This notion of set recursion is then used by Rathjen to develop an inner model for ***CZF*** extended by constructively acceptable choice principles within a variant of ***CZF*** itself with *only* exponentiation in place of subset collection.[9]

We now show that we can capture $E_{\mathcal{P}}$–computability quite naturally in our (much weaker) theory $\boldsymbol{COST}_b$.

LEMMA 4.1 ($\boldsymbol{COST}^-$). *There exist operations* ***int***, ***prod***, ***dom***, ***ran***, ***opair***, $\boldsymbol{proj}_i$ $(i = 0, 1)$, *satisfying* (*in the given order*):

(i) *if a is inhabited, then **int** a is a set such that*

$$\boldsymbol{int}\, a = \bigcap a;$$

(ii) *if F is a set theoretic function, **dom** F, **ran** F are sets such that **dom** F = $Dom(F)$, **ran** F = $Ran(F)$;*

(iii) ***opair*** $xy = \langle x, y\rangle = \{\{x\}, \{x, y\}\}$ *and, if x, y are sets,*

$$\boldsymbol{proj_0}(\langle x, y\rangle) = x \wedge \boldsymbol{proj_1}(\langle x, y\rangle) = y;$$

(iv) *if a, b are sets, so is **prod** $ab := a \times b$, the cartesian product of a, b.*

PROOF. Apply the uniform versions of union, bounded separation, pairing and direct image. In particular, we can choose

$$\boldsymbol{int}\, a := \Big\{x \in \bigcup a : \forall v \in a\,(x \in v)\Big\};$$

$$\boldsymbol{dom}\ F := \Big\{x \in \bigcup\bigcup F : (\exists y \in \bigcup\bigcup F)(\langle x, y\rangle \in F)\Big\};$$

$$\boldsymbol{ran}\, F := \Big\{y \in \bigcup\bigcup F : (\exists x \in \bigcup\bigcup F)(\langle x, y\rangle \in F)\Big\}.$$

Concerning projections[10], let

$$\boldsymbol{proj_0} z := (z)_0 = \bigcup\bigcap z;$$

$$\boldsymbol{proj_1} z := (z)_1 = \Big\{w \in \bigcup\bigcup z : \forall y \in \bigcup z\,(z = \langle (z)_0, y\rangle \to w \in y)\Big\}.$$

The cartesian product is defined by

$$a \times b := \bigcup \mathsf{im}\,(\lambda y.\mathsf{im}\,(\lambda x.\,\boldsymbol{opair}\,xy)a)b. \qquad \dashv$$

[9]See Rathjen and Tupailo [32] for further metamathematical applications of the formula-as-classes interpretation like, for example, to a study of the existence property for ***CZF*** with constructive choice principles.

[10]The projections we define below are well-behaved only on sets. Of course, it is possible to define a non-set theoretic ordered pairing operation with projections over the whole universe by means of the applicative structure; e.g. choose $\mathsf{P}xy := \lambda u.uxy$, $\mathsf{P}_0 z := zK$, $\mathsf{P}_1 z := zK\mathsf{I}$, where I is the term $\lambda x.x$.

DEFINITION 4.2. (i) Let

$$\Pi_{x\in a}b_x := \left\{F : a \to \bigcup_{x\in a} b_x : \forall x \in a\,(F(x) \in b_x)\right\};$$

$$\Sigma_{x\in a}b_x := \left\{\langle x, y\rangle \in a \times \bigcup_{x\in a} b_x : x \in a \wedge y \in b_x\right\}.$$

(ii) Let $a \dotplus b := \Sigma_{x\in \mathbf{2}}C_x$, with $\mathbf{2} := \{0, 1\}$, $C_0 = a$ and $C_1 = b$.

PROPOSITION 4.3 ($\boldsymbol{COST}_b$). *There exist application terms satisfying the following schemata*:

(a) $Kxy \simeq x$;
(b) $Sxyz \simeq xz(yz)$;
(c) $\boldsymbol{opair}\,xy \simeq \langle x, y\rangle$;
(d) $\boldsymbol{proj_0}x \simeq (x)_0$;
(e) $\boldsymbol{proj_1}x \simeq (x)_1$;
(f) $SUCn \simeq n + 1$;
(g) $PR(n + 1) \simeq n$;
(h) $Dxynn \simeq x$;
(i) $Dxynm \simeq y$, *provided* $n \neq m$;
(j) $\overline{0}x \simeq 0$;
(k) $\overline{\omega}x \simeq Nat$;
(l) $\overline{\boldsymbol{fa}}\,Gx \simeq G(x)$, *if* $Fun(G)$ *and* $x \in Dom(G)$;
(m) $\overline{\boldsymbol{ab}}\,ea \simeq H$, *if* $Fun(H)$, $Dom(H) = a$ *and* $\forall u \in a\,(eu \simeq H(u))$;
(n) $\sigma aG \simeq \Sigma_{z\in a}G(z)$, *if* $Fun(G)$ *and* $Dom(G) = a$;
(o) $\pi aG \simeq \Pi_{z\in a}G(z)$, *if* $Fun(G)$ *and* $Dom(G) = a$;
(p) $\overline{pl}\,xy \simeq x \dotplus y$;
(q) $Ixyz \simeq \{e \in \top : x = y \wedge x, y \in z\}$.

PROOF. (a)-(k), (p)-(q) are either immediate or follow directly from lemma 4.1.

(l) Function application can be expressed as $\bigcup\{y \in Ran(G) : \langle x, y\rangle \in G\}$ and hence as an operation $\overline{\boldsymbol{fa}}Gx$ of G and x (apply uniform union, pairing and separation).

(m) The graph of the operation e on the set a can be expressed as $\{z \in a \times \mathsf{im}\, ea : \mathsf{eq}(e(z)_0)(z)_1 \simeq \top\}$ and hence as an operation of e and a (apply the lemma, uniform direct image and bounded separation, see remark 3.5).

(n) σaG is defined by the application term

$$\{u \in a \times c : \exists x \in a\, \exists y \in c(u = \langle x, y\rangle \in G)\}$$

where $c = \bigcup(\mathsf{im}\ (\overline{\boldsymbol{fa}}G)a)$ and $G : a \to \boldsymbol{V}$.

(o) It is enough to choose the operation $\pi a G$ specified by the application term

$$\Big\{F : F \in \exp a\Big(\bigcup(\mathrm{im}\,(\overline{\boldsymbol{fa}}G)a)\Big) \wedge \forall z\big(z \in F \rightarrow z \in \sigma(\overline{\boldsymbol{fa}}G)a\big)\Big\}.$$

Then apply remark 3.5, uniform direct image and exponentiation and the previous point. ⊣

Application terms generated according to the equations of 4.3 define the partial $E_{\mathcal{P}}$-recursive functions of [31].

Then it is possible to prove the following version of Rathjen's theorems 4.13, 4.26, 4.31 in [31], with $\boldsymbol{V}(\boldsymbol{Y}^*)$ defined as in [31, Definition 3.1] and representing an internal analogue of Aczel's $\boldsymbol{V}$ type.

THEOREM 4.4. *If* $\boldsymbol{CZF}$[11] $+ \Pi\Sigma I - AC + \boldsymbol{RDC} \vdash \varphi$, *then we can find g such that*:

(i) $\boldsymbol{COST} \vdash \forall \vec{x} \in \boldsymbol{V}(\boldsymbol{Y}^*)\,(g\vec{x} \in \|\varphi\|_{\vec{x}})$;
(ii) $\|\varphi\|_{\vec{x}}$ *is a class which is a set in* $\boldsymbol{V}(\boldsymbol{Y}^*)$ *whenever* φ *is bounded*;
(iii) g *is* $E_{\mathcal{P}}$*-recursive.*

$\Pi\Sigma I - AC$ is the axiom of choice in the form: every set belonging to the smallest class closed under Π, Σ and I, has a base (for undefined notions, we refer the reader to [31]). Recall that $\boldsymbol{RDC}$, the schema of *relativized dependent choice* has the form:

$$(\forall x)(\psi(x) \rightarrow (\exists y)(\psi(y) \wedge \varphi(x,y))) \rightarrow (\forall x_0)(\psi(x_0) \rightarrow (\exists F)(F(0) = \\ = x_0 \wedge (\forall n \in Nat)(\psi(F(n)) \wedge \varphi(F(n), F(n+1)))))$$

for every formula ψ, φ.

We again stress that theorem 4.4 does not require the scheme of subset collection, but only exponentiation. Moreover, strong collection is only applied for the inductive definition of $\boldsymbol{V}(\boldsymbol{Y}^*)$.

§5. Operations vs. functions.

In $\boldsymbol{COST}$ we have set theoretic functions and operations. The question naturally arises of which relationship exists between them. According to 4.3, to each set theoretic function F with empty support[12] there corresponds an operation $\lambda x.(\overline{\boldsymbol{fa}}F)x$ which coincides with F on the common domain.

We can even consistently achieve a sort of "harmony" between functions and operations by assuming Beeson's axiom, $\boldsymbol{FO}$ (see [8]). $\boldsymbol{FO}$ asserts that every set theoretic function is an operation, that is

$$(\boldsymbol{FO}) \quad \forall z(Fun(z) \wedge Dom(z){=}a \wedge Ran(z) \subseteq b \rightarrow \forall x \in a \exists y \in b zx \simeq y).$$

[11] We leave the case of $\boldsymbol{CZF}_b^{op}$ for further investigations.

[12] F is a pure set; else the definition of $\overline{\boldsymbol{fa}}$ does not work.

Clearly, ***FO*** implies that every element of the set $\mathsf{exp}\,ab$ is an operation from a to b, that is

$$f \in \mathsf{exp}\,ab \rightarrow \forall x \in a\, \exists y \in b\, (fx \simeq y).$$

One might now wonder if it is consistent to assume the existence of the set

$$\mathsf{op}ab := \{\, f : \forall x \in a\, \exists y \in b\, (fx \simeq y)\}$$

of *all operations* from a to b. Pierluigi Minari has observed that, if $\mathsf{op}ab$ is a set, then one can reproduce the fixed point argument of proposition 3.15:

LEMMA 5.1. $\boldsymbol{COST}^{-} + \forall a \forall b\, \exists c (\mathsf{op}\,ab = c)$ *is inconsistent.*

PROOF. Assume $\mathsf{op}ab$ is a set, uniformly in a, b. By pairing and separation, there exists a total operation g such that

$$gfz = \{x \in \top : fz \simeq x\} = \{x \in \top : f \in \mathsf{op}\,\{z\}\{x\}\}. \tag{5.1}$$

Exactly as in proposition 3.15, there exists e satisfying $ee = \{x \in \top : ee = x\}$ and a contradiction arises.[13] ⊣

Even though we cannot assume the existence a set of all operations from a set a to a set b, we know that to each f such that $\forall x \in a\, \exists y \in b\, (fx \simeq y)$ we can associate a set theoretic function, say $F : a \rightarrow b$ such that $\forall x \in a\, F(x) \simeq fx$. That is, $\mathsf{exp}\,ab$ contains a representative of each operation $f : a \rightarrow b$. More generally, we can prove:

LEMMA 5.2 (Function existence, in $\boldsymbol{COST}^{-}$).

$$\forall x \in a \exists! y\, \varphi(x, y) \rightarrow \exists! F (Fun(F) \wedge \forall x \in a\, \varphi(x, F(x))).$$

PROOF. The proof makes essential use of replacement. ⊣

If in addition we allow for ***FO***, then we can state the existence of an operation:

PROPOSITION 5.3 (Operation existence). $\boldsymbol{COST}^{-} + \boldsymbol{FO}$ *proves for arbitrary* φ:

$$\forall x \in a\, \exists! y\, \varphi(x, y) \rightarrow \exists f\, \forall x \in a\, \varphi(x, fx)$$

PROOF. Simply use lemma 5.2 and ***FO***. ⊣

This shows that locally (on a given set) each implicitly defined function is witnessed by an element of the underlying combinatory structure.

---

[13]An argument by the present authors, which does not rely upon self-application, but uses ***FO*** is based on the idea that $\mathsf{op}\,ab$ is too large to be a set. Indeed, consider the collection $C(x, y)$ of all functions F satisfying $F(x) = y$: this is provably not a set. But ***FO*** implies that $C(x, y)$ is a subcollection of $\mathsf{op}\{x\}\{y\}$ and hence a set, if $\mathsf{op}\{x\}\{y\}$ is a set: absurd.

5.1. Appendix: choice for operations? The axiom of choice in the usual form[14] is not constructively acceptable, since it implies the law of excluded middle by a well known argument (see [15]).

The situation does not change substantially if we consider forms of *the axiom of choice for operations* in the form ***OAC***

$$\forall x \in a\, \exists y\, \varphi(x, y) \to \exists f\, \forall x \in a\, \varphi(x, fx), \tag{5.2}$$

or in the generalized class form ***GAC***:

$$\forall x\, (\varphi(x) \to \exists y\, \psi(x, y)) \to \exists f\, \forall x\, (\varphi(x) \to \psi(x, fx)) \tag{5.3}$$

GAC! is ***GAC*** with the uniqueness restriction on the quantifier $\exists y$ in the antecedent of (5.3).

LEMMA 5.4. *$\boldsymbol{COST}^-$ + $\boldsymbol{OAC}$ proves $\varphi \vee \neg\varphi$ for arbitrary bounded formulas. Moreover, $\boldsymbol{COST}^-$ + $\boldsymbol{GAC}$ and $\boldsymbol{COST}^-$ + $\boldsymbol{GAC}$! are inconsistent.*

PROOF. The argument by Goodman and Myhill - as given for example in [6] - goes through (almost) literally unchanged and justifies the first claim. Concerning the inconsistencies, as ***GAC*** trivially implies ***GAC***!, it is enough to refute ***GAC***!. Following Beeson [7, p. 228], let $\gamma(x, y)$ be the formula

$$(xx = 0 \to y = 1) \wedge (xx \in Nat \wedge xx \neq 0 \to y = 0) \wedge$$
$$\wedge\, (xx \notin Nat \to y = 0) \wedge y \in Nat$$

Then one can verify, for $\delta(x) := \exists y\, \gamma(x, y)$:

$$\forall x(\delta(x) \to \exists!\, y \in Nat\, \gamma(x, y))$$

Hence by ***GAC***! there is an operation f such that

$$\forall x(\delta(x) \to (fx \in Nat \wedge \gamma(x, fx))). \tag{5.4}$$

Then one shows $\delta(f) \leftrightarrow \neg\delta(f)$: contradiction. ⊣

Note that $\exists$ occurs in δ. Both arguments use decidability of equality on *Nat*; on the other hand, while Diaconescu's argument relies upon extensionality, pairing and bounded separation, the tricky proof of the lemma hinges upon the combinatory structure and exploits self-application of operations.

§6. Assigning a combinatory structure to the universe of constructive sets.

Informally, the content of this section can be summarized as follows: we start with a constructive universe of sets, which has a set of urelements endowed with a structure of combinatory algebra. We then show how to lift the applicative structure to the whole universe. By the results of 3.1, we cannot hope to get a total application relation: however, the argument turns out to be constructive and susceptible of being rephrased as a suitable interpretation theorem.

[14]If $\{F_x\}_{x \in a}$ is a family of inhabited sets indexed by a, there exists a function G defined on a such that $G(x) \in F_x$.

Formally speaking, we first describe a system $\boldsymbol{CZF^{op}}$ (respectively $\boldsymbol{CZF_b^{op}}$), which roughly describes a universe of sets on top of a ground combinatory algebra. Even if this is not necessary, we assume that *the underlying combinatory algebra of urelements is total and also extensional*, so that the final conservation result will be stronger. Then we build a model for $\boldsymbol{COST}$ ($\boldsymbol{COST_b}$) in the system $\boldsymbol{CZF^{op}}$ (respectively $\boldsymbol{CZF_b^{op}}$).

6.1. Step 1: the theory $\boldsymbol{CZF^{op}}$.

6.1.1. *The language of $\boldsymbol{CZF^{op}}$.* The language of $\boldsymbol{CZF^{op}}$ is the applicative extension, $\mathcal{L}^{op}$, of the usual first order language of Zermelo Fraenkel set theory, $\mathcal{L}$, by the individual constants K, S, $PAIR$, $LEFT$, $RIGHT$, 0, SUC, PR, D, the unary predicates $\mathcal{S}, \mathcal{N}, \mathcal{U}$ and the ternary application symbol App, now to be understood *as restricted to urelements*. Terms and formulas are defined as usual. We still use the notation of section 2.1 for application terms (which will now denote urelements). We also assume pairing with projections on urelements as primitive; thus the expressions $(t, s) := PAIRts$, $(t)_1 := RIGHTt$, $(t)_0 := LEFTt$ stand for the application terms designating the ordered pair composed by t and s, together with its left and right projections.

As for $\mathcal{L}^R$ also in this case we make use of the same conventions for variables for sets, natural numbers and urelements. (Having no operation on sets in $\boldsymbol{CZF^{op}}$, we drop for this system the convention of using capital letters for functions).

6.1.2. *The system $\boldsymbol{CZF^{op}}$.*

DEFINITION 6.1. The system $\boldsymbol{CZF^{op}}$ is the $\mathcal{L}^{op}$-theory, whose axioms are the usual axioms for intuitionistic first order logic *with equality* together with the following applicative and set theoretic axioms and axiom schemes:

(1) *Ontological axioms and extensionality*: as in definition 2.2, group (1);
(2) *Combinators and natural numbers*: as in definition 2.2, group (2), except that now all basic combinatory axioms are relativized to $\mathcal{U}$; in addition we have strictness and pairing axioms:

$$App(x, y, z) \to \mathcal{U}(x) \wedge \mathcal{U}(y) \wedge \mathcal{U}(z);$$

$$\mathcal{U}(x) \wedge \mathcal{U}(y) \to (x, y)_0 = x \wedge (x, y)_1 = y;$$

$$\mathcal{U}(PAIR) \wedge \mathcal{U}(RIGHT) \wedge \mathcal{U}(LEFT).$$

(3) *Set existence axioms*: non-uniform versions of empty set, pairing, union, bounded separation, existence of the set of the urelements:
   (a) *Emptyset*

$$\exists c\, \forall y\, \neg(y \in c)$$

   (b) *Pair*

$$\exists c\, (y \in c \wedge z \in c)$$

   (c) *Union*

$$\exists c\, \forall y\, (y \in c \leftrightarrow \exists z \in a\, (y \in z))$$

(d) *Infinity*

$$\exists c\, \forall y\, (y \in c \leftrightarrow \mathcal{U}(y))$$

(e) Δ_0 - *Separation scheme*. For every Δ_0 formula $\varphi(y)$,[15]

$$\exists c\, \forall y\, (y \in c \leftrightarrow y \in a \wedge \varphi(y))$$

(4) *Subset Collection-Strong Collection*: as in definition 2.2, group (6);
(5) *Induction*: the schemata of $\mathcal{N}$–induction and $\in$–induction, as in definition 2.2, group (5).

Depending on the amount of induction one allows for sets and for natural numbers one obtains various subsystems of $\boldsymbol{CZF}^{op}$. In the following we shall be interested in the system $\boldsymbol{CZF}^{op}_b$, or *bounded* $\boldsymbol{CZF}^{op}$.

DEFINITION 6.2. $\boldsymbol{CZF}^{op}_b$ is the system obtained from $\boldsymbol{CZF}^{op}$ by replacing the induction scheme on $\mathcal{N}$ by the induction axiom, i.e. the following principle

$$0 \in a \wedge \forall n(n \in a \rightarrow SUCn \in a) \rightarrow \forall n(n \in a).$$

6.2. Step 2: the interpretation theorem. We now prove:

THEOREM 6.3. *$\boldsymbol{COST}_b$ is interpretable in $\boldsymbol{CZF}^{op}_b$ and $\boldsymbol{COST}$ is interpretable in $\boldsymbol{CZF}^{op}$.*

PROOF. We work informally in $\boldsymbol{CZF}^{op}_b$ thus providing a sketch of the proof in the case of the weaker systems. It should be clear how to obtain a proof for the stronger systems. First of all, we interpret the basic predicates $\mathcal{U}$, $\mathcal{N}$, $\mathcal{S}$, $=$ onto themselves and we associate suitable terms $c^\diamond$ in $\boldsymbol{CZF}^{op}_b$ to the basic constants c of $\boldsymbol{COST}_b$. We choose $c{=}c^\diamond$ if c is a constant among K, S, 0, SUC, PR, D. As to the remaining constants c for set operations, the associated terms $c^\diamond$ of $\boldsymbol{CZF}^{op}_b$ are fixed provably pairwise distinct elements (for instance, numerals corresponding to the odd numbers starting with, say, 5, so that el $\equiv$ 5, etc ... [16]). The interpretation of the set constants of $\boldsymbol{COST}_b$ and their axioms is straightforward by making use of the empty set axiom, the infinity axiom, bounded separation (which allows for application on urelements) and the existence of the cartesian product in $\boldsymbol{CZF}^{op}_b$.

We then assign a combinatory structure to the universe of constructive sets; this is achieved by defining a global application relation as the least fixed point

$$\simeq^\infty = \Phi(\simeq^\infty) \tag{6.1}$$

of a suitable operator Φ, which embodies the clauses corresponding to (i) combinators, (ii) primitive predicate functions and (iii) set constructors.

[15] We stress that now bounded formulas φ may contain application.

[16] In 6.2 we adopt $t \equiv s$ as a notation for definitional equality.

The definition of $\simeq^\infty$ falls under Aczel's treatment of inductive definitions in ***CZF*** (see [5], [6, Chapter 8]). More precisely, $\simeq^\infty$ satisfies the equation

$$\simeq^\infty = \bigcup_a \simeq^a \tag{6.2}$$

where

$$\simeq^a = \Phi\Big(\bigcup_{x\in a} \simeq^x\Big) \tag{6.3}$$

As a consequence, the verification of the various applicative axioms amounts to applying (unbounded) instances of $\in$–induction. For instance, in order to check uniqueness for application, it is enough to verify, for arbitrary a,

$$\forall x\,\forall y\,\forall u\,\forall v\,(xy \simeq^a u \wedge xy \simeq^\infty v \to u = v) \tag{6.4}$$

under the induction hypothesis

$$\forall c \in a\,(\forall x\,\forall y\,\forall u\,\forall v\,(xy \simeq^c u \wedge xy \simeq^\infty v \to u = v)). \tag{6.5}$$

For the sake of clarity, whenever we specify each single clause $\mathcal{I}$ of Φ, we check (i) that $\mathcal{I}$ preserves uniqueness (6.4) under assumption (6.5); (ii) that the appropriate existence conditions are met.

The specification of all clauses (and of the formula defining Φ) is very tedious, so that we give here only some examples. First of all, when we deal with operations under which $\mathcal{U}$ is closed, we make $\simeq^\infty$ coincide with application on the ground algebra:

($\sharp$) *Extension clause*:

$$x \in \mathcal{U} \wedge y \in \mathcal{U} \wedge z \in \mathcal{U} \wedge App(x,y,z) \Rightarrow xy \simeq^\infty z$$

This takes care of the cases of SUC, PR, D and of the behaviour of K and S on $\mathcal{U}$. Uniqueness and existence conditions are trivially met.

Then we have to extend K, S in the case of arguments which are not urelements; in this case we make use of set theoretic pairing in order to choose values Kx, Sx, Sxy, namely:

(1) $y \notin \mathcal{U} \Rightarrow Ky \simeq^\infty \langle K, y\rangle \equiv K_y \wedge K_y x \simeq^\infty y$;
(2) $y \in \mathcal{U} \wedge Ky \equiv K_y \wedge x \notin \mathcal{U} \Rightarrow (K_y)x \simeq^\infty y$[17];
(3) $x \notin \mathcal{U} \Rightarrow Sx \simeq^\infty \langle S, x\rangle \equiv S_x \wedge S_x y \simeq^\infty \langle S_x, y\rangle \equiv S_{xy}$;
(4) $x \in \mathcal{U} \wedge Sx \equiv S_x \wedge y \notin \mathcal{U} \Rightarrow S_x y \simeq^\infty \langle S_x, y\rangle \equiv S_{xy}$;
(5) $xz \simeq^\infty u \wedge yz \simeq^\infty v \wedge uv \simeq^\infty w \Rightarrow S_{xy} z \simeq^\infty w$.

Uniqueness and existence are obvious for all items, except for uniqueness in the last case. Assume

$$Sxyz \simeq^a p_1 \wedge Sxyz \simeq^\infty p_2.$$

[17] Note that in this case K_y is the unique z such that $App(K,y,z)$; the case $x \in \mathcal{U}$ is taken care of by the extension clause.

By (6.3) and (6.1), there are $i \in a$, $j \in a$, $k \in a$, u_1, u_2, v_1, v_2 such that

$$xz \simeq^i u_1 \wedge yz \simeq^j v_1 \wedge u_1 v_1 \simeq^k p_1;$$
$$xz \simeq^\infty u_2 \wedge yz \simeq^\infty v_2 \wedge u_2 v_2 \simeq^\infty p_2.$$

By induction hypothesis (6.5), $u_1 = u_2$, $v_1 = v_2$, whence also $p_1 = p_2$.

We are then left with clauses for el, pair, un, im, sep, exp. Again we explicitly give only a few instances.

(a) *Membership operation:*

(i) $\mathsf{el}a \simeq^\infty \langle \mathsf{el}, a \rangle \equiv \mathsf{el}_a$;

(ii) $\dfrac{\forall u\,(u \in z \leftrightarrow u = \bot \wedge x \in y)}{\mathsf{el}_x y \simeq^\infty z}$.

Uniqueness: trivial by extensionality.

Existence: by bounded separation, $\{u \in \top : x \in y\}$ is a set.

(b) *Set theoretic constructors*

◇ Bounded Separation:

(i) $\mathsf{sep} f \simeq^\infty \langle \mathsf{sep}, f \rangle \equiv \mathsf{sep}_f$;

(ii) from

$$f : a \to \Omega$$
$$\forall u\,(u \in z \to fu \simeq^\infty \top \wedge u \in a)$$
$$\forall u \in a\,(\forall y\,(y \simeq^\infty fu \to y = \top) \to u \in z)$$

infer $\mathsf{sep}_f a \simeq^\infty z$.

Uniqueness is easy to check. To see that $\mathsf{sep} f a \downarrow$, choose

$$\mathsf{sep} f a = \{x \in a : \langle x, y \rangle \in F_f\}$$

where F_f is the set corresponding to the graph of f. This set exists since by assumption $(\forall x \in a)(\exists! y \subseteq \top)(fx \simeq^\infty y)$ and hence we can bound $\exists y$ by replacement which is a consequence of the schema of strong collection of $\boldsymbol{CZF}_b^{op}$.[18]

◇ Direct Image:

(i) $\mathsf{im} f \simeq^\infty \langle \mathsf{im}, f \rangle \equiv \mathsf{im}_f$;

(ii) from

$$f : a \to V$$
$$\forall x \in a\, \exists y \in z\,(fx \simeq^\infty y)$$
$$\forall y \in z\, \exists x \in a\,(fx \simeq^\infty y)$$

infer $\mathsf{im}_f a \simeq^\infty z$.

---

[18]We stress that the instance of replacement we apply in this step is not bounded, as it refers to the unbounded condition defining application. A similar definition would not be allowed in admissible set theory.

To see that $\mathsf{im} fa \downarrow$ and it is a set, choose

$$\mathsf{im} fa = \{y : \exists x \in a\,(fx \simeq^{\infty} y)\}$$

and apply replacement.
Uniqueness: assume

(6.6) $$\mathsf{im}_f a \simeq^u z_1 \wedge \mathsf{im}_f a \simeq^{\infty} z_2.$$

Then by (6.3) and the first member of (6.6):

(6.7) $$\forall x \in a\, \exists y \in z_1\, \exists i \in u\,(y \simeq^i fx);$$

(6.8) $$\forall y \in z_1\, \exists x \in a\, \exists j \in u(y \simeq^j fx).$$

By (6.1) and the second member of (6.6):

(6.9) $$\forall x \in a\, \exists y \in z_2\,(y \simeq^{\infty} fx);$$

(6.10) $$\forall y \in z_2\, \exists x \in a\,(y \simeq^{\infty} fx).$$

We now check $z_1 \subseteq z_2$. Let $y_1 \in z_1$. Then by (6.8), for some $x \in a$ and some $j \in u$, $y_1 \simeq^j fx$, whence there exists $y_2 \in z_2$ such that $y_2 \simeq^{\infty} fx$ (by (6.9)). By induction hypothesis it follows that $y_1 = y_2$; so $y_1 \in z_2$.
We prove $z_2 \subseteq z_1$. Let $y_2 \in z_2$. Then by (6.10), for some $x \in a$, $y_2 \simeq^{\infty} fx$, whence there exist $i \in u$ and $y_1 \in z_1$ such that $y_1 \simeq^i fx$ (by (6.7)). By induction hypothesis it follows that $y_1 = y_2$; so $y_2 \in z_1$.
Therefore we have proved that $z_1 = z_2$.

◇ Exponentiation:
(i) $\mathsf{exp} a \simeq^{\infty} \langle \mathsf{exp}, a \rangle \equiv \mathsf{exp}_a$;
(ii) $$\frac{\forall f\,(f \in z \leftrightarrow \forall x \in a\, \exists! y \in b\,(\langle x, y \rangle \in f))}{\mathsf{exp}_a b \simeq^{\infty} z}.$$
Uniqueness: trivial by extensionality
Existence: apply exponentiation, which is a consequence of subset collection.

We finally must check that restricted *Nat*–induction is validated by our interpretation, provably in $\boldsymbol{CZF}_b^{op}$, i.e. by using number theoretic induction only for sets. Assume that $f : Nat \to \Omega$: by the function existence lemma 5.2, there exists a function F which is pointwise equal with f on the set Nat. By separation, there exists a set

$$b = \{x \in Nat : \langle x, \top \rangle \in F\}$$

But by assumption on f, b contains 0 and is closed under successor; hence $Nat \subseteq b$. ⊣

We now observe that also the principle ***FO*** from section 5 can be interpreted in $\boldsymbol{CZF}_b^{op}$.

THEOREM 6.4. *$\boldsymbol{COST}_b + \boldsymbol{FO}$ is interpretable in $\boldsymbol{CZF}_b^{op}$.*

PROOF. It is enough to extend the inductive clauses for App with the following rule

$$\frac{Fun(F) \qquad \langle x, y\rangle \in F}{Fx \simeq y}. \qquad \dashv$$

REMARK 6.5. If we omit subset collection from $\boldsymbol{COST}_b$ and $\boldsymbol{COST}$ the interpretation theorem 6.3 still works for the corresponding subsystems of $\boldsymbol{CZF}_b^{op}$ and $\boldsymbol{CZF}^{op}$ without subset collection; this is because the crucial theorem on inductively defined classes requires only the strong collection schema.

§7. Proof theoretic reduction of $\boldsymbol{CZF}_b^{op}$ to $\boldsymbol{PA}$: the auxiliary theory. We now come to the third part of the paper where we investigate the proof theoretic strength of the theory $\boldsymbol{CZF}_b^{op}$. The strength of the theory will be determined by interpreting it in the auxiliary theory $\boldsymbol{T_c}$ to be introduced below. This can be regarded as a theory of partial (non-extensional) classes, which subsumes a classical version of the weakest system, EM_0, of Feferman's explicit mathematics (see [17, 7])[19]. It turns out that $\boldsymbol{T_c}$ has the same arithmetical content as $\boldsymbol{PA}$.

7.1. The theory $\boldsymbol{T_c}$. The basic first order language $\mathcal{L}_T$ of $\boldsymbol{T_c}$ is essentially the language of $\boldsymbol{CZF}^{op}$ without the predicates $\in$, $\mathcal{S}$, $\mathcal{U}$ and App but with an additional predicate $\mathcal{T}$ for truth, i.e. it comprises the predicate symbols $=$, $\mathcal{T}$, $\mathcal{N}$, the binary function symbol ap (application), combinators K, S, successor, predecessor, definition by cases on numbers, pairing with projections. Terms are inductively generated from variables and individual constants via application. As usual $ts := ap(t, s)$; missing brackets are restored by associating to the left. Formulas are inductively generated from atoms of the form $t = s$, $\mathcal{T}(t)$, $\mathcal{N}(t)$ by means of sentential operations and quantifiers. We adopt the following conventions:

(i) By $[\varphi]$ we denote a term representing the propositional function associated with φ and such that $\boldsymbol{FV}([\varphi]) = \boldsymbol{FV}(\varphi)$. We fix distinct closed *terms* $\hat{\forall}$, $\hat{\exists}$, $\hat{\neg}$, $\hat{\wedge}, \ldots$, naming the *logical constants*, and $\hat{=}$, $\hat{N}$ naming the equality predicate and the number predicate. Then $[\varphi]$ is inductively defined by stipulating $[t = s] = (\hat{=}\, ts)$, $[\mathcal{N}(s)] = \hat{\mathcal{N}}s$, $[\mathcal{T}(s)] = s$ and closing under application of the "small hat" operations, with the warning that $[\forall x\varphi] = \hat{\forall}(\lambda x[\varphi])$, $[\exists x\varphi] = \hat{\exists}(\lambda x[\varphi])$.

(ii) Given a formula φ we define abstraction by letting $\{x : \varphi\} := \lambda x.[\varphi]$.

---

[19]Alternatively, you may regard it as an abstract version of the Kripke-Feferman system for self-referential truth, based on combinatory logic (for additional information, see [11]); related systems go back to Aczel [1, 3].

(iii) We define intensional membership, η, as follows:

$$x\,\eta\,a := \mathcal{T}(ax);$$
$$x\,\bar{\eta}\,a := \mathcal{T}(\hat{\neg}(ax)).$$

(iv) The notion of class (or classification) is so specified:

$$\boldsymbol{Cl}(a) := \forall x\,(x\,\eta\,a \vee x\,\bar{\eta}\,a).$$

(v) A formula φ is *$\mathcal{T}$–positive* iff φ is inductively generated from prime formulas of the form $\mathcal{T}(t)$, $t = s$, $\neg t = s$, $\mathcal{N}(t)$, $\neg\mathcal{N}(t)$ by means of $\vee$, $\wedge$, $\forall$, $\exists$.

(vi) A formula φ is *$\mathcal{T}$–positive operative in v* (in short, a *positive operator*) iff φ belongs to the smallest class of formulas inductively generated from prime formulas of the form $\mathcal{T}(t)$, $s\,\eta\,v$, $t = s$, $\neg t = s$, $\mathcal{N}(t)$, $\neg\mathcal{N}(t)$ by means of $\vee$, $\wedge$, $\forall y$, $\exists y$, where y is distinct from v and v does not occur in t, s.

(vii) For each formula φ, fixed points are defined by letting:

$$\boldsymbol{I}(\varphi) := \boldsymbol{Y}(\lambda v.\{x : \varphi(x, v)\})$$

where $\boldsymbol{Y}$ is Curry's fixed point combinator.

The system $\boldsymbol{T_c}$ comprises, besides *classical* predicate calculus with equality:

(1) The base theory $\boldsymbol{TON}^-$ (see [33, 22]), which formalises the notion of total extensional combinatory algebra expanded with natural numbers. This includes the obvious axioms on combinators, pairing, projections. In addition, closure axioms for the predicate $\mathcal{N}$ defining a copy of the natural numbers, together with number theoretic conditions on the basic operations of successor SUC, predecessor $PRED$, 0, definition by cases on N.

(2) A fixed point axiom ($\boldsymbol{Tr}$) for abstract truth

$$\boldsymbol{Tr}(x, \mathcal{T}) \leftrightarrow \mathcal{T}(x);$$

here $\boldsymbol{Tr}(x, \mathcal{T})$ is a formula encoding the closure properties:

$$\frac{a = b}{\mathcal{T}[a = b]} \qquad \frac{\neg(a = b)}{\mathcal{T}[\neg(a = b)]} \qquad \frac{\mathcal{N}(a)}{\mathcal{T}[\mathcal{N}(a)]} \qquad \frac{\neg\mathcal{N}(a)}{\mathcal{T}[\neg\mathcal{N}(a)]}$$

for the basic atomic formulas with $=$ and $\mathcal{N}$, and the following additional clauses for the compound formulas:

$$\frac{\mathcal{T}(a)}{\mathcal{T}(\hat{\neg}\hat{\neg}a)} \qquad \frac{\mathcal{T}a \quad \mathcal{T}b}{\mathcal{T}(a\hat{\wedge}b)} \qquad \frac{\mathcal{T}(\hat{\neg}a)\ [\ or\ \mathcal{T}\hat{\neg}b]}{\mathcal{T}(\hat{\neg}(a\hat{\wedge}b))}$$

$$\frac{\forall x\,\mathcal{T}(ax)}{\mathcal{T}(\hat{\forall}a)} \qquad \frac{\exists x\,\mathcal{T}\hat{\neg}ax}{\mathcal{T}(\hat{\neg}\hat{\forall}a)}$$

(3) Consistency axiom: $\neg(\mathcal{T}x \wedge \mathcal{T}\hat{\neg}x)$.

(4) Induction on natural numbers $\mathcal{N}$ for *classes*:

$$\boldsymbol{Cl}(a) \wedge \boldsymbol{Clos}_{\mathcal{N}}(a) \rightarrow \forall x(\mathcal{N}(x) \rightarrow x\,\eta\,a)$$

with $\boldsymbol{Clos}_{\mathcal{N}}(a) := 0\eta a \wedge \forall x\,(x\eta a \rightarrow (SUCx)\eta a)$.

(5) the principle ***GID***, ensuring the minimality of the fixed points: if $\varphi(x, v)$ is a positive operator,

$$\boldsymbol{Clos}_{\varphi}(\psi) \rightarrow \forall x\,(x\eta \boldsymbol{I}(\varphi) \rightarrow \psi(x))$$

with $\boldsymbol{Clos}_{\varphi}(\psi) := \forall x\,(\varphi(x, \psi) \rightarrow \psi(x))$.[20]

By $\boldsymbol{T_f}$ we understand the theory which results from $\boldsymbol{T_c}$ by allowing full induction on natural numbers, i.e. for *arbitrary formulas*. $\boldsymbol{T}^-$ is the theory $\boldsymbol{T_c}$ without number theoretic induction.

Observe that $\boldsymbol{T_c}$ and $\boldsymbol{T_f}$ are classical theories and this seems necessary to make the interpretation below work.

Let ***CL*** be $\{x : \boldsymbol{Cl}(x)\}$ (which turns out to be provably not a class). Then we can show that ***CL*** has natural closure conditions (loosely analogous to those of the universe of small types in Martin-Löf's type theory); also the existence of least fixed points for inductive definitions is granted in $\boldsymbol{T_c}$.

Definition 7.1. If $\vec{a}$ is a finite list $a_0, \dots a_n$ of variables distinct from x, we say that $\varphi(x, \vec{a})$, where $FV(\varphi) \subseteq \vec{a}, x$, is *elementary in* $\vec{a}$, if φ is inductively generated from atomic formulas of the form $t = s, \mathcal{N}(t), t\,\eta\,a_i$, and their negations (with $0 \leq i \leq n$) by means of sentential connectives and quantification on variables not occurring in the list $\vec{a}$.

Proposition 7.2 ($\boldsymbol{T}^-$). (i) ***CL*** *is closed under elementary comprehension, generalized disjoint union* (Σ), *generalized disjoint product* (Π).[21] *Moreover* ***CL*** *contains* $V = \{x : x = x\}$ *and* $N = \{x : \mathcal{N}(x)\}$.

(ii) *If* φ *is* $\mathcal{T}$*–positive, then* $\mathcal{T}[\varphi] \leftrightarrow \varphi$ *and positive comprehension holds*

$$\forall x\,(x\eta\{u : \varphi\} \leftrightarrow \varphi[u := x])$$

(iii) *A version of the second recursion theorem holds. That is if* φ *is positive*

$$\forall x\,(x\eta \boldsymbol{I}(\varphi) \leftrightarrow \varphi(x, \boldsymbol{I}(\varphi))).$$

Proof. See [11], II.9B, II.10A. ⊣

Theorem 7.3. $\boldsymbol{T_c}$ ($\boldsymbol{T_f}$) *is proof theoretically equivalent to* ***PA*** ($\boldsymbol{ID_1}$[22]).

---

[20] $\varphi(x, \psi)$ is the formula obtained by replacing each occurrence of the formula $t\,\eta\,v$ in $\varphi(x, v)$ by means of $\psi(t)$.

[21] Elementary comprehension means that, if $\varphi(x, \vec{a})$ is elementary in $\vec{a}$ and each component of $\vec{a}$ is a class, then $\{x : \varphi(x, \vec{a})\}$ is a class and it satisfies $\forall u(u\,\eta\,\{x : \varphi(x, \vec{a})\} \leftrightarrow \varphi(u, \vec{a}))$. Generalized union ($\Sigma$) and generalized product (Π) correspond to the axioms (in the given order): if a is a class, and f is an operation mapping elements of a into classes, there exist (i) the class $\Sigma(a, f)$ of all pairs (x, y) with $x\,\eta\,a$ and $y\,\eta\,fx$; (ii) the class $\Pi(a, f)$ of all operations g such that for every $x\,\eta\,a$, then $gx\,\eta\,fx$.

[22] This is the theory of non-iterated inductive definitions.

PROOF. A quick argument runs as follows. First of all, it is enough to prove that $\boldsymbol{T_c}$ is conservative over ***TON*** for formulas not containing $\mathcal{T}$ (since ***TON*** has an inner model in ***PA***). Given an arbitrary countable model $\mathcal{M}$ of ***TON***, choose a recursively saturated elementary extension $\mathcal{M}'$ of $\mathcal{M}$ and consider the least fixed point model of the truth axioms with ground applicative structure $\mathcal{M}'$. This is a model of $\boldsymbol{T_c}$, except possibly class induction on numbers. Now classes in minimal models exactly represent the hyperelementary subsets of $\mathcal{M}'$, and these sets are indeed elementary as $\mathcal{M}'$ is recursively saturated [23]. But induction on numbers for $\mathcal{T}$-free formulas holds by assumption in $\mathcal{M}$ and hence in $\mathcal{M}'$. For a proof–theoretic argument, see [10]. The second part (related to $\boldsymbol{T_f}$) is also standard, making use of an inner model construction. ⊣

§8. The model construction in $\boldsymbol{T_c}$. We define a suitable counterpart of a universe $V_{\mathcal{U}}$ of sets above the collection of urelements $\mathcal{U}$, $\mathcal{U}$ being an extensional combinatory algebra with embedded natural numbers. Let (x, y) denote the basic pairing operation which is built-in the axioms of $\boldsymbol{T_c}$; (x, y, z) stands for $(x, (y, z))$, and, if $a = (x, y, z)$, $a_0 = x$, $a_1 = y$ and $a_2 = z$. Intuitively, the universe of sets $V_{\mathcal{U}}$ is defined by means of two rules, one for urelements and one for sets:

$$\overline{(0, x)\; \eta\; V_{\mathcal{U}}}$$

and

$$\frac{\boldsymbol{Cl}(a) \qquad \forall u\, \eta\, a\, (f u\; \eta\; V_{\mathcal{U}})}{(1, a, f)\; \eta\; V_{\mathcal{U}}}.$$

Urelements of $\boldsymbol{CZF}_b^{op}$ are here represented by pairs of the form $(0, x)$, where x ranges over the elements of the underlying applicative algebra. Sets are instead inductively built in a similar way as in Aczel's type theoretic interpretation, $(1, a, f)$ corresponding to Aczel's $\sup(a, f)$. Here classifications play a similar role as small types in the same interpretation.[24]

In the following, unless otherwise stated, we work in the theory $\boldsymbol{T}^-$. We also write $\sup(a, f)$ for $(1, a, f)$.

The intuitive rules above may be formalized via an inductive definition and we easily obtain the following proposition.

PROPOSITION 8.1. *There exists a closed term $V_{\mathcal{U}}$ such that*

---

[23]For the notion of hyperelementary set, see [25]; the result we refer to is a generalization of a well-known theorem of Barwise and Schlipf; see [11].

[24]Classifications *should not be assimilated to types*, though they can be used to *simulate* types and type constructions. Classifications can be generated also according to procedures that are not accepted according to the constructive discipline of types (in the sense of Martin-Löf). To mark the difference with type theory, there exists a universal classification with a global equality relation, and the applicative structure allows untyped operations. Of course, it is possible to develop a formal theory of classifications with an underlying constructive logic, and this leads exactly to constructive explicit mathematics in the sense of Feferman [17], Beeson [7].

(i) $a\,\eta\,V_{\mathcal{U}} \leftrightarrow (a = (0, a_1)) \vee (a = (1, a_1, a_2) \wedge \boldsymbol{Cl}(a_1) \wedge (\forall u\,\eta\,a_1)((a_2 u)\,\eta\,V_{\mathcal{U}}))$;
(ii) $\boldsymbol{Clos}_{V_{\mathcal{U}}}(\varphi) \rightarrow \forall x\,(x\eta V_{\mathcal{U}} \rightarrow \varphi(x))$,

where φ is an arbitrary formula, $\boldsymbol{Clos}_{V_{\mathcal{U}}}(\varphi) := \forall x(\mathcal{V}(x,\varphi) \rightarrow \varphi(x))$, and $\mathcal{V}(x,\varphi)$ is an abbreviation of $(x = (0, x_1)) \vee (x = (1, x_1, x_2) \wedge \boldsymbol{Cl}(x_1) \wedge (\forall u\,\eta\,x_1)(\varphi(x_2 u)))$.

Proof. As to (i), let $\mathcal{V}(x, a)$ be $\mathcal{V}(x,\varphi)$ where $\varphi(x)$ is $x\,\eta\,a$. Then $\mathcal{V}(x,a) \leftrightarrow T[\mathcal{V}(x, a)]$ and the claim follows by proposition 7.2. (ii) is an application of ***GID***. ⊣

In the following, applications of proposition 8.1 (ii) will be simply referred to as *proofs by induction on $V_{\mathcal{U}}$*.

Proposition 8.2. *There are operations assigning $\bar{a}$ and $\tilde{a}$ to each $a\,\eta\,V_{\mathcal{U}}$ and such that $\boldsymbol{Cl}(\bar{a})$ and $\tilde{a} : \bar{a} \rightarrow V_{\mathcal{U}}$ (that is $\forall x\eta\bar{a}\,(\tilde{a}x\eta V_{\mathcal{U}})$), in particular*

$$\text{(i)}\quad \bar{a} = \begin{cases} \emptyset, & \textit{if } a_0 = 0, \\ a_1, & \textit{if } a_0 = 1; \end{cases}$$

$$\text{(ii)}\quad \tilde{a} = \begin{cases} \lambda x.0, & \textit{if } a_0 = 0, \\ a_2, & \textit{if } a_0 = 1. \end{cases}$$

Here $\emptyset$ is the empty class $\{x : \neg x = x\}$.

Proof. By induction on $V_{\mathcal{U}}$, using the recursion theorem and definition by cases on natural numbers. ⊣

For $a\,\eta\,V_{\mathcal{U}}$, we write

$$\mathcal{S}^*(a) := a_0 = 1; \tag{8.1}$$

$$\mathcal{U}^*(a) := a_0 = 0; \tag{8.2}$$

$$\mathcal{N}^*(a) := a_0 = 0 \wedge \mathcal{N}(a_1). \tag{8.3}$$

We interchangeably write $x\,\eta\,\mathcal{S}^*$, $x\,\eta\,\mathcal{U}^*$, $x\,\eta\,\mathcal{N}^*$ for $x\,\eta\,\{u : \mathcal{S}^*(u)\}$, $x\,\eta\,\{u : \mathcal{U}^*(u)\}$, $x\,\eta\,\{u : \mathcal{N}^*(u)\}$ (in the given order). If $a\,\eta\,V_{\mathcal{U}}$, $b\,\eta\,V_{\mathcal{U}}$, $c\,\eta\,V_{\mathcal{U}}$, we further write:

$$App^*(a,b,c) := a_0 = b_0 = c_0 = 0 \wedge a_1 b_1 = c_1 \tag{8.4}$$

Lemma 8.3. *If $a\,\eta\,V_{\mathcal{U}}$ then*

$$(a\,\eta\,\mathcal{S}^* \vee a\,\eta\,\mathcal{U}^*) \wedge \neg(a\,\eta\,\mathcal{S}^* \wedge a\,\eta\,\mathcal{U}^*).$$

Proof. By induction on $V_{\mathcal{U}}$. ⊣

We next define recursively (i.e. by (7.2) (ii)–(iii)) an equivalence relation, $\doteq$, on $V_{\mathcal{U}}$.

LEMMA 8.4. *There exists a term* $\doteq$ *such that*

$$
\begin{aligned}
a \doteq b \leftrightarrow a\,\eta\, V_{\mathcal{U}} \wedge b\,\eta\, V_{\mathcal{U}} \wedge ((a_0 = b_0 = 0 \wedge a_1 = b_1) \vee \\
\vee (a_0 = b_0 = 1 \wedge \\
\wedge \forall x\,\eta\,\bar{a}\,\exists y\,\eta\,\bar{b}\,(\tilde{a}x \doteq \tilde{b}y) \wedge \\
\wedge \forall y\,\eta\,\bar{b}\,\exists x\,\eta\,\bar{a}\,(\tilde{a}x \doteq \tilde{b}y))).
\end{aligned}
$$

LEMMA 8.5. *For* $a, b, c\,\eta\, V_{\mathcal{U}}$ *the following holds*

(i) $a \doteq a$

(ii) $a \doteq b \rightarrow b \doteq a$

(iii) $a \doteq b \wedge b \doteq c \rightarrow a \doteq c$.

PROOF. By induction on $V_{\mathcal{U}}$. ⊣

Membership is then defined in the obvious way by means of $\doteq$.

DEFINITION 8.6. For $a\eta V_{\mathcal{U}}$ and $b\eta V_{\mathcal{U}}$, let

$$a \dot{\in} b := \exists x\,\eta\,\bar{b}\,(a \doteq \tilde{b}x) \wedge b_0 = 1.$$

LEMMA 8.7. *Let* $a, b\,\eta\, V_{\mathcal{U}}$. *If* $\mathcal{S}^*(a)$ *and* $\mathcal{S}^*(b)$ *then*

$$\forall x\,\eta\, V_{\mathcal{U}}\,(x \dot{\in} a \leftrightarrow x \dot{\in} b) \rightarrow a \doteq b.$$

PROOF. Assume $\mathcal{S}^*(a)$, $\mathcal{S}^*(b)$ and $\forall x\,\eta\, V_{\mathcal{U}}\,(x \dot{\in} a \leftrightarrow x \dot{\in} b)$. Suppose $z\,\eta\,\bar{a}$. Then $\tilde{a}z\,\eta\, V_{\mathcal{U}}$ and $\tilde{a}z \dot{\in} a$ (apply (8.5)(i)), so that by hypothesis, also $\tilde{a}z \dot{\in} b$. Then $\exists y\,\eta\,\bar{b}$ such that $\tilde{a}z \doteq \tilde{a}y$. Similarly one proves the other conjunct in the definition of $a \doteq b$. ⊣

We now define a dual of $\doteq$ which is used, for example, in verifying Δ_0–separation.

LEMMA 8.8. *There exists a term* $\dot{\neq}$ *such that*

$$
\begin{aligned}
a \dot{\neq} b \leftrightarrow\ & a\eta V_{\mathcal{U}} \wedge b\eta V_{\mathcal{U}} \wedge (a_0 \neq b_0 \vee (a_0 = b_0 = 0 \wedge a_1 \neq b_1) \vee \\
& \vee (a_0 = b_0 = 1 \wedge (\exists x\,\eta\,\bar{a}\,\forall y\,(y\,\bar{\eta}\,\bar{b} \vee \tilde{a}x \dot{\neq} \tilde{b}y)) \vee \\
& \vee (\exists y\,\eta\,\bar{b}\,\forall x\,(x\,\bar{\eta}\,\bar{a} \vee \tilde{a}x \dot{\neq} \tilde{b}y)))).
\end{aligned}
$$

LEMMA 8.9. *For* $a, b\,\eta\, V_{\mathcal{U}}$

(i) $\neg(a \doteq b \wedge a \dot{\neq} b) \wedge (a \doteq b \vee a \dot{\neq} b)$;

(ii) $\neg a \dot{\neq} a$;

(iii) $a \dot{\neq} b \rightarrow b \dot{\neq} a$;

(iv) $a \doteq b \wedge b \dot{\neq} c \rightarrow a \dot{\neq} c$.

PROOF. Use ***GID*** and lemmata 8.4, 8.8. ⊣

We define a dual $\dot{\notin}$ of $\dot{\in}$ as follows.

DEFINITION 8.10. For $a, b\,\eta\, V_{\mathcal{U}}$ let

$$a \dot{\notin} b := \forall x\,(x\,\bar{\eta}\,\bar{b} \vee \tilde{b}x \dot{\neq} a) \vee b_0 \neq 1.$$

LEMMA 8.11. *For* $a, b, c\,\eta\, V_{\mathcal{U}}$

(i) $\neg(a \dot{\in} b \wedge a \dot{\notin} b)$;
(ii) $a \dot{\in} b \vee a \dot{\notin} b$;
(iii) $\dot{\in}, \dot{\notin}$ *are preserved by* $\doteq$, *that is*:

$$(a \doteq b \wedge b \dot{\in} c \rightarrow a \dot{\in} c) \wedge (a \doteq b \wedge c \dot{\in} b \rightarrow c \dot{\in} a);$$
$$(a \doteq b \wedge b \dot{\notin} c \rightarrow a \dot{\notin} c) \wedge (a \doteq b \wedge c \dot{\notin} b \rightarrow c \dot{\notin} a).$$

PROOF. Apply the consistency axiom for $\mathcal{T}$ and lemmata 8.4, 8.5, 8.8, 8.9. ⊣

LEMMA 8.12. *For* $a, b \,\eta\, \boldsymbol{V}_{\mathcal{U}}$

$$\mathcal{T}[a \doteq b] \vee \mathcal{T}[\neg a \doteq b];$$
$$\mathcal{T}[a \dot{\in} b] \vee \mathcal{T}[\neg a \dot{\in} b].$$

PROOF. By proposition 7.2 (ii) we obtain:

$$(a \doteq b \leftrightarrow \mathcal{T}[a \doteq b]) \wedge (a \not\doteq b \leftrightarrow \mathcal{T}[a \not\doteq b]);$$
$$(a \dot{\in} b \leftrightarrow \mathcal{T}[a \dot{\in} b]) \wedge (a \dot{\notin} b \leftrightarrow \mathcal{T}[a \dot{\notin} b]).$$

Assume $\neg\mathcal{T}[a \doteq b]$. Then by lemma 8.9 (i), $a \not\doteq b$. But we can check by induction on $\boldsymbol{V}_{\mathcal{U}}$ and the closure axioms of $\boldsymbol{T}^-$: $a \not\doteq b \rightarrow \mathcal{T}[\neg a \doteq b]$.

Assume $\neg\mathcal{T}[a \dot{\in} b]$: by lemma 8.11(ii), $a \dot{\notin} b$. By the previous fact, we easily derive that $a \dot{\notin} b \rightarrow \mathcal{T}[\neg a \dot{\in} b]$. ⊣

DEFINITION 8.13. (i) To each term t of $\boldsymbol{CZF}_b^{op}$ we assign a term t^* as follows: we let $c^* = (0, c)$ if c is an individual constant and $(x)^* = x$, if x is a variable.

(ii) Let φ be a Δ_0 formula of $\boldsymbol{CZF}_b^{op}$; we inductively define φ^* in $\boldsymbol{T}^-$ as follows:

$$(\bot)^* := 0 = 1$$
$$(App(t, r, s))^* := App^*(t^*, s^*, r^*)$$
$$(\mathcal{S}(t))^* := \mathcal{S}^*(t^*)$$
$$(\mathcal{U}(t))^* := \mathcal{U}^*(t^*)$$
$$(\mathcal{N}(t))^* := \mathcal{N}^*(t^*)$$
$$(t = s)^* := t^* \doteq s^*$$
$$(t \in s)^* := t^* \dot{\in} s^*$$
$$(\psi \wedge \chi)^* := (\psi)^* \wedge (\chi)^*$$
$$(\psi \vee \chi)^* := (\psi)^* \vee (\chi)^*$$

$$(\psi \to \chi)^* :=(\psi)^* \to (\chi)^*$$
$$(\exists x \in t\, \psi(x))^* :=\exists x(x \dot{\in} t^* \wedge \psi^*(x))$$
$$(\forall x \in t\, \psi(x))^* :=\forall x(x \dot{\in} t^* \to \psi^*(x)).$$

(iii) Let φ be Δ_0; then φ^*_+ is obtained from (ii) above by replacing the clauses for the bounded quantifiers with the following

$$(\exists x \in t\, \psi(x))^*_+ :=\exists x\, \eta\, \overline{t^*}\, .\psi^*_+(\widetilde{t^*}x)$$
$$(\forall x \in t\, \psi(x))^*_+ :=\forall x\, \eta\, \overline{t^*}\, .\psi^*_+(\widetilde{t^*}x).$$

LEMMA 8.14 ($\boldsymbol{T_c}$). (i) *If φ is an arbitrary formula of $\boldsymbol{CZF}_b^{op}$,*

$$a\, \eta\, V_{\mathcal{U}} \wedge b\, \eta\, V_{\mathcal{U}} \wedge a \doteq b \wedge \varphi^*(a) \to \varphi^*(b);$$

(ii) *Let φ be a Δ_0 formula of set theory; if $\vec{a}\, \eta\, V_{\mathcal{U}}$*[25]*, then*

$$\varphi^*(\vec{a}) \leftrightarrow \varphi^*_+(\vec{a}); \tag{8.5}$$

$$\mathcal{T}[\varphi^*_+(\vec{a})] \vee \mathcal{T}[\neg\varphi^*_+(\vec{a})]. \tag{8.6}$$

PROOF. (i) The atomic cases follow by Lemmata 8.5 and 8.11, the other cases by induction. (ii) is also proved by induction on the formula φ using (i). In particular, (8.6) requires lemmata 8.11 and 8.12 for the atomic case (besides the fact that $\mathcal{T}[x = y \vee \neg x = y]$). ⊣

8.1. The set theoretic structure on $V_{\mathcal{U}}$.

PROPOSITION 8.15. *The structure $\langle V_{\mathcal{U}}, \doteq, \dot{\in}, \mathcal{U}^*, \mathcal{N}^*, \mathcal{S}^*, App^*\rangle$ is a model of the theory $\boldsymbol{CZF}_b^{op}$, ($\boldsymbol{CZF}^{op}$), except strong collection and subset collection, provably in $\boldsymbol{T_c}$ ($\boldsymbol{T_f}$).*

PROOF. (Hint) The ontological and combinatory axioms, the extensionality and natural numbers axioms of $\boldsymbol{CZF}_b^{op}$ are immediately verified by definition of the basic predicates and application ((8.1) - (8.4)) and the lemmata 8.3, 8.7. The schema of $\in$-induction follows by ***GID*** on $V_{\mathcal{U}}$ (see proposition 8.1). Concerning set induction on numbers, if $\mathcal{S}^*(a)$, then $a^* := \{x : \mathcal{N}(x) \wedge \exists v\, \eta\, \bar{a}\, .((\tilde{a}v)_1 = x \wedge (\tilde{a}v)_0 = 0)\}$ is a class, and by class induction on numbers it follows that $\forall x(\mathcal{N}(x) \to x\, \eta\, a^*)$, which readily yields the required conclusion $\forall x\, \eta\, V_{\mathcal{U}}(\mathcal{N}^*(x) \to x \dot{\in} a)$.

Emptyset: A witness is $\sup(N_0, h_0)$, where $N_0 := \{x : x \neq x\}$ and $h_0 := \lambda x.0$.

Pair: A witness for the pair of $a, b\, \eta\, V_{\mathcal{U}}$ is $\sup(\mathbf{2}, \lambda z.Dabz0)$, where $\mathbf{2} := \{x : Nat(x) \wedge (x = 0 \vee x = 1)\}$.

Union: A witness for the union over the set $a\, \eta\, V_{\mathcal{U}}$ is $\sup(c, \lambda z.\widetilde{(\tilde{a}z_0)}z_1)$, where $c := \{z : z_0\, \eta\, \bar{a} \wedge z_1\, \eta\, \overline{\tilde{a}(z_0)}\}$.

Infinity: Take $\sup(\mathcal{U}^*, \lambda x.x)$, with $\mathcal{U}^* := \{u : \exists y\, (u = (0, y))\}$. As $\mathcal{U}^*$ is a class, $\sup(\mathcal{U}^*, \lambda x.x)$ is in $V_{\mathcal{U}}$.

[25] If $\vec{a} := a_0, \ldots, a_n$, $\vec{a}\, \eta\, V_{\mathcal{U}}$ shortens $a_0\, \eta\, V_{\mathcal{U}}, \ldots, a_n\, \eta\, V_{\mathcal{U}}$.

Δ_0*–separation:* A witness for $a\,\eta\, V_{\mathcal{U}}$ and a Δ_0 formula φ is $b := \sup(c, \lambda z.\tilde{a}z)$ with $c := \{z : z\,\eta\,\bar{a} \wedge \varphi^*_+(\tilde{a}z)\}$. Then $\mathcal{S}^*(b)$, since c is a class (lemma 8.14) and $\tilde{a}$ maps elements of $\bar{a}$ into $V_{\mathcal{U}}$. Assume $u \dot{\in} a$ and $\varphi^*(u)$. Then $u \doteq \tilde{a}x$ for some $x\,\eta\,\bar{a}$; hence $\varphi^*(\tilde{a}x)$ and $\varphi^*_+(\tilde{a}x)$ again by lemma 8.14, whence $x\,\eta\,\bar{b}$ and $u \doteq \tilde{b}x$, i.e. $u \dot{\in} b$. The converse direction is similar. ⊣

It is possible to extend $\boldsymbol{T_c}$ conservatively, so that *also bounded collection* is validated in $\langle V_{\mathcal{U}}, \doteq, \dot{\in}, \mathcal{U}^*, \mathcal{N}^*, \mathcal{S}^*, App^*\rangle$, and hence the structure is a model of *admissible set theory with urelements*. This fact ultimately depends upon the recursion theoretic structure built in the minimal fixed point models of $\boldsymbol{T_f}$ (or $\boldsymbol{T_c}$) and more precisely upon (a uniform version of) the so-called *ordinal comparison theorem* (see [10, 11]).

As to the *full* schemata of strong collection and subset collection, the problem is overcome by switching to a non-classical constructive semantics, i.e. showing that $\langle V_{\mathcal{U}}, \doteq, \dot{\in}, \mathcal{U}^*, \mathcal{N}^*, \mathcal{S}^*, App^*\rangle$ is *a realizability model* for $\boldsymbol{CZF^{op}}$.

Definition 8.16. $a + b := \{u : u = (u_0, u_1) \wedge ((u_0 = 0 \wedge u_1\,\eta\,a) \vee (u_0 = 1 \wedge u_1\,\eta\,b))\}$ represents the direct sum of a, b.

We adopt the set theoretic notation $\boldsymbol{1}$ also for denoting the classification which only has the empty classification as element. If φ is a bounded formula of $\boldsymbol{CZF^{op}_b}$ we inductively define a map $\varphi \mapsto \|\varphi\|$, where (roughly) $\|\varphi\|$ collects the proof objects for φ, provided the parameters range over $V_{\mathcal{U}}$.

$$\begin{aligned}
\|\bot\| &= \{e\,\eta\,\boldsymbol{1} : 0 = 1\};\\
\|\mathcal{U}(a)\| &= \{e\,\eta\,\boldsymbol{1} : \mathcal{U}^*(a)\};\\
\|\mathcal{S}(a)\| &= \{e\,\eta\,\boldsymbol{1} : \mathcal{S}^*(a)\};\\
\|\mathcal{N}(a)\| &= \{e\,\eta\,\boldsymbol{1} : \mathcal{N}^*(a)\};\\
\|App(a,b,c)\| &= \{e\,\eta\,\boldsymbol{1} : App^*(a,b,c)\};\\
\|a = b\| &= \{e : e = 0 \wedge a_0 = b_0 = 0 \wedge a_1 = b_1\}\\
&\quad + \{e : a_0 = b_0 = 1 \wedge e = (e_0, e_1)\\
&\quad \wedge \forall u\,\eta\,\bar{a}\,(e_0u)_0\,\eta\,\bar{b} \wedge (e_0u)_1\,\eta\,\|\tilde{a}u = \tilde{b}(e_0u)_0\|\\
&\quad \wedge \forall v\,\eta\,\bar{b}\,(e_1v)_0\,\eta\,\bar{a} \wedge (e_1v)_1\,\eta\,\|\tilde{a}(e_1v)_0 = \tilde{b}v\|\};\\
\|a \in b\| &= \{e : e = (e_0, e_1) \wedge e_0\,\eta\,\bar{b} \wedge e_1\,\eta\,\|a = \tilde{b}e_0\| \wedge b_0 = 1\};\\
\|\varphi \wedge \psi\| &= \{e : e = (e_0, e_1) \wedge (e_0)\,\eta\,\|\varphi\| \wedge (e_1)\,\eta\,\|\psi\|\};\\
\|\varphi \vee \psi\| &= \|\varphi\| + \|\psi\|;\\
\|\varphi \to \psi\| &= \{e : \forall q\,\eta\,\|\varphi\|(eq\,\eta\,\|\psi\|)\};\\
\|\exists x \in a\,\varphi(x)\| &= \{e : (e_0, e_1) \wedge (e_0)\,\eta\,\bar{a} \wedge e_1\,\eta\,\|\varphi(\tilde{a}e_0)\|\};\\
\|\forall x \in a\,\varphi(x)\| &= \{e : \forall u\,\eta\,\bar{a}\,(eu\,\eta\,\|\varphi(\tilde{a}u)\|)\}.
\end{aligned}$$

Note that, formally speaking, the definition of $\|\varphi\|$ above makes sense only after one shows by a fixed point argument in $\boldsymbol{T}^-$ that there exists an operation $H(a,b)$ satisfying the equation for $\|a = b\|$ (hence the definition inductively extends H to arbitrary bounded conditions).

8.1.1. *The main theorem.*

DEFINITION 8.17. Let φ be an arbitrary formula of $\boldsymbol{CZF}_b^{op}$; we inductively define a formula $e \Vdash \varphi$ of $\boldsymbol{T_c}$ with the same free variables of φ and a fresh variable e:

if φ is bounded, $e \Vdash \varphi$ iff $e\,\eta\,\|\varphi\|$; else:

$$e \Vdash \varphi \to \psi \text{ iff } \forall f (f \Vdash \varphi \to ef \Vdash \psi);$$

$$e \Vdash \varphi \wedge \psi \text{ iff } e = (e_0, e_1) \wedge e_0 \Vdash \varphi \wedge e_1 \Vdash \psi;$$

$$e \Vdash \varphi \vee \psi \text{ iff } (e = (0, e_1) \wedge e_1 \Vdash \varphi) \vee (e = (1, e_1) \wedge e_1 \Vdash \psi);$$

$$e \Vdash \forall x \in a\, \varphi(x) \text{ iff } \forall x\, \eta\, \tilde{a}\, (ex \Vdash \varphi(\tilde{a}x));$$

$$e \Vdash \exists x \in a\, \varphi(x) \text{ iff } e = (e_0, e_1) \wedge e_0\, \eta\, \tilde{a} \wedge e_1 \Vdash \varphi(\tilde{a}e_0);$$

$$e \Vdash \exists x\, \varphi \text{ iff } e = (e_0, e_1) \wedge e_0\, \eta\, V_{\mathcal{U}} \wedge e_1 \Vdash \varphi(e_0);$$

$$e \Vdash \forall x\, \varphi \text{ iff } \forall x\, \eta\, V_{\mathcal{U}}\, (ex \Vdash \varphi(x)).$$

LEMMA 8.18. *Let φ be a bounded formula of $\boldsymbol{CZF}_b^{op}$. Then $\boldsymbol{T}^-$ proves*

$$\vec{x} \in V_{\mathcal{U}} \to Cl(\|\varphi(\vec{x})\|); \tag{8.7}$$

$$e \Vdash \varphi(\vec{x}) \text{ iff } e\,\eta\,\|\varphi(\vec{x})\|. \tag{8.8}$$

PROOF. Ad (8.7): apply ***GID*** (in the form of induction on $V_{\mathcal{U}}$) in the atomic $=$-case. Then use closure of ***CL*** under direct sum, generalized disjoint sum, Σ, product, Π, and elementary comprehension. As to the second item (8.8), proceed by induction on the definition of realizability. ⊣

We say that a formula $\varphi(\vec{x})$ is *realized in a theory* $\boldsymbol{T}$, if there exists a term e such that $\boldsymbol{T}$ proves $\vec{a}\,\eta\, V_{\mathcal{U}} \to e\vec{a} \Vdash \varphi(\vec{a})$.

LEMMA 8.19. *Every sentence of intuitionistic predicate logic (in the language of $\boldsymbol{CZF}_b^{op}$) is realized in $\boldsymbol{T}^-$.*

The proof is standard (see [33]).

LEMMA 8.20. *Every equality axiom of $\boldsymbol{CZF}_b^{op}$ is realized in $\boldsymbol{T}^-$.*

PROOF. Consider $\doteq$–reflexivity: if a is an urelement, simply let $\boldsymbol{ref}\,a = 0$; else, if a is a set, choose by fixed point theorem $\boldsymbol{ref}\,a = (\lambda u.(u, \boldsymbol{ref}(\tilde{a}u)), \lambda u.(u, \boldsymbol{ref}(\tilde{a}u)))$ and apply $V_{\mathcal{U}}$–induction in order to check that $\boldsymbol{ref}\,a$ realizes $a \doteq a$. Since definition by cases is admissible, as a_0 is a natural number, $\boldsymbol{ref}$ is well-defined on $V_{\mathcal{U}}$. The other cases are similar (e.g. one recursively defines operations $\boldsymbol{sym}$, $\boldsymbol{tr}$, yielding the realizers for symmetry and transitivity). Preservation of $\dot{\in}$ under $\doteq$ easily follows with $\boldsymbol{tr}$, $\boldsymbol{sym}$. ⊣

LEMMA 8.21. *There are realizers* e, f *such that provably in* $\boldsymbol{T}^-$:

$$e \Vdash \exists x \in a\, \varphi(x) \leftrightarrow \exists x\, (x \in a \wedge \varphi(x));$$
$$e \Vdash \forall x \in a\, \varphi(x) \leftrightarrow \forall x\, (x \in a \rightarrow \varphi(x)).$$

PROOF. Apply lemma 8.20 (see [13] for details). ⊣

THEOREM 8.22. *Every theorem of* $\boldsymbol{CZF}_b^{op}$ *is realized in* $\boldsymbol{T_c}$, *i.e. if* $\boldsymbol{CZF}_b^{op} \vdash \varphi(\vec{x})$, *then there exists a closed term e such that, provably in* $\boldsymbol{T_c}$,

$$\vec{a} \in V_{\mathcal{U}} \rightarrow e\vec{a} \Vdash \varphi(\vec{a}).$$

PROOF. We only point out the main steps (the verification, though quite lengthy, is more or less analogous to the corresponding steps in [13]). As to the logical part, this is taken care of by lemmata 8.19 and 8.20.

(a) *Ontological axioms*. Choose $\lambda x.\emptyset$ and $\lambda x.Dx_0 0(0,\emptyset)(1,\emptyset)$, $\lambda x.x$, $\lambda x.\emptyset$ for the ontological axioms (a)–(d) of 2.2 (in the given order). As to extensionality, the realizer is obtained by using projections and an operation **ref** such that **ref** $x \Vdash x \doteq x$ (granted by lemma 8.20).

(b) *Applicative axioms*. The verification is essentially straightforward. As to the axiom of totality of application over the urelements, choose $\lambda x \lambda y.((0, x_1 y_1), (\emptyset, \emptyset))$. Concerning uniqueness of application, choose $\lambda x.0$.

(c) *Strong collection*. Assume $q \Vdash \forall x \in a \exists y \varphi(x, y)$ where $a_0 = 1$ (a is a set). We want r_0, $r_1 = (r_{10}, r_{11})$ such that $r_0\, \eta\, V_{\mathcal{U}}$ and

$$r_{10} \Vdash \forall x \in a \exists y \in r_0.\varphi(x, y);$$
$$r_{11} \Vdash \forall y \in r_0 \exists x \in a.\varphi(x, y)$$

Choose $r_0 = sup(\bar{a}, \lambda z.(qz)_0)$; clearly by assumption on a, $r_0\, \eta\, V_{\mathcal{U}}$. Then let

$$r_{10} := \lambda x.(x, (qx)_1) = r_{11}.$$

(d) *Subset collection*. Assume that a, b are sets. We need a set $(eab)_0$ such that if $u\, \eta\, V_{\mathcal{U}}$

$$(eab)_1 u \Vdash \forall x \in a \exists y \in b\, \varphi(x, y, u) \rightarrow$$
$$\rightarrow \exists d \in (eab)_0 (\forall x \in a \exists y \in d\, \varphi(x, y, u) \wedge$$
$$\wedge \forall y \in d \exists x \in a\, \varphi(x, y, u))$$

Observe that $\Pi(\bar{a}, \lambda x.\bar{b})$ is a class, since $\bar{a}$ is a class and $(\lambda x.\bar{b})u = \bar{b}$ is a class, for every u. Hence choose

$$(eab)_0 = \sup(\Pi(\bar{a}, \lambda x.\bar{b}), \lambda z. \sup(\bar{a}, \lambda y.\bar{b}(zy)))$$
$$(eab)_1 = \lambda u \lambda q.(\lambda y.(qy)_0, (\lambda y.(y, (qy)_1), \lambda y.(y, (qy)_1))).$$

(e) *Bounded separation.* Let $\varphi(x)$ be a bounded formula and assume $a \,\eta\, V_{\mathcal{U}}$. We need a pair ea such that

$$(ea)_1 \Vdash \forall y(y \in (ea)_0 \leftrightarrow y \in a \wedge \varphi(y)) \tag{8.9}$$

Observe that $\Sigma(\bar{a}, \lambda w.\|\varphi(\tilde{a}w)\|)$ is a class, as $\bar{a}$ is a class by assumption on a and $\|\varphi(\tilde{a}w)\|$ is a class for every $w \,\eta\, \bar{a}$ (by lemma 8.18). Hence $(ea)_0 = \sup(\Sigma(\bar{a}, \lambda w.\|\varphi(\tilde{a}w)\|), \lambda z.\tilde{a}(z_0))$ is in $V_{\mathcal{U}}$. In order to define $(ea)_1$, proceed as follows. If $q \Vdash u \in (ea)_0$, then $q_{00} \,\eta\, \bar{a}$, $q_{01} \,\eta\, \|\varphi(\tilde{a}(q_{00})\|$ and $q_1 \Vdash u \doteq \tilde{a}(q_{00})$. By lemma 8.20, there exists an operation H such that $H(q_1, q_{01})$ is a realizer of $\varphi(u)$. Hence $u \in a \wedge \varphi(u)$ is realized by the pair $((q_{00}, q_1), H(q_1, q_{01}))$. This essentially fixes up (8.9) from left to right (uniformly in a, u). Conversely, if $q_0 \Vdash u \in a$ and $q_1 \Vdash \varphi(u)$, $((q_{00}, J(q_1, q_{01})), q_{01})$ is a realizer of $u \in (ea)_0$ ($J(q_1, q_{01})$ being a realizer of $\varphi(\tilde{a}(q_0))$ which exists again by lemma 8.20).

(f) *Union* and *Pairing*. If a is a set, pairing is realized by $eab = (\Sigma(\mathbf{2}, f), \lambda y \lambda x.x))$, where $\mathbf{2}$ is the class $\{u : u = 0 \vee u = 1\}$, $fz = a$ if $z = 0$, $fz = b$ if $z = 1$. As to union, if a is a set, choose

$$ea = \left(\sup\left(\Sigma(\bar{a}, \lambda w.\overline{\tilde{a}w}), \lambda z.\widetilde{\tilde{a}(z_0)}z_1\right), \lambda u \lambda x.((x_0, x_{10}), x_{11})\right)$$

(g) *Existence of the set of all urelements.* Choose $e_0 = \sup(\mathcal{U}^*, \lambda x.x)$, where $\mathcal{U}^*$ is the class $\{u : \exists y\,(u = (0, y))\}$. Then e_0 is in $V_{\mathcal{U}}$; if we let $e_1 = \lambda u.\langle K\emptyset, K(u, 0)\rangle$, then $e_1 u$ realizes $u \in e_0 \leftrightarrow \mathcal{U}(u)$.

(h) *Full set induction.* By fixed point theorem there is a solution to the equation $e = \lambda x \lambda u.xu(\lambda y.ex(\tilde{u}y))$. Then apply induction on $V_{\mathcal{U}}$ to check that e is a sound realizer.

(i) *N-induction on sets*: As in the previous case, the realizer is defined by recursion (on numbers); that it is sound, follows by class induction on numbers ($\|x \in a\|$ is a class if a is in $V_{\mathcal{U}}$ by lemma 8.18). ⊣

REFERENCES

[1] P. Aczel, *The strength of Martin Löf type theory with one universe*, Tech. rep., Department of Philosophy, University of Helsinki, 1977.

[2] ———, *The type theoretic interpretation of constructive set theory*, ***Logic Colloquium '77 (Proc. Conf., Wrocław, 1977)*** (L. Pacholski and J. Paris, editors), vol. 96, North-Holland, Amsterdam, 1978, pp. 55–66.

[3] ———, *Frege structures and the notions of proposition, truth and set*, ***The Kleene Symposium (Proc. Sympos., Univ. Wisconsin, Madison, Wis., 1978)*** (J. Barwise et al., editors), vol. 101, North-Holland, Amsterdam, 1980, pp. 31–59.

[4] ———, *The type theoretic interpretation of constructive set theory: choice principles*, ***The L.E.J. Brouwer Centenary Symposium (Noordwijkerhout, 1981)*** (A. S. Troelstra and D. Van Dalen, editors), vol. 110, North-Holland, Amsterdam, 1982, pp. 1–40.

[5] ———, *The type theoretic interpretation of constructive set theory: inductive definitions*, ***Logic, Methodology and Philosophy of Science, VII (Salzburg, 1983)*** (R. B. Marcus et al., editors), vol. 114, North-Holland, Amsterdam, 1986, pp. 17–49.

[6] P. ACZEL and M. RATHJEN, *Notes on Constructive Set Theory*, Institut Mittag-Leffler, preprint, 40, 2000.

[7] M. BEESON, ***Foundations of Constructive Mathematics***, vol. 6, Springer-Verlag, Berlin, 1985.

[8] ———, *Towards a computation system based on set theory*, ***Theoretical Computer Science***, vol. 60 (1988), no. 3, pp. 297–340.

[9] E. BISHOP, ***Foundations of Constructive Analysis***, McGraw-Hill, New York, 1967.

[10] A. CANTINI, *Levels of implication and type free theories of partial classifications with approximation operator*, ***Zeitschrift für Mathematische Logik und Grundlagen der Mathematik***, vol. 38 (1992), no. 2, pp. 107–141.

[11] ———, ***Logical Frameworks for Truth and Abstraction***, Studies in Logic and the Foundations of Mathematics, vol. 135, North-Holland, Amsterdam, 1996.

[12] A. CANTINI and P. MINARI, *Uniform inseparability in explicit mathematics*, ***The Journal of Symbolic Logic***, vol. 64 (1999), no. 1, pp. 313–326.

[13] L. CROSILLA, ***Realizability Models for Constructive Set Theories with Restricted Induction Principles***, Ph.D. thesis, Department of Pure Mathematics, University of Leeds, September 2000.

[14] L. CROSILLA and M. RATHJEN, *Inaccessible set axioms may have little consistency strength*, ***Annals of Pure and Applied Logic***, vol. 115 (2002), no. 1-3, pp. 33–70.

[15] R. DIACONESCU, *Axiom of choice and complementation*, ***Proceedings of the American Mathematical Society***, vol. 51 (1975), pp. 176–178.

[16] S. FEFERMAN, *A language and axioms for explicit mathematics*, ***Algebra and Logic (Fourteenth Summer Res. Inst., Austral. Math. Soc., Monash Univ., Clayton, 1974)*** (J. Crossley, editor), Lecture Notes in Mathematics, vol. 450, Springer, Berlin, 1975, pp. 87–139.

[17] ———, *Constructive theories of functions and classes*, ***Logic Colloquium '78 (Mons, 1978)*** (M. Boffa, D. van Dalen, and K. McAloon, editors), vol. 97, North-Holland, Amsterdam, 1979, pp. 159–224.

[18] ———, *Notes on Operational Set Theory I*, unpublished, Stanford University (2001) pp. 1-10.

[19] R. FLAGG and J. MYHILL, *Implication and analysis in classical Frege structures*, ***Annals of Pure and Applied Logic***, vol. 34 (1987), no. 1, pp. 33–85.

[20] H. FRIEDMAN, *The consistency of classical set theory relative to a set theory with intuitionistic logic*, ***The Journal of Symbolic Logic***, vol. 38 (1973), pp. 315–319.

[21] ———, *Set theoretic foundations for constructive analysis*, ***Annals of Mathematics. Second Series***, vol. 105 (1977), no. 1, pp. 1–28.

[22] G. JÄGER and T. STRAHM, *Totality in applicative theories*, ***Annals of Pure and Applied Logic***, vol. 74 (1995), no. 2, pp. 105–120.

[23] P. MARTIN-LÖF, *An intuitionistic theory of types: predicative part*, ***Logic Colloquium '73 (Bristol, 1973)*** (H. E. Rose and J. C. Sheperdson, editors), Studies in Logic and the Foundations of Mathematics, vol. 80, North-Holland, Amsterdam, 1975, pp. 73–118.

[24] ———, ***Intuitionistic Type Theory***, Studies in Proof Theory. Lecture Notes, vol. 1, Bibliopolis, Naples, 1984, Notes by Giovanni Sambin.

[25] Y. MOSCHOVAKIS, ***Elementary Induction on Abstract Structures***, North-Holland, Amsterdam, 1984.

[26] L. S. MOSS, *Power set recursion*, ***Annals of Pure and Applied Logic***, vol. 71 (1995), no. 3, pp. 247–306.

[27] J. MYHILL, *Some properties of intuitionistic Zermelo-Frankel set theory*, ***Cambridge Summer School in Mathematical Logic (1971)***, Lecture Notes in Mathematics, vol. 337, Springer, Berlin, 1973, pp. 206–231.

[28] ———, *Constructive set theory*, ***The Journal of Symbolic Logic***, vol. 40 (1975), no. 3, pp. 347–382.

[29] D. Normann, *Set recursion*, ***Generalized Recursion Theory, II (Proc. Second Sympos., Univ. Oslo, Oslo, 1977)***, Studies in Logic and the Foundations of Mathematics, vol. 94, North-Holland, Amsterdam, 1978, pp. 303–320.

[30] M. Rathjen, *Replacement versus collection and related topics in constructive Zermelo-Fraenkel set theory*, ***Annals of Pure and Applied Logic***, vol. 136 (2005), no. 1-2, pp. 156–174.

[31] ———, *The formulae as classes interpretation*, ***Proof Technology and Computation, Proceedings of the International Summer School Markoberdorf 2003***, IOS press, Amsterdam, 2006.

[32] M. Rathjen and S. Tupailo, *Characterizing the interpretation of set theory in Martin-Löf type theory*, ***Annals of Pure and Applied Logic***, vol. 141 (2006), no. 3, pp. 442–471.

[33] A. S. Troelstra and D. van Dalen, ***Constructivism in Mathematics: An introduction, Vols. I and II***, Studies in Logic and the Foundations of Mathematics, vol. 121, North-Holland, Amsterdam, 1988.

DIPARTIMENTO DI FILOSOFIA
UNIVERSITÀ DEGLI STUDI DI FIRENZE
VIA BOLOGNESE, 52
50139 FIRENZE, ITALY
E-mail: Cantini@philos.unifi.it
E-mail: Laura.Crosilla@unifi.it

PARAMETRIZED LOCAL ZETA FUNCTIONS

RAF CLUCKERS

Abstract. We extend further the theory of parameterized p-adic integrals, using the recently proven cell decomposition for subanalytic sets and functions. We construct a framework of (analytic) constructible functions (including parameterized local zeta functions) which is stable under integration. We bound the multiplicity of the complex poles of these integrals, uniform in the parameters. Finally, following Denef, we give an application to Poincaré series attached to the counting of well determined points modulo p^n for natural numbers n.

§1. introduction. Parametrized p-adic integrals are being better understood recently by the work of Denef in an algebraic context and the work of the author in an analytic context (and, by the theory of motivic integrals, but we will not treat uniformity questions here). In this paper we give a contribution to the theory of p-adic integrals, based on the Cell Decomposition Theorem of [3]. In [3], a framework of constructible functions which is stable under integration is introduced. Here we extend the notion of constructible functions such that parametrized (possibly twisted) local zeta functions, which are p-adic integrals of quite a general form, are included, and we prove that it is stable under p-adic integration. We control the possible order of poles of the zeta functions, uniform in the parameters.

After introducing this framework, we can use it to describe parametrized variants of (what is often called) the Igusa-Poincaré series and the Denef-Serre-Poincaré series of an analytic subset of $\mathbb{Z}_p^m$, which are both built up by counting well determined points modulo $p^n\mathbb{Z}_p$.

1.1. Let us first give an example of the type of integrals we study. Under the supposition of integrability for all $\lambda \in \mathbb{Q}_p^r$ and all big enough real

2000 *Mathematics Subject Classification.* Primary 03C98, 11S80, 11S40; Secondary 03C10, 11U09.

Key words and phrases. p-adic integrals, Igusa's local zeta functions, cell decomposition, Tate series, restricted power series, subanalytic p-adic sets, analytic functions.

The author is a postdoctoral fellow of the Fund for Scientific Research – Flanders (Belgium)(F.W.O.)

Logic Colloquium '04
Edited by A. Andretta, K. Kearnes, and D. Zambella
Lecture Notes in Logic, 29

numbers $s \gg 0$, we study parametrized p-adic integrals of the form

$$I(s,\lambda) := \int_{\mathbb{Q}_p^m} |f(\lambda,x)|^s\, g(\lambda,x)\, |dx| \tag{1.1}$$

against the normalized Haar measure $|dx|$, where f is a subanalytic function from $\mathbb{Q}_p^{r+m}$ to $\mathbb{Q}_p$ and $g : \mathbb{Q}_p^{r+m} \to \mathbb{Q}$ a constructible function[1] on $\mathbb{Q}_p^{r+m}$. When g is a Schwartz-Bruhat function[2] and f is a polynomial, I is nothing else than a parameterized local zeta function as studied by Weil, Igusa, and many others; see [13] for an excellent introduction to local zeta functions. In Theorem 2.5 we prove that $I(s,\lambda)$ is a finite sum of terms of a similar form as the integrand in (1.1) (up to some extra rational factors in p^s), and we also give upperbounds for the multiplicity of the (complex) poles in s, see Theorem 2.5. Similar integrals are described by Denef [7, Theorem 3.1] in an algebraic context, namely, in the case that f is a semialgebraic function and $g = 1$.

If $\chi : \mathbb{Z}_p^\times \to \mathbb{C}^\times$ is a multiplicative character extended by zero outside $\mathbb{Z}_p^\times$, and $\mathrm{ac} : \mathbb{Q}_p \to \mathbb{Z}_p$ is the map $x \mapsto xp^{-v(x)}$ for $x \neq 0$ and $\mathrm{ac}(0) = 0$, we pursue a similar study for integrals of the form

$$I(s,\lambda,\chi) := \int_{\mathbb{Q}_p^m} |f(\lambda,x)|^s\, g(\lambda,x)\, \chi(\mathrm{ac}(h(\lambda,x)))|dx|, \tag{1.2}$$

where f and g are as in (1.1) and $h : \mathbb{Q}_p^{r+m} \to \mathbb{Q}_p$ is a subanalytic function, cf. Lemma 2.10 and Theorem 2.5.

Terminology and notation. Let p denote a fixed prime number, $\mathbb{Q}_p$ the field of p-adic numbers and K a fixed finite field extension of $\mathbb{Q}_p$ with valuation ring R. For $x \in K^\times$ let $v(x) \in \mathbb{Z}$ denote the p-adic valuation of x and $|x| = q^{-v(x)}$ the p-adic norm, with q the cardinality of the residue field. For $n > 0$, let P_n be the multiplicative group of the n-th powers in $K^\times$ and let λP_n denote $\{\lambda x \mid x \in P_n\}$ for $\lambda \in K$. With π a fixed uniformizer of R, define $\mathrm{ac} : K \to R$ as the map sending $x \neq 0$ to $x\pi^{-v(x)}$ and 0 to 0.

For $X = (X_1, \ldots, X_m)$, let $R\langle X\rangle$ be the ring of restricted power series over R in the variables X; it is the ring of power series $\sum a_i X^i$ in $R[[X]]$ such that $|a_i|$ tends to 0 as $|i| \to \infty$. (Here, we use the multi-index notation where $i = (i_1, \ldots, i_m)$, $|i| = i_1 + \cdots + i_m$ and $X^i = X_1^{i_1} \ldots X_m^{i_m}$.) For $x \in R^m$ and $f = \sum a_i X^i$ in $R\langle X\rangle$, the series $\sum a_i x^i$ converges to a limit in K, thus, one

[1]The definition of subanalytic functions is given in the section on terminology below; the definition of constructible functions is given in section 2.

[2]A Schwartz-Bruhat function on $\mathbb{Q}_p^{r+m}$ is a locally constant function (into $\mathbb{C}$) with compact support.

can associate to f a *restricted analytic function* given by

$$f : K^m \to K : x \mapsto \begin{cases} \sum_i a_i x^i & \text{if } x \in R^m, \\ 0 & \text{else.} \end{cases}$$

By a p-adic manifold we mean a p-adic analytic manifold which is Hausdorff and everywhere of the same dimension; see [1].

A key notion is the following:

DEFINITION 1.1. A subset of K^n is called (globally) *subanalytic* if it can be obtained in finitely many steps by taking unions, intersections, complements and linear projections of the zero locus of polynomials in K^{n+e} and of zerosets of restricted analytic functions in $K^{n+e'}$ for some e and e'. A function $f : X \subset K^m \to K^n$ is called subanalytic if its graph is a subanalytic set.

A subset of K^n is called *semialgebraic* if it can be obtained in finitely many steps by taking unions, intersections, complements and linear projections of the zero locus of polynomials in K^{n+e} for some e. A function $f : X \subset K^m \to K^n$ is called semialgebraic if its graph is a semialgebraic set.

We refer to [9, 10], [11], and [3] for the basic theory of subanalytic sets.

§2. Integration of constructible functions. We will use the notion of cells and the Cell Decomposition Theorem of [3] in order to integrate constructible functions. Our usage of this cell decomposition is concentrated in Proposition 2.3.

DEFINITION 2.1. [3] A cell $A \subset K$ is a (nonempty) set of the form

$$\{t \in K \mid |\alpha| \square_1 |t - c| \square_2 |\beta|,\ t - c \in \lambda P_n\}, \tag{2.1}$$

with constants $n > 0$, $\lambda, c \in K$, $\alpha, \beta \in K^\times$, and $\square_i$ either $<$ or no condition. A cell $A \subset K^{m+1}$, $m \geq 0$, is a set of the form

$$\begin{aligned} \{(x,t) \in K^{m+1} \mid \quad & x \in D,\ |\alpha(x)| \square_1 |t - c(x)| \square_2 |\beta(x)|, \\ & t - c(x) \in \lambda P_n\}, \end{aligned} \tag{2.2}$$

with $(x,t) = (x_1, \dots, x_m, t)$, $n > 0$, $\lambda \in K$, $D = \pi_m(A)$ a cell where π_m is the projection $K^{m+1} \to K^m$, subanalytic functions $\alpha, \beta : K^m \to K^\times$, and $c : K^m \to K$, and the $\square_i$ either $<$ or no condition, such that the functions α, β, and c are analytic on D. We call c the center of the cell A and λP_n the coset of the cell.

THEOREM 2.2 (Subanalytic p-adic Cell Decomposition, [3]). *Let $X \subset K^{m+1}$ be a subanalytic set and $f_j : X \to K$ subanalytic functions for $j = 1, \dots, r$. Then there exists a finite partition of X into cells A with center c and coset λP_n such that*

$$|f_j(x,t)| = |\delta_j(x)| \cdot |(t - c(x))^{a_j} \lambda^{-a_j}|^{\frac{1}{n}}, \quad \textit{for each } (x,t) \in A, \tag{2.3}$$

with $(x,t)=(x_1,\dots,x_m,t)$, *integers* a_j, *and* $\delta_j : K^m \to K$ *subanalytic functions, analytic on* $\pi_m(A)$, *for* $j=1,\dots,r$. *If* $\lambda = 0$, *we use the convention that* $a_j = 0$ *and* $0^0 = 1$.

Theorem 2.2 is a generalisation of the cell decomposition Theorem for polynomial maps by Denef [6], [8].

2.1. Simple sets. A set $X \subset \mathbb{Z}^n \times K^m$ is called (subanalytically) *simple* if

$$\left\{(\lambda,x)\in K^{n+m} \mid (v(\lambda_1),\dots,v(\lambda_n),x)\in X \;\&\; \prod_{i=1\dots,n} \lambda_i \neq 0\right\}$$

is a subanalytic set. A function $h : X \to Y$ with X and Y simple sets is called *simple* if its graph is simple[3].

We recall the cell decomposition for simple subsets of $\mathbb{Z}^{k+1}$ of [2]. A simple set $A \subset \mathbb{Z}^{k+1}$, $k \geq 0$, is called a 1-cell if there are simple functions $a_1, a_2, d : \mathbb{Z}^k \to \mathbb{Z}$ and integers $0 \leq c < n$ such that A equals

$$\{(x_1,\dots,x_k,t)\in\mathbb{Z}^{k+1} \mid x\in C,\quad a_1(x)\,\square_1\, t\,\square_2\, a_2(x),\quad t\equiv c \bmod n\},$$

with $\square_i$ either $<$ or no condition for $i = 1, 2$, and C a simple set which is equal to the projection of A on $\mathbb{Z}^k$ and such that the number of elements in the fiber $A_x := \{t \in \mathbb{Z} \mid (x,t) \in A\}$ gets arbitrarily large when x runs through C. A simple set $A \subset \mathbb{Z}^{k+1}$, $k \geq 0$, is called a 0-cell if there is a simple set $C \subset \mathbb{Z}^k$ and a simple function $a : C \to \mathbb{Z}$ such that $A = \{(x, a(x)) \mid x \in C\}$. We call a function $f : X \subset \mathbb{Z}^k \to \mathbb{Z}$ linear if there exists a polynomial p over $\mathbb{Z}$ of degree ≤ 1 such that $f(x) = p(x)$ for each $x \in X$.

PROPOSITION 2.3. [2] *Let* $X \subset \mathbb{Z}^{k+1}$ *be simple and* $f_i : X \to \mathbb{Z}$ *be simple functions for* $i = 1,\dots,r$. *Then there exists a finite partition of* X *into cells* A_j *such that each of the restrictions* $f_i|_{A_j}$ *is linear.*

2.2. Constructible functions. Let $\mathbb{R}_{\gg 0}$ be the germ of real numbers around $+\infty$, hence, $s \in \mathbb{R}_{\gg 0}$ means that s is sufficiently big. Make the conventions $(\mathbb{R}_{\gg 0})^0 = \mathbb{R}^0_{\gg 0} = \{+\infty\}$, $q^{-\infty} = 0$, $1/(a+\infty) = 1/(a-\infty) = 0$ for real a, and so on. Recall that q is the residue field cardinality of K. Let A_q be the ring

$$A_q := \mathbb{Z}\left[q^{-1}, \left(\frac{1}{1-q^{-i}}\right)_{i\in\mathbb{N}\setminus\{0\}}\right].$$

For any sets X and S, write X^S for the set of functions $X \to S$.

Let

$$\mathcal{C}(\mathbb{R}^n_{\gg 0} \times X)$$

be the A_q-subalgebra of $(\mathbb{R}^n_{\gg 0} \times X)^{\mathbb{R}}$ generated by all functions $(s,x) \mapsto q^{\alpha(x)\cdot s+\beta(x)}$, $(s,x) \mapsto \beta(x)$, and $(s,x) \mapsto \frac{1}{1-q^{a\cdot s+b}}$, for $\alpha : X \to \mathbb{Z}^n$ and

[3]This notion generalizes and is compatible with the analytic notion of simple functions in [9].

$\beta : X \to \mathbb{Z}$ simple functions, with $a \in \mathbb{Z}^n \setminus \{(0,\ldots,0)\}$, $b \in \mathbb{Z}$, $a \cdot s = \sum_{i=1}^n a_i s_i$, and $\alpha(x) \cdot s = \sum_{i=1}^n \alpha_i(x) s_i$. We call elements of $\mathcal{C}(\mathbb{R}^n_{\gg 0} \times X)$ (subanalytically) *constructible functions* on $\mathbb{R}^n_{\gg 0} \times X$.

DEFINITION 2.4. Let f be a function in $\mathcal{C}(\mathbb{R}^n_{\gg 0} \times X)$. Put

$$S(f) = \min\big(\max_{i\in I,\ \ell \in L_i} (\sharp J_i + c_{i\ell})\big),$$

where the minimum is taken over all possible ways of writing f as

$$f(s,x) = \sum_{i\in I} \frac{r_i q^{\alpha_i \cdot s + \beta_i} \prod_{j\in J_i} \gamma_{ij}}{\prod_{\ell \in L_i} (1 - q^{a_{i\ell}\cdot s + b_{i\ell}})^{c_{i\ell}}}, \tag{2.4}$$

with I and the J_i, L_i finite sets, $r_i \in A_q$, $a_{i\ell} \in \mathbb{Z}^n \setminus \{(0,\ldots,0)\}$, $b_{i\ell} \in \mathbb{Z}$, $c_{i\ell} \in \mathbb{N}$, and $\gamma_{ij}, \beta_i : X \to \mathbb{Z}$ and $\alpha_i : X \to \mathbb{Z}^n$ simple functions, and where for all i and different ℓ_0, ℓ_1 in L_i the tuples $(a_{i\ell_0}, b_{i\ell_0})$ and $(a_{i\ell_1}, b_{i\ell_1})$ are linearly independent in the module $\mathbb{Z}^{n+1}$.

Note that any f in $\mathcal{C}(\mathbb{R}^n_{\gg 0} \times X)$ can be written as in (2.4) since

$$\frac{1}{(1 - q^{a\cdot s + b})} = \frac{Q(q^{a\cdot s+b})}{(1 - q^{n(a\cdot s) + nb})},$$

with Q a polynomial over $\mathbb{Z}$. The number $S(f)$ is a kind of uniform upper bound for the poles in s of f, along hyperplanes.

2.3. The following theorem is a generalisation of a conjecture of Denef (see [10] for a statement of the conjecture and [3] for its proof). It is the main result of this paper.

THEOREM 2.5 (Basic Theorem on p-adic Analytic Integrals). *Let X be a simple set, k, m, n natural numbers, and let f be in $\mathcal{C}(\mathbb{R}^n_{\gg 0} \times X \times \mathbb{Z}^k \times K^m)$. Suppose that the function $f(s, x, \cdot, \cdot) : \mathbb{Z}^k \times K^m \to \mathbb{R} : (z, y) \mapsto f(s, x, z, y)$ is absolutely integrable for all $(s, x) \in \mathbb{R}^n_{\gg 0} \times X$ with respect to the discrete measure on $\mathbb{Z}^k$ and the Haar measure on K^m and put*

$$g(s,x) := \sum_{z\in\mathbb{Z}^k} \int_{K^m} f(s,x,z,y)|dy|.$$

Then, g is in $\mathcal{C}(\mathbb{R}^n_{\gg 0} \times X)$. Moreover, $S(g) \le S(f) + m + k$.

Further, in the case that $X = \mathbb{Z}^\ell \times K^r$ and $f(s, x, \cdot, \cdot)$ is absolutely integrable for almost all $x \in X$ (that is, up to a measure zero set) and all $s \in \mathbb{R}^n_{\gg 0}$, there exists a simple set $X' \subset X$ such that $X \setminus X'$ has measure zero and such that $f(s, x, \cdot, \cdot)$ is absolutely integrable for all $(s, x) \in \mathbb{R}^n_{\gg 0} \times X'$.

PROOF. We may suppose that $X = K^r \times \mathbb{Z}^\ell$. By applying Theorem 2.2 in a nowadays standard way, we can write parameterized p-adic integrals over K^m of constructible functions on $\mathbb{R}^n_{\gg 0} \times \mathbb{Z}^{\ell+k} \times K^{r+m}$ as sums of constructible functions on $\mathbb{R}^n_{\gg 0} \times \mathbb{Z}^{\ell+k+m} \times K^r$ over $\mathbb{Z}^m$. Again by Cell Decomposition and by ignoring cells of small dimension, it is enough to consider the case that $r = m = 0$. The third statement then becomes trivial since we use the counting measure on $\mathbb{Z}^\ell$ (see [4] for more details of this reduction).

We prove the first two statements. By Fubini's Theorem and by induction, it is enough to consider the case that $k = 1$. Let $f_i : \mathbb{Z}^{\ell+1} \to \mathbb{Z}$ be the simple functions which occur in a writing of f as a sum $\sum_i \prod_j G_{ij}(s, x)$ of products of generators of $\mathcal{C}(\mathbb{R}^n_{\gg 0} \times \mathbb{Z}^{\ell+1})$.[4] Apply Proposition 2.3 to the functions f_i. One sees that summing over the so-obtained cells essentially amounts to summing (derivatives) of geometric series of monomials in q and q^{s_i}, $i = 1, \dots, n$, along arithmetic progressions. More precisely, $g(s, x)$ is a finite A_q-linear combination of sums of the form

$$\sum_{z \in I_x} q^{(\frac{z-c}{e})(a \cdot s + b)} \left(\frac{z - c}{e} \right)^d h(s, x), \tag{2.5}$$

where $a \in \mathbb{Z}^n$, $b \in \mathbb{Z}$, $d \in \mathbb{N}$, $(s, x) \in \mathbb{R}^n_{\gg 0} \times C$ with $C \subset \mathbb{Z}^\ell$ simple, $a : C \to \mathbb{Z}^n$ simple, and $h \in \mathcal{C}(\mathbb{R}^n_{\gg 0} \times \mathbb{Z}^\ell)$, and

$$I_x = \{z \in \mathbb{Z} \mid \alpha(x) \square_1 z \square_2 \beta(x),\ z \equiv c \bmod e\}, \tag{2.6}$$

with $0 \leq c < e$, $\alpha, \beta : C \to \mathbb{Z}$ simple functions, and $\square_i$ either $<$ or no condition. If we are working on a 0-cell, such a term is automatically integrable over I_x and its sum over I_x is of the right form. Suppose that we are working on a 1-cell (that is, the cardinality of I_x is not bounded when $x \in C$ varies). The only cases that there can be nonsummable terms among the terms (2.5) is when at least one of the $\square_i$ is no condition; in such a case, one sees that there can be no cancellation between unsummable terms and summable terms (cf. Lemma 2.6), by comparing the asymptotics of the terms for z going to $\pm\infty$. Hence, we may suppose that the terms (2.5) are summable over I_x for all $x \in C$. This also shows that we may suppose that $S(f)$ can be calculated on the terms (2.5). If both $\square_1$ and $\square_2$ are no condition, only the zero term is summable. Hence, by symmetry, we may suppose that $\square_1$ is $<$. The first two statements now follow easily from Lemma 2.8 when $a = (0, \dots, 0)$, and, when $a \neq (0, \dots, 0)$, they follow from Lemma 2.7. ⊣

The following Lemma is proven by comparing the asymptotics of the terms for i going to $+\infty$.

LEMMA 2.6. *Let r_j and $t > 1$ be real numbers, and a_j, d_j integers with $d_j \geq 0$, $j = 1, \dots, n$. For each integer $i \geq 0$ let v_i be*

$$v_i := \sum_{j=1}^n r_j\, t^{a_j i}\, i^{d_j}$$

[4]By this, we mean all simple functions, also those which occur as an exponent of p or q^s.

and suppose that the series $\sum_{i\geq 0} v_i$ *is summable. Then, for each* i,

$$v_i = \sum_{\substack{j \\ a_j<-1}} r_j\, t^{a_j i}\, i^{d_j}.$$

LEMMA 2.7. [5] *Let* P *be a degree* d *polynomial* P *in* $\mathbb{Z}[T]$. *The equality*

$$\sum_{n\geq a} P(n)T^n = \sum_{i=0}^{d} \frac{[\Delta^i P(a)]T^{a+i}}{(1-T)^{i+1}} \tag{2.7}$$

holds in $\mathbb{Z}[[T]]$ *for all* $a \in \mathbb{N}$. *Here* Δ^i *is the* i*-th iterate of the difference operator* $P \mapsto P(X+1) - P(X)$ *with the convention* $\Delta^0 P = P$.

The next Lemma is a refinement of the claim in the proof of Lemma 3.2 of [7].

LEMMA 2.8. [5] *Let* P *be a degree* d *polynomial* P *in* $\mathbb{Z}[T]$. *There exist Presburger functions* $\gamma_{ij} : \mathbb{Z} \to \mathbb{Z}$ *for* i *in a finite set* I *and* $j = 0 \ldots d$ *such that*

$$\sum_{0\leq n\leq a} P(n) = \sum_{i\in I} \prod_{j=0,\ldots,d} \gamma_i(a),$$

for each $a > 0$.

REMARK 2.9. Probably one can deduce from the proof of Theorem 2.5 that

$$S(g) \leq S(f) + m + k$$

in Theorem 2.5

2.4. Recall that ac : $K \to R$ is the map $x \mapsto x\pi^{-v(x)}$ for $x \neq 0$ and $\mathrm{ac}(0) = 0$, with π a fixed uniformizer. Theorem 2.5 can be applied as well to describe integrals of the form

$$I(s,x) = \int_{K^m} \sum_{i=1}^{r} \chi(\mathrm{ac}(f_i(x,y)))h_i(s,x,y)|dy|,$$

with X and the $f_i : X \times K^m \to K$ (subanalytically) simple, the h_i in $\mathcal{C}(\mathbb{R}^n_{\gg 0} \times X \times K^m)$, and $\chi : R^\times \to \mathbb{C}^\times$ a multiplicative character extended by zero, when suitable integrability conditions are met for the h_i. This follows from the following lemma.

Let B_q be the image of $R^\times$ under all multiplicative characters $R^\times \to \mathbb{C}^\times$ and let C_q be $B_q \otimes A_q$.

LEMMA 2.10. *For a simple set* $X \subset \mathbb{Z}^n \times K^m$, *the* C_q*-algebra* $\mathcal{C}(\mathbb{R}^n_{\gg 0} \times X) \otimes C_q$ *of functions* $\mathbb{R}^n_{\gg 0} \times X \to \mathbb{C}$ *is equal (as* C_q*-algebra of functions) to the* C_q*-algebra generated by* $\mathcal{C}(\mathbb{R}^n_{\gg 0} \times X)$ *and the functions* $(s,x) \mapsto \chi(\mathrm{ac}(f(x)))$ *with* $f : X \to K$ *subanalytic and where* $\chi : R^\times \to \mathbb{C}^\times$ *runs over all multiplicative characters.*

PROOF. The Lemma follows at once by the fact that, for any $\xi \in C_p$ and any χ, the set $\{x \in R^\times \mid \chi(x) = \xi\}$ is a simple set, cf. [8]. ⊣

§3. Poincaré series. To end the paper we give an application of Theorem 2.5 to Poincaré series based on counting the number of points modulo powers of a uniformizer π, following Denef [7]. Let $f : K^{r+m} \to K$ be a subanalytic function in the variables (λ, x). For each $n \in \mathbb{N}$ and $\lambda \in K^r$, let

$$N_n(\lambda) = \sharp\{x \bmod \pi^n \mid x \in R^m \text{ and } f(\lambda, x) \equiv 0 \bmod \pi^n\}, \quad \text{and}$$
$$\tilde{N}_n(\lambda) = \sharp\{x \bmod \pi^n \mid x \in R^m \text{ and } f(\lambda, x) = 0\}.$$

To these data we associate Poincaré series parametrized by λ:

$$P(T, \lambda) = \sum_{n \in \mathbb{N}} N_n(\lambda) T^n \quad \text{and} \quad \tilde{P}(T, \lambda) = \sum_{n \in \mathbb{N}} \tilde{N}_n(\lambda) T^n.$$

THEOREM 3.1. *There exist numbers $c_i \in \mathbb{N}$, $a_i \in \mathbb{N} \setminus \{0\}$, $b_i \in \mathbb{Z}$, simple functions $\alpha_i : K^r \to \mathbb{N}$, and γ_i in $\mathcal{C}(K^r)$, such that $S(\gamma_i) + c_i \leq m$ for each i and*

$$P(T, \lambda) = \sum_{i \in I} \frac{\gamma_i(\lambda) T^{\alpha_i(\lambda)}}{(1 - q^{b_i} T^{a_i})^{c_i}} \quad \textit{for all } \lambda \in R^s.$$

The same statement holds for $\tilde{P}(T, \lambda)$.

PROOF. The Theorem follows from Theorem 2.5 by using partial fractions. This is because there is an easy functional equation between $P(p^{-m-s}, \lambda)$, resp. $\tilde{P}_\lambda(p^{-m-s}, \lambda)$, and an integral over K^m of some g in $\mathcal{C}(\mathbb{R}_{\gg 0} \times K^{r+m})$ with $S(g) = 0$ (see [7] or [9] for these functional equations). ⊣

Note that many other Poincaré series can be built up in a similar way and can be described by the previous theorem. See [12] for a general context, uniformity results in p when f is semialgebraic, and motivic variants.

REFERENCES

[1] N. BOURBAKI, ***Variétés différentielles et analytiques. Fascicule de résultats***, Hermann, Paris, 1967, (French).

[2] R. CLUCKERS, *Presburger sets and p-minimal fields*, ***The Journal of Symbolic Logic***, vol. 68 (2003), pp. 153–162, arXiv:math.LO/0206197.

[3] ———, *Analytic p-adic cell decomposition and integrals*, ***Transactions of the American Mathematical Society***, vol. 356 (2004), no. 4, pp. 1489–1499, arXiv:math.NT/0206161.

[4] ———, *Multi-variate Igusa theory: Decay rates of p-adic exponential sums*, ***IMRN***, (2004), no. 76, pp. 4093–4108, arXiv:math.NT/0306351.

[5] R. CLUCKERS and F. LOESER, *Constructible motivic functions and motivic integration*, arxiv:math.AG/0410203.

[6] J. DENEF, *The rationality of the Poincaré series associated to the p-adic points on a variety*, ***Inventiones Mathematicae***, vol. 77 (1984), pp. 1–23.

[7] ———, *On the evaluation of certain p-adic integrals*, ***Théorie des nombres, sémin. delange-pisot-poitou 1983–84***, vol. 59, 1985, pp. 25–47.

[8] ———, *p-adic semialgebraic sets and cell decomposition*, ***Journal für die reine und angewandte Mathematik***, vol. 369 (1986), pp. 154–166.

[9] ———, *Multiplicity of the poles of the poincaré series of a p-adic subanalytic set*, ***Sém. Th. Nombres Bordeaux***, vol. 43 (1987-1988), pp. 1–8.

[10] ———, *Arithmetic and geometric applications of quantifier elimination for valued fields*, ***Model Theory, Algebra, and Geometry*** (D. Haskell, A. Pillay, and C. Steinhorn, editors), MSRI Publications, vol. 39, Cambridge University Press, 2000, pp. 173–198.

[11] J. Denef and L. van den Dries, *p-adic and real subanalytic sets*, ***Annals of Mathematics***, vol. 128 (1988), no. 1, pp. 79–138.

[12] J. Denef and F. Loeser, *On some rational generating series occuring in arithmetic geometry*, ***Geometric Aspects of Dwork Theory***, vol. 1, de Gruyter, 2004, arXiv:math.NT/0212202, pp. 509–526.

[13] J. Igusa, ***An Introduction to the Theory of Local Zeta Functions***, Studies in advanced mahtematics, AMS, 2000.

KATHOLIEKE UNIVERSITEIT LEUVEN
DEPARTEMENT WISKUNDE, CELESTIJNENLAAN 200B
B-3001 LEUVEN, BELGIUM

Current address: École Normale Supérieure, Département de mathématiques et applications, 45 rue d'Ulm, 75230 Paris Cedex 05, France

E-mail: cluckers@ens.fr

URL: www.wis.kuleuven.ac.be/algebra/raf/

COLOURINGS OF HYPERGRAPHS, PERMUTATION GROUPS AND CSP'S

BENOIT LAROSE AND LUCIEN HADDAD

Abstract. We investigate the complexity of strong colouring problems of hypergraphs associated to groups of permutations initiated in [13]. By reformulating these as Constraint Satisfaction Problems (CSP's) we are able to exploit recent algebraic results to answer various questions posed by Haddad and Rödl [13]. In particular, we show that all known tractable cases are explained by the presence of a Mal'tsev operation, and that all known **NP**-complete cases correspond to algebras that admit no Taylor operation, thus confirming the dichotomy conjecture by Bulatov et al. [7, 8]. We classify completely the case where the group consists of all affine transformations on a ring of matrices over a commutative ring.

§1. Introduction. We start by briefly describing the *strong colouring problem for permutation groups* that was investigated in [13] and [12]; this problem has its origins in the classification of maximal partial clones over a finite non-empty set [14, 15, 16]. We shall immediately reformulate this decision problem as a Constraint Satisfaction Problem (CSP) in order to exploit various universal algebraic tools to study its algorithmic complexity.

As usual, let S_h denote the symmetric group on h letters, which we consider acting on the set $\{1, 2, \ldots, h\}$. Let G be a subgroup of S_h and let ρ be an h-ary relation on a finite set X. We say that G is the *symmetry group* of ρ if, for every $\pi \in G$ and every $(x_1, \ldots, x_h) \in X^h$, we have that $(x_1, \ldots, x_h) \in \rho$ if and only if $(x_{\pi(1)}, \ldots, x_{\pi(h)}) \in \rho$. Define an h-ary relation ρ_G on $\{1, \ldots, h\}$ as follows:

$$\rho_G = \{(\pi(1), \ldots, \pi(h)) : \pi \in G\}.$$

A map $\phi : X \to \{1, \ldots, h\}$ is a *strong h-colouring* of ρ if it is a homomorphism from $\langle X; \rho\rangle$ to $\langle\{1, \ldots, h\}; \rho_G\rangle$, i.e. if $(\phi(x_1), \ldots, \phi(x_h)) \in \rho_G$ whenever

---

2000 *Mathematics Subject Classification.* 05C15, 68R10, 68Q25.

Key words and phrases. Hypergraph colouring, Permutation groups, Constraint Satisfaction Problems, Taylor operations, Mal'tsev operations.

Both authors' research is supported by a grant from NSERC

Part of this research was conducted while the first author was visiting RMC Kingston

Logic Colloquium '04
Edited by A. Andretta, K. Kearnes, and D. Zambella
Lecture Notes in Logic, 29

$(x_1, \ldots, x_h) \in \rho$. The relation ρ is *areflexive* if for all $(x_1, \ldots, x_h) \in \rho$ and all $1 \leq i < j \leq h$ we have $x_i \neq x_j$.

- *Strong Colouring Problem for G*
  Input: an areflexive relation ρ with symmetry group G;
  Question: does ρ have a strong h-colouring ?

It is immediate that this problem is in the complexity class **NP**. In [13] Haddad and Rödl ask for which groups is the problem in **P** or **NP**-complete. They give various examples of both types of behaviour, and in [12] Haddad, Hell and Mendelsohn investigate this problem further.

We now reformulate this decision problem in a more convenient form. Let $A = \{1, \ldots, h\}$.

- $CSP(\rho_G)$
  Input: a relational structure $\langle X; \mu \rangle$ where μ is h-ary;
  Question: is there a homomorphism from $\langle X; \mu \rangle$ to $\langle A; \rho_G \rangle$?

It is easy to see that in fact these problems are equivalent. Indeed, it is clear that if the input structure for $CSP(\rho_G)$ is not areflexive then there cannot be a homomorphism. Furthermore, it is easy to verify that if a homomorphism exists, and if both $(x_1, \ldots, x_h)$ and $(x_{\pi(1)}, \ldots, x_{\pi(h)})$ are in μ then $\pi \in G$. Finally, we can always add to μ all permutation of its tuples by permutations in G without changing the set of solutions. Obviously all these steps can be done in time polynomial in the size of the structure $\langle X; \mu \rangle$ and thus we may safely restrict the inputs to the problem $CSP(\rho_G)$ (abbreviated from now as $CSP(G)$) to areflexive structures with symmetry group G.

More generally, for any finite set of relations $R = \{\rho_1, \ldots, \rho_k\}$ on A where ρ_i has arity d_i, let $CSP(R)$ denote the following decision problem:

- $CSP(R)$
  Input: a relational structure $\langle X; \mu_1, \ldots, \mu_k \rangle$ where μ_i is d_i-ary;
  Question: is there a homomorphism from $\langle X; \mu_1, \ldots, \mu_k \rangle$ to $\langle A; \rho_1, \ldots, \rho_k \rangle$?

In 1993, Feder and Vardi [10] conjectured that, depending on the constraint relations, the problem $CSP(\rho_1, \ldots, \rho_k)$ should be either in **P** or **NP**-complete. This *dichotomy conjecture* has attracted a great deal of attention lately (see for example [4, 2, 3, 7, 11, 21, 22, 23, 25, 30]), and the conjecture has been settled in various special cases. A deep connection with universal algebra was uncovered by P. Jeavons [21] and further refined in collaboration with Bulatov and Krokhin [7, 8]. They have stated a precise conjecture predicting which sets of relations should give rise to tractable problems and which are **NP**-complete (a conjecture along the same lines is sketched in [11].) We now state the precise conjecture and briefly outline the necessary background (we refer the reader to [19, 31] and [33] for basic results in universal algebra.)

Let f be an n-ary operation on a set A and let θ be a k-ary relation on A. We say that f *preserves* θ or that θ is *invariant under* f if the following holds: given any matrix M of size $k \times n$ whose columns are in θ, applying f to the rows of M will produce a k-tuple in θ. Given a set R of relations on A, we define $Pol(R)$ to be the set of all operations on A that preserve all relations in R. Jeavons observed that the complexity of the problem $CSP(R)$ is essentially determined by $Pol(R)$. Bulatov, Jeavons and Krokhin prove that, to settle the dichotomy conjecture, it is sufficient to consider sets of relations R such that every member of $Pol(R)$ is surjective [7]. When a set of relations has this property we say that it is a *core*.[1]

An operation f on A of arity at least 2 is *idempotent* if it satisfies the identity $f(x, \ldots, x) \approx x$ (where $\approx$ indicates the equality holds with all variables universally quantified). Equivalently, an operation on A is idempotent if it preserves every unary relation of the form $\{a\}$ for $a \in A$. An n-ary idempotent operation f is a *Taylor operation* if it satisfies, for every $1 \leq i \leq n$ an identity of the form

$$f(x_1, \ldots, x_{i-1}, x, x_{i+1}, \ldots, x_n) \approx f(y_1, \ldots, y_{i-1}, y, y_{i+1}, \ldots, y_n)$$

where $x_j, y_j \in \{x, y\}$ for all $1 \leq j \leq n$ (see [34, 19]). For instance, a binary operation is a Taylor operation if and only if it is idempotent and commutative; in particular, semilattice operations are Taylor operations. Here are other common instances of Taylor operations:

- a 3-ary operation M is a *majority* operation if it satisfies the identities

$$M(x, x, y) \approx M(x, y, x) \approx M(y, x, x) \approx x.$$

- a 3-ary operation m is a *Mal'tsev* operation if it satisfies

$$m(x, x, y) \approx m(y, x, x) \approx y.$$

The following hardness criterion was first proved in [7] in a different formulation, but may be found in the present form in [30].

THEOREM 1.1. *Let R be a core. If $Pol(R)$ contains no Taylor operation then the problem $CSP(R)$ is* **NP**-*complete.*

It is conjectured that, in essence, this is the only reason why a constraint satisfaction problem is hard:

DICHOTOMY CONJECTURE. [7] Let R be a core. If $Pol(R)$ contains a Taylor operation then $CSP(R)$ is in **P**; otherwise it is **NP**-complete.

---

[1]More precisely, we should say that the relational structure $\langle A; R\rangle$ is a core, i.e. that all its endomorphisms are automorphisms, see for instance [11] or [18]. Every finite relational structure has, up to isomorphism, a unique core, and the CSP problems for the structure and its core are poly-time equivalent [7].

We view our investigation of the decision problems $CSP(G)$ as an opportunity to pursue the study of this conjecture in a context with a rather different flavour from the cases that have been studied so far. Secondly, the clone $Pol(R)$ is of independent interest because it is the largest element of the so-called monoidal interval determined by the group G; these intervals have attracted some attention in the recent literature (see for example [24, 27, 26, 32]) and we shall feel free to analyse in more detail some of the idempotent terms of these algebras.

We now briefly outline the contents of the paper. We shall prove that in all known cases where $CSP(G)$ is in **P**, there is a Taylor operation that preserves the relation ρ_G; in fact, we shall exhibit a Mal'tsev operation in every case. This includes the cases of regular groups, primitive groups and wreath products of these (Proposition 3.1, Corollary 3.4 and Theorem 3.2). Furthermore, we prove that in all known **NP**-complete cases there is no Taylor operation, confirming the dichotomy conjecture, and further proving that the problem cannot be of so-called *bounded width* (see [29]). We shall also present some evidence that the classification of the tractable cases cannot avoid the intransitive groups, see subsection 3.3. In particular, we answer in section 4 the following question posed by Haddad and Rödl [13]: is there a group G for which $CSP(G)$ is **NP**-complete but such that its actions on orbits are tractable? We construct an infinite family of such examples. We shall also answer another question posed in the same paper, namely, we provide an infinite family of tractable $CSP(G)$ such that G is transitive and has a (normal) transitive subgroup H such that $CSP(H)$ is **NP**-complete. Finally, we shall completely classify the complexity of the problem when the permutation group G consists of all transformations of the form $x \mapsto ax+b$ where a and b are $n \times n$ matrices over some commutative ring S and a is invertible (Theorem 4.5).

§2. **Preliminaries.** Unless otherwise specified, the groups we consider in the sequel are concrete groups of permutations. Recall that for G a group of permutations on A and $x \in A$, the set of all $\sigma(x)$ with $\sigma \in G$ is called the *orbit of x under G*. The group G is *transitive* if it has a single orbit, i.e. for every $x, y \in A$ there exists $\sigma \in G$ such that $\sigma(x) = y$. For $x \in A$ the *stabiliser* of x is the subgroup $G_x = \{\sigma \in G : \sigma(x) = x\}$; for $S \subseteq A$ the (set-wise) stabiliser of S is $G_S = \{\sigma \in G : \sigma(S) \subseteq S\}$. A transitive group G is *regular* if for each x and y in A the permutation taking x into y is unique; equivalently, if every non-identity permutation in G is fixed-point free, i.e. $|G_x| = 1$ for all $x \in A$. A group G is *primitive* if there is no non-trivial equivalence relation θ on the set A which is invariant under the permutations of G (viewed as unary operations on A).

It will be convenient (and interesting) from time to time to consider our problem from an algebraic point of view. For any group of permutations G

on a set A let $\mathbb{A}_G$ denote the (non-indexed) algebra $\langle A; Pol\ \rho_G\rangle$. A set of operations on A which contains all projections and is closed under composition is called a *clone*. It is easy to verify that $Pol\ \rho_G$ is a clone; its members are called the *terms* of the algebra $\mathbb{A}_G$. The *unary part* of a clone C is the set of its unary operations. It is a simple exercise to verify the following fact: a clone C has unary part equal to G if and only if it is contained in the clone $Pol\ \rho_G$. In particular, the only unary operations that preserve ρ_G are the members of G, and hence the relational structure $\langle A; \rho_G\rangle$ is a core. Thus Theorem 1.1 above can be invoked to prove **NP**-hardness.

Since a Taylor operation preserving the relation ρ_G is an idempotent term of the algebra $\mathbb{A}_G$, we may use the following technique to produce "obstructions" to the existence of such a term. A k-ary relation μ on A is an *idempotent k-subalgebra of* $\mathbb{A}_G$ if it is preserved by every idempotent term of $\mathbb{A}_G$. (In other words, these are the subalgebras of the algebra $\mathbb{B}^k$ where $\mathbb{B}$ is the idempotent reduct of $\mathbb{A}$.) When the arity of the relation is clear we shall simply say *idempotent subalgebra.*

Alternatively, these relations can be described as follows. A first-order formula ϕ in the language of ρ_G (with equality and constants) is *primitive positive* if it is built up using only the existential quantifier and conjunction, i.e. of the form

$$\phi \equiv \exists x_1 \cdots \exists x_m \psi$$

where ψ is a conjunction of atomic formulas involving ρ_G, equality and the unary relations $\{a\}$ for every $a \in A$.

LEMMA 2.1. [1] *Let μ be a k-ary relation on A. Then μ is an idempotent k-subalgebra if and only if there exists a primitive positive first-order formula $\phi(x_1, \ldots, x_k)$ in the language of ρ_G (with equality and constants) with free variables $x_1, \ldots, x_k$ such that*

$$\mu = \{(a_1, \ldots, a_k) : \phi(a_1, \ldots, a_k) \text{ holds}\}.$$

Here are examples of such constructions. Let θ be a k-ary relation on A. For each sequence of indices $I = (i_1, \ldots, i_j)$ where $1 \leq i_s \leq k$ for all s, we define the *projection of θ onto I* by

$$\theta_I = \{(x_1, x_2, \ldots, x_j) : \exists (y_1, \ldots, y_k) \in \theta \text{ such that } y_{i_1} = x_1, \ldots, y_{i_j} = x_j\}.$$

If θ is an idempotent k-subalgebra of $\mathbb{A}_G$ then θ_I is an idempotent j-subalgebra. We are also allowed to use constants, since we are considering idempotent terms. For instance, if θ is a k-ary idempotent subalgebra and $a_{j+1}, \ldots, a_k \in A$ we may build the idempotent j-subalgebra

$$\begin{aligned}\{(x_1, x_2, \ldots, x_j) :& \exists \overline{y} \in \theta \text{ such that } y_1 = x_1, \ldots, y_j = x_j, y_{j+1}\\ &= a_{j+1}, \ldots, y_k = a_k\}\end{aligned}$$

where $\overline{y} = (y_1, \ldots, y_k)$.

The following simple application will illustrate the method. Suppose the group G is doubly transitive on A, i.e. for every pairs $(a,b),(c,d) \in A^2$ there exists a permutation $\sigma \in G$ such that $\sigma(a) = c$ and $\sigma(b) = d$. Consider the projection of ρ_G onto $I = (1,2)$; it is obviously the disequality relation $\{(a,b) : a \neq b\}$, in other words, the adjacency relation of the complete graph on A. For $|A| > 2$, this graph is invariant under no Taylor operation; in fact, this result holds for all non-bipartite, symmetric, areflexive graphs (see Lemma 4.2 below). It follows that the relation ρ_G is invariant under no Taylor operation, and hence that $CSP(G)$ is **NP**-complete, by Theorem 1.1.

On the other hand, the Dichotomy Conjecture has been verified for various special cases of Taylor terms. We mention here the results we will require in the sequel. It is easy to verify that on any group $\langle A; \cdot \rangle$, the term operation $m(x,y,z) = xy^{-1}z$ is a Mal'tsev operation; and a k-ary relation on A is invariant under m if and only if it is a coset of a subgroup of A^k. The following result is due to Feder and Vardi:

THEOREM 2.2. [11] *Let $\langle A; \rho_1, \dots, \rho_k \rangle$ be a relational structure such that the relations $\rho_1, \dots, \rho_k$ are invariant under the operation $m(x,y,z) = xy^{-1}z$ for some group operation on A. Then the problem $CSP(\rho_1, \dots, \rho_k)$ is in* **P**.

More generally, we have the following result:

THEOREM 2.3. [6] *Let $\langle A; \rho_1, \dots, \rho_k \rangle$ be a relational structure such that the relations $\rho_1, \dots, \rho_k$ are invariant under some Mal'tsev operation on A. Then the problem $CSP(\rho_1, \dots, \rho_k)$ is in* **P**.

The *discriminator* on the set A is the 3-ary operation defined by

$$t(x,y,z) = \begin{cases} z, & \text{if } x = y, \\ x, & \text{otherwise.} \end{cases}$$

It is easy to see that t is a Mal'tsev operation; it also satisfies the identity $t(x,y,x) \approx x$, and the composition $t(x, t(x,y,z), z)$ is a majority operation. Thus if ρ_G is invariant under t the problem $CSP(G)$ is in **P**, and in fact is of bounded width, by a result of Jeavons, Cohen and Cooper [22].

§3. Regular groups, wreath products and simple algebras.

3.1. Regular groups. In [13] it is shown that if G is regular then $CSP(G)$ is tractable. We generalise this slightly, with an altogether different proof.

PROPOSITION 3.1. *If G is a group of permutations on A such that $|G_x| = 1$ for all $x \in A$, then the discriminator t preserves ρ_G, and hence $CSP(G)$ is in* **P**. *If G is transitive, then the discriminator preserves ρ_G if and only if G is regular.*

PROOF. Suppose that $|G_x| = 1$ for all $x \in A$: if $\alpha, \beta, \gamma \in G$ are such that $\alpha(x) = \beta(x)$ for some $x \in A$, then $\alpha = \beta$, and hence $t(\alpha(x), \beta(x), \gamma(x)) = \gamma(x)$ for all $x \in A$. Otherwise, $t(\alpha(x), \beta(x), \gamma(x)) = \alpha(x)$ for all $x \in A$ and so t preserves ρ_G.

Now suppose that G is transitive. If G is not regular, we can find $\alpha, \beta \in G$ and $x, y \in A$ such that $\alpha(x) = \beta(x)$ and $\alpha(y) \neq \beta(y)$. Let $\gamma \in G$ such that $\gamma(x) = \alpha(y)$. Apply the discriminator t to the tuples of ρ_G corresponding to α, β and γ: we get that $t(\alpha(x), \beta(x), \gamma(x)) = \gamma(x)$ and $t(\alpha(y), \beta(y), \gamma(y)) = \alpha(y)$; since these values are equal t cannot preserve ρ_G. ⊣

We note in passing that there is another Mal'tsev operation that preserves the relation ρ_G in the case where $|G_x| = 1$ for all $x \in A$. Indeed, suppose that G has k orbits. Consider the action of G on the set $G \times \{1, \dots, k\}$ defined by $(g', i)^g = (gg', i)$ (i.e. G acts by left translations on each separate fiber $\{(x, i) : x \in G\}$.) It is easy to see that this action is equivalent to the action of G on A, and that the operation $m(x, y, z) = xy^{-1}z$ of the group $G \times \mathbb{Z}/k\mathbb{Z}$ preserves ρ_G.

3.2. Wreath products. We expand on a result of [12] concerning wreath products of transitive permutation groups. Let H be a group of permutations on A an K a group of permutations on B. The *wreath product* of H and K, denoted by $H \wr K$, is the group of permutations on $A \times B$ that consists of all permutations of the form

$$(a, b) \longmapsto (\alpha_b(a), \beta(b))$$

where $\beta \in K$ and $\alpha_b \in H$ for each $b \in B$ (see [17]).

Without loss of generality let us assume that $A = \{1, \dots, h\}$ and $B = \{1, \dots, k\}$ for some integers $h, k \geq 2$. Then the relation $\rho_{H \wr K}$ is, up to permutations of indices, the hk-ary relation containing all tuples of the form $((\alpha_1(1), \beta(1)), \dots, (\alpha_1(h), \beta(1)), (\alpha_2(1), \beta(2)), \dots, \dots, (\alpha_k(1), \beta(k)),$
$\dots, (\alpha_k(h), \beta(k)))$
where the α_i are elements of H and β is in K.

THEOREM 3.2. *Let $h, k \geq 2$, let $A = \{1, \dots, h\}$ and $B = \{1, \dots, k\}$, and let H and K be transitive groups of permutations on A and B respectively.*

(1) *the relation $\rho_{H \wr K}$ is invariant under a Taylor operation if and only if the same holds for ρ_H and ρ_K;*
(2) *each of the problems $CSP(\rho_H)$ and $CSP(\rho_K)$ reduces to $CSP(\rho_{H \wr K})$ in polynomial time;*
(3) *if both the problems $CSP(\rho_H)$ and $CSP(\rho_K)$ are in* **P** *then so is $CSP(\rho_{H \wr K})$.*

PROOF. Part (3) was proved in [12].

It is easy to verify the following: $(*)$ if f and g are n-ary operations preserving ρ_H and ρ_K respectively then the operation

$$(f, g)((x_1, y_1), \dots, (x_n, y_n)) = (f(x_1, \dots, x_n), g(y_1, \dots, y_n))$$

preserves $\rho_{H \wr K}$.

Now we prove a claim which we'll use to prove both (1) and (2): consider the binary relation θ on $A \times B$ defined as follows: θ consists of all pairs $((x_1, y_1), (x_1', y_1'))$ such that there exist $(x_i, y_i), (x_i', y_i')$, $i = 1, \ldots, hk$, such that $((x_1, y_1), (x_2, y_2), \ldots)$ and $((x_1', y_1'), (x_2', y_2'), \ldots)$ are in $\rho_{H \wr K}$ and $(x_i, y_i) = (x_i', y_i')$ for all $i \geq h + 1$. It is not difficult to see that $((x_1, y_1), (x_1', y_1')) \in \theta$ if and only if $y_1 = y_1'$. Thus θ is a congruence of the algebra $\mathbb{A}_{H \wr K}$, and we may safely identify the blocks of θ with the elements of B. In particular, every term of $\mathbb{A}_{H \wr K}$ has the form

$$\Phi((x_1, y_1), \ldots, (x_n, y_n)) = (\ldots, g(y_1, \ldots, y_n))$$

for some operation g on K. In fact, the quotient algebra is term-equivalent to $\mathbb{A}_K$. Indeed, every term operation of $\mathbb{A}_K$ is obtained as the "action" of a term of $\mathbb{A}_{H \wr K}$ on the blocks of θ: if g is the term of $\mathbb{A}_K$ take (π_1, g) as the term of $\mathbb{A}_{H \wr K}$. On the other hand, the unary part of the clone of the quotient algebra is the action of $H \wr K$ on the blocks of θ, which is easily seen to be that of K. Since $\mathbb{A}_K$ is the algebra whose clone is the largest with this property, we get the result.

Now we prove that the subset $A' = \{(a, 1) : a \in A\}$ is an idempotent subalgebra of $\mathbb{A}_{H \wr K}$; we may identify it in the obvious way to A. If Φ is an idempotent term of $\mathbb{A}_{H \wr K}$ then the associated term g (see above) of $\mathbb{A}_K$ is clearly idempotent, and the result follows immediately.

If Φ is an idempotent term of $\mathbb{A}_{H \wr K}$ then it must preserve the following relation:

$$\{((x_1, y_1), \ldots, (x_h, y_h)) : ((x_1, y_1), \ldots, (x_h, y_h), (1, 2), \ldots, (h, 2), \ldots, (1, k), \ldots, (h, k)) \in \rho_{H \wr K}\}$$

which is easily seen to be equal to

$$\{(\alpha(1), 1), \ldots, (\alpha(h), 1) : \alpha \in H\}.$$

This shows that the idempotent subalgebra of $\mathbb{A}_{H \wr K}$ with universe A' is a reduct of the algebra $\mathbb{A}_H$. Now if f is an idempotent term of $\mathbb{A}_H$ then the operation (f, π_1) is a term of $\mathbb{A}_{H \wr K}$ so the idempotent subalgebra is actually term-equivalent to the idempotent reduct of the algebra $\mathbb{A}_H$.

This shows that if $\mathbb{A}_{H \wr K}$ has a Taylor operation then so do $\mathbb{A}_H$ and $\mathbb{A}_K$, and by Theorem 7 of [30] that (2) holds.

Conversely, suppose that f is an n-ary term of $\mathbb{A}_H$ and g is a m-ary term of $\mathbb{A}_K$. Define an $(n + m)$-ary term of $\mathbb{A}_{H \wr K}$ by

$$\Phi((x_1, y_1), \ldots, (x_{n+m}, y_{n+m})) = (f(x_1, \ldots, x_n), g(y_{n+1}, \ldots, y_{n+m}));$$

notice that it is a term of $\mathbb{A}_{H \wr K}$ by $(*)$ because it is of the form (f', g') where f' and g' are obtained from f and g respectively by adding fictitious variables.

It is easy to verify the following: if f satisfies the identity

$$f(u_1, \ldots, u_n) \approx f(v_1, \ldots, v_n)$$

then Φ satisfies the identity

$$\Phi(u_1, \ldots, u_n, x, x, \ldots, x) \approx \phi(v_1, \ldots, v_n, x, x, \ldots, x)$$

and similarly for g; in particular, Φ is a Taylor operation if both f and g are Taylor, which proves (1). ⊣

REMARK. If $\mathbb{A}_H$ and $\mathbb{A}_K$ each have Taylor terms f and g of the same arity that obey the same linear identities, then $\mathbb{A}_{H \wr K}$ will have a term obeying those identities. Indeed, it suffices to identify variables in the term defined in the last proof, as follows:

$$\Psi((x_1, y_1), \ldots, (x_n, y_n)) = \Phi((x_1, y_1), \ldots, (x_n, y_n), (x_1, y_1), \ldots, (x_n, y_n)).$$

In particular, $\mathbb{A}_{H \wr K}$ admits a Mal'tsev term or majority term if and only if both $\mathbb{A}_H$ and $\mathbb{A}_K$ do.

3.3. Simple algebras. If we wish to determine whether there exists a Taylor term that preserves the relation ρ_G, it seems natural to try to break down the problem into smaller parts, as for instance, if G can be decomposed as a wreath product of smaller groups. In another vein, if G is not transitive on A, then we may consider its actions on orbits: for each orbit $A' \subseteq A$, we have a group of permutations G' on A' consisting of the restrictions of the members of G to A'. In [13], Haddad and Rödl ask whether there exists a group G such that for every group G' obtained this way, the problem $CSP(G')$ is in **P**, but $CSP(G)$ is **NP**-complete; we shall provide examples in the next section. Unfortunately, it seems unlikely that a characterisation of the tractable cases can be obtained in the transitive case only, as the following discussion will show.

Another approach to break down our problem is to use congruences of the algebra $\mathbb{A}_G$, and analyse the quotients. We require the following result from [32] (see also Lemma 3.5 of [24]):

LEMMA 3.3. *Let G be a transitive group of permutations on A and let $\mathbb{A}$ be a simple algebra whose clone of term operations has unary part equal to G. Then either G is regular or $\mathbb{A}$ is essentially unary.*

COROLLARY 3.4. *Let G be a transitive group of permutations on A. If the algebra $\mathbb{A}_G$ is simple, in particular if the group G is primitive, then $\mathbb{A}_G$ has a Taylor term if and only if the discriminator is a term of $\mathbb{A}_G$. If this holds $CSP(G)$ is in* **P**, *otherwise it is* **NP**-*complete.*

PROOF. If the algebra $\mathbb{A}_G$ has a Taylor term then it is not essentially unary; the result then follows immediately from Lemma 3.3 and Proposition 3.1. ⊣

Suppose that G is transitive and that $\mathbb{A}_G$ is not simple. If α is a maximal congruence of $\mathbb{A}_G$ then the quotient algebra $\mathbb{A}_G/\alpha$ is simple; furthermore the unary part of its clone of term operations consists of the permutations of the blocks of α induced by the elements of G; this is a transitive group, and thus by Lemma 3.3 either the quotient algebra is essentially unary, and hence $\mathbb{A}_G$ has no Taylor term, or else the action of G on the blocks of α is regular. Then the following is a term operation of $\mathbb{A}_G$:

$$T(x, y, z) = \begin{cases} z, & \text{if } (x, y) \in \alpha, \\ x, & \text{otherwise.} \end{cases}$$

Indeed, if $\sigma, \tau, \mu \in G$ are such that $(\sigma(x), \tau(x)) \in \alpha$ for some $x \in A$, then the same holds for every $x \in A$ by regularity on the blocks. Thus $T(\sigma(x), \tau(x), \mu(x)) = \mu(x)$ for all $x \in A$. Otherwise we get $T(\sigma(x), \tau(x), \mu(x)) = \sigma(x)$ for all $x \in A$ and so T preserves ρ_G.

Hence we obtain the following:

PROPOSITION 3.5. *Let G be a transitive group of permutations on A. If $\mathbb{A}_G$ has a Taylor term then it has a quotient with a discriminator term.*

LEMMA 3.6. *Let G be a transitive group of permutations on A, let α be a congruence of $\mathbb{A}_G$ such that G acts regularly on the blocks of α; let B be a block of α and let $H = G_B$. If the relation ρ_G is invariant under a Taylor operation then so is ρ_H. If the relation ρ_H is invariant under a Mal'tsev operation then so is ρ_G.*

PROOF. As we remarked earlier, the fact that G acts regularly on the blocks of α translates as follows: for any $\sigma, \tau \in G$, if there exists some $x \in A$ such that $\sigma(x)$ and $\tau(x)$ are in the same block of α then the same holds for every $x \in A$. In particular, the relation ρ_H is an idempotent k-subalgebra since it is equal to

$$\rho_H = \{(x_1, \ldots, x_k) \in \rho_G : (x_1, 1) \in \alpha\}$$

(and α, being a congruence, is itself an idempotent subalgebra.) It follows that if f is a Taylor operation preserving ρ_G, it also preserves ρ_H.

Now suppose that ρ_H is invariant under a Mal'tsev operation f. We define an operation on A as follows: let $\phi(x, y, z) = f(x, y, z)$ whenever x, y, z lie in the same block of α. If x, y, z are not in the same block, let $\phi(x, y, z) = z$ if $(x, y) \in \alpha$ and $\phi(x, y, z) = x$ otherwise.

It is easy to verify that ϕ is a Mal'tsev operation; we must now show that it preserves ρ_G. Let $\sigma, \tau, \mu \in G$. First suppose that $\sigma(x), \tau(x), \mu(x)$ are all in the same block of α for some x: this will then hold for all $x \in A$. In particular, we get that the tuples corresponding to the identity permutation, $\tau \circ \sigma^{-1}$ and $\mu \circ \sigma^{-1}$ are all in ρ_H.

Notice that if $(x_1, \ldots, x_k) \in \rho_G$, then $(x_{\nu(1)}, \ldots, x_{\nu(k)}) \in \rho_G$ if and only if $\nu \in G$. Indeed, if $(x_1, \ldots, x_k)$ corresponds to $\pi \in G$ and $(x_{\nu(1)}, \ldots, x_{\nu(k)})$ corresponds to $\xi \in G$ then we have that $\pi \circ \nu = \xi$. Thus we proceed as follows: if M is the $|G| \times 3$ matrix whose rows are the tuples corresponding to σ, τ, μ, permute the rows using σ^{-1} to obtain columns that are members of ρ_H. Apply ϕ, which in this case amounts to applying f, and hence the resulting column is in ρ_H. Now reorder the rows using σ to obtain $\phi(\sigma, \tau, \mu)$ which is in ρ_G.

Now suppose that $\sigma(x), \tau(x), \mu(x)$ do not lie all in the same block of α but the first two do; then this holds for all $x \in A$. Hence $\phi(\sigma(x), \tau(x), \mu(x)) = \mu(x)$ for all x. The remaining case is similar. ⊣

The second statement of the last lemma can be extended (for instance to majority operations). But is it true in general that if ρ_H admits a Taylor operation then so does ρ_G ?

§4. Dihedral groups and Affine transformations. In this section we consider various examples of permutation groups and analyse which admit Taylor terms or not.

4.1. Dihedral groups. In the following, we let D_n ($n \geq 3$) denote the $2n$-element permutation group on $\{0, 1, \ldots, n-1\}$ which is the automorphism group of the n-cycle, i.e. which is generated by the permutations $(0\,1\,\ldots\,n-1)$ and $(0\,n-1)(1\,n-2)\cdots$. The complexity of $CSP(D_n)$ was first determined in [12] by different methods.

LEMMA 4.1. *The relation ρ_{D_4} is invariant under a Mal'tsev operation. In particular, the problem $CSP(\rho_{D_4})$ is in* **P**.

PROOF. There are in fact many different Mal'tsev operations preserving this relation. Notice first that D_4 is the wreath product of two copies of the 2-element group, and hence admits the Mal'tsev operation

$$((x, x'), (y, y'), (z, z')) \mapsto (x + y + z, x' + y' + z')$$

where the sum is that of the group $\mathbb{Z}/2\mathbb{Z}$ (see the comment following Theorem 3.2). Alternatively, one may invoke the results on affine transformations (see below), by noticing that D_4 consists of all affine transformations $x \mapsto ax + b$ with $a, b \in \mathbb{Z}/4\mathbb{Z}$ and a invertible. It follows that the operation $M(x, y, z) = x - y + z$ preserves ρ_{D_4} (where the sum is in $\mathbb{Z}/4\mathbb{Z}$.) ⊣

We shall now prove that no Taylor operation preserves the relation ρ_{D_n} for $n \neq 4$. For this we require the following results on graphs, one of which was mentioned earlier in section 2.

LEMMA 4.2. [5, 29] *Let θ be a binary, areflexive, symmetric relation on a set A and let Γ be the associated simple graph. If Γ is not bipartite then θ is invariant under no Taylor operation.*

Let θ be a binary, reflexive relation. We'll say that θ is *intransitive* if whenever $(a, b), (b, c), (a, c) \in \theta$ then $|\{a, b, c\}| \leq 2$. We'll say that θ is *not acyclic* if its symmetric closure (with all loops removed) is not a tree.

LEMMA 4.3. [28] *Let θ be a binary, reflexive relation on A which is intransitive. If θ is not acyclic then it is invariant under no Taylor operation.*

THEOREM 4.4. *If $n \neq 4$ then no Taylor operation preserves the relation ρ_{D_n}; in particular the problem $CSP(D_n)$ is* **NP**-*complete.*

PROOF. We divide the proof in 3 cases. Suppose first that n is odd: consider the projection θ of ρ_{D_n} onto the coordinates 0 and 1 (an edge of the n-gon). It is easy to see that θ is symmetric, areflexive, and is in fact the n-cycle; by Lemma 4.2 we are done.

Now suppose that n is even. Consider the relation α that consists of all pairs (x_0, y_0) such that there exist $(x_0, \ldots, x_{n-1}), (y_0, \ldots, y_{n-1}) \in \rho_{D_n}$ with $x_1 = y_1$. It is easy to see that

$$\alpha = \{(x, y) : x - y = \pm 2\}$$

where the sum is taken modulo n. Thus α is a reflexive, symmetric relation. If $n \geq 8$ then this is an intransitive cycle so we apply Lemma 4.3; if $n = 6$, then α is a congruence of $\mathbb{A}_{D_6}$ (on whose blocks D_6 acts regularly), and we define the idempotent subalgebra γ as the set of all pairs (x_0, x_2) such that there exists $(x_0, \ldots, x_5) \in \rho_{D_6}$ such that $(x_0, 0) \in \theta$. It is easy to see that γ is an areflexive, symmetric cycle of length 3, and hence Lemma 4.2 applies. ⊣

4.2. Affine transformations. In this section we completely determine the complexity of $CSP(G)$ in the case where G is the group of all (bijective) affine transformations on a matrix ring.[2] (See [27] for a study of monoidal intervals related to affine transformations.) More precisely, let S be a (finite) commutative ring with 1, let n be a positive integer and let $R = M_n(S)$ the ring of $n \times n$ matrices over S. Let $U = GL_n(S)$ denote the group of invertible matrices in $M_n(S)$. The group $G = G_{n,R}$ will consist of all transformations on R of the form $x \mapsto ax + b$ where $a \in U$ and $b \in R$. It is clear that G acts transitively on R.

THEOREM 4.5. *Let $G = G_{n,R}$. Then the following conditions are equivalent*:

(1) *$\mathbb{A}_G$ admits a Taylor term;*
(2) *$n = 1$ and the operation $m(x, y, z) = x - y + z$ is a term of $\mathbb{A}_G$.*

If one of these conditions holds then $CSP(G)$ is in **P**; *otherwise it is* **NP**-*complete.*

PROOF. It will suffice to prove that (1) implies (2). We start by showing that if $\mathbb{A}_G$ has a Taylor term then $n = 1$. For any non-zero $c \in R$, the orbit of $(0, c)$ under G is invariant under the Taylor term (since it is a projection of ρ_G on two indices), and it is the edge relation of a simple graph. Indeed, it is easy to see that it consists of all pairs (x, y) in R^2 such that $x - y \in Uc$; since $-U = U$ and $0 \notin U$ it follows that the binary relation is symmetric and areflexive. It follows from Lemma 4.2 that the graph in question must be bipartite.

Consider the special case where $c = 1$. The neighbourhood of 0 in this graph is the set U; since the graph contains no 3-cycle, there cannot be an edge between two elements of U. However, in the case $n \geq 2$ consider

[2]For basic results on rings we refer the reader to [9] and [20].

the matrix

$$B = \begin{bmatrix} 0 & \cdots & 0 & 1 \\ \cdots & \cdots & \cdots & 0 \\ \cdots & I_{n-1} & \cdots & \vdots \\ \cdots & \cdots & \cdots & 0 \\ \cdots & \cdots & \cdots & 1 \end{bmatrix}$$

where as usual I_n denotes the identity matrix of order n; in the matrix B the lower left hand corner consists of the identity matrix of order $n-1$. The matrix B has determinant $(-1)^{n+1}$ and hence is in U. Furthermore, if we expand the determinant of the matrix

$$B - I_n = \begin{bmatrix} -1 & 0 & \cdots & \cdots & \cdots & 0 & 1 \\ 1 & -1 & \cdots & \cdots & \cdots & 0 & 0 \\ 0 & 1 & -1 & 0 & \cdots & 0 & 0 \\ \vdots & \ddots & \ddots & \cdots & \cdots & \vdots & \vdots \\ \vdots & \cdots & \cdots & \cdots & 1 & -1 & 0 \\ 0 & \cdots & \cdots & \cdots & 0 & 1 & 0 \end{bmatrix}$$

following the first row we see that it is equal to $(-1)^{n+1}$ and hence is in U. Thus B and I_n are adjacent elements of U, proving that if $\mathbb{A}_G$ has a Taylor term then $n = 1$.

So we may now assume that R $(= S)$ is a commutative ring. Since it is finite, R is a product of local rings, i.e. $R = R_1 \times \cdots \times R_v$ where each R_i has a unique maximal ideal M_i; in particular, the group of units U_i of R_i consists of those elements not in M_i. Also, note that $U = U_1 \times \cdots \times U_v$.

Suppose that for some i there exist $u, v \in U_i$ such that $u - v \in U_i$: we shall deduce a contradiction along the lines of the previous arguments. Let $c = (0, \ldots, 0, 1, 0, \ldots, 0)$ where 1 appears in the i-th position, and consider the graph with set of edges equal to the orbit of $(0, c)$. It is easy to see that (x, y) is an edge if and only if $x - y = (0, \ldots, 0, u_i, 0, \ldots, 0)$ for some $u_i \in U_i$. But then we have a 3-cycle formed of the elements $\{0, (0, \ldots, 0, u, 0, \ldots, 0), (0, \ldots, 0, v, 0, \ldots, 0)\}$.

Fix $1 \leq i \leq v$. By the last paragraph, we have that $u - v \in M_i$ for all $u, v \in U_i$. In particular, the (additive) factor group R_i/M_i has 2 elements. It follows that for each i, U_i is a coset of M_i and thus the set of units U is a coset of a subgroup of R. In particular U is closed under the operation $m(x, y, z) = x - y + z$. It follows that the relation ρ_G is invariant under this operation: indeed, if $\alpha_i \in G$ with $\alpha_i(x) = a_i x + b_i$ for $i = 1, \ldots, 3$ then we have that $m(\alpha_1, \alpha_2, \alpha_3) = \gamma$ where $\gamma(x) = (a_1 - a_2 + a_3)x + (b_1 - b_2 + b_3)$ for all $x \in R$. ⊣

Observe that in the above proof we have a Taylor term if and only if R is a product of local rings $R_1, \ldots, R_s$ with maximal ideal of size $|R_i|/2$. For instance one may take products of the rings $\mathbb{Z}/k\mathbb{Z}$ where $k = 2^n$. We use these groups to answer two questions posed in [13].

EXAMPLE 1. A transitive group G with a normal, transitive subgroup H such that $CSP(G)$ is in **P** but $CSP(H)$ is **NP**-complete.

The example is $G_{1,R}$ where $R = \mathbb{Z}/k\mathbb{Z}$ where $k = 2^m$ with $m \geq 3$. We know that it admits a Mal'tsev operation. Consider the subgroup H defined by

$$H = \{\sigma : \sigma(x) = ax + b \text{ where } a \in \{-1, 1\}\}.$$

Notice that H is the dihedral group D_{2^m} and we've seen above that the associated problem is **NP**-complete if $m \geq 3$. It is a simple exercise to verify that H is normal in G.

EXAMPLE 2. A group G such that $CSP(G)$ is **NP**-complete, but $CSP(G')$ is in **P** for every G' which is the action of G on an orbit.

For simplicity we consider the case of two orbits only. Let $m \geq 3$ and let $k = 2^m$. For $i = 1, 2$ let

$$A_i = \{(x, i) : x \in \mathbb{Z}/k\mathbb{Z}\}.$$

The group G consists of all permutations of $A_1 \cup A_2$ of the following form: for every elements a, b, c, d in $\mathbb{Z}/k\mathbb{Z}$ with a and c odd and $a = \pm c$, define

$$\sigma_{(a,b,c,d)}(x, i) = \begin{cases} (ax + b, 1), & \text{if } i = 1, \\ (cx + d, 2), & \text{if } i = 2. \end{cases}$$

One verifies easily that G is closed under composition, and it is clear that A_1 and A_2 are the orbits of G. It is also immediate to verify that the restriction G_i of G to the orbit A_i is equivalent to the group $G_{1,R}$ with $R = \mathbb{Z}/k\mathbb{Z}$; thus $CSP(G_i)$ is in **P** for $i = 1, 2$. However, we prove that ρ_G is invariant under no Taylor operation. Assume that the indexing is chosen as follows:

$$\rho_G = \{(\sigma(0, 1), \ldots, \sigma(k-1, 1), \sigma(0, 2)), \ldots, \sigma(k-1, 2)\}.$$

Consider the following idempotent subalgebra:

$$\theta = \{(x_0, \ldots, x_{k-1}) : (x_0, \ldots, x_{k-1}, (0, 2), \ldots, (k-1, 2)) \in \rho_G\}.$$

Since $cx + d = x$ for all x implies that $c = 1$ and $d = 0$, it follows that we have

$$\theta = \{((a0 + b, 1), \ldots, (a(k-1) + b, 1)) : a = \pm 1\}$$

which we recognise to be ρ_{D_k}, which admits no Taylor operation.

§5. Conclusion. We have shown that all known **NP**-complete cases of $CSP(G)$ are explained by the absence of a Taylor operation. On the other hand, we have verified that all known tractable cases of $CSP(G)$ admit a Taylor operation, and in fact, a Mal'tsev operation. This prompts the following question:

PROBLEM. Is there a (transitive) group G such that $CSP(G)$ admits a Taylor operation but no Mal'tsev operation?

In view of Example 2 above and Proposition 3.6, it seems likely that an eventual classification of the tractable cases of $CSP(G)$ will involve a somewhat detailed inspection of how subgroups of wreath products of tractable groups behave.

Acknowledgment. The authors wish to thank Ágnes Szendrei and Keith Kearnes for fruitful discussions.

REFERENCES

[1] V. G. BODNARČHUK, L. A. KALUZHNIN, V. N. KOTOV, and B. A. ROMOV, *Galois theory for post algebras I-II (russian)*, ***Kibernetica***, vol. 3 (1969), pp. 1–10, and **5** (1969), 1–9. English version: ***Cybernetics*** (1969), 243–252 and 531–539.

[2] A. A. BULATOV, *A graph of a relational structure and constraint satisfaction problems*, LICS 2004: 448–457.

[3] ———, *Tractable conservative constraint satisfaction problems*, LICS 2003: 321–330.

[4] ———, *A dichotomy theorem for constraints on a three-element set*, ***Proceedings of 43rd IEEE Symposium on Foundations of Computer Science (FOCS'02)***, Vancouver, Canada, 2002, pp. 649–658.

[5] ———, *H-coloring dichotomy revisited*, ***Theoretical Computer Science***, vol. 349 (2005), no. 1, pp. 31–39.

[6] A. A. BULATOV and V. DALMAU, *A simple algorithm for Mal'tsev constraints*, ***SIAM Journal on Computing***, vol. 36 (2006), no. 1, pp. 16–27.

[7] A. A. BULATOV, A. KROKHIN, and P. JEAVONS, *Constraint satisfaction problems and finite algebras*, ***Proceedings of 27th International Colloquim on Automata, Languages and Programming ((ICALP'00), Geneva, Switzerland***, Lecture Notes in Computer Science, vol. 1853, Springer, Berlin, 2000, pp. 272–282.

[8] ———, *Classifying the complexity of constraints using finite algebras*, submitted, 2003, 30 pp.

[9] D. S. DUMMIT and R. M. FOOTE, ***Abstract Algebra***, 3rd ed., Wiley, 2004.

[10] T. FEDER and M. Y. VARDI, *The computational structure of monotone monadic SNP and constraint satisfaction: a study through datalog and group theory*, ***Proc. 25th Ann. ACM Symp. on Theory of Computing***, 1993, pp. 612–Ű622.

[11] ———, *The computational structure of monotone monadic SNP and constraint satisfaction: a study through Datalog and group theory*, ***SIAM Journal on Computing***, vol. 28 (1999), no. 1, pp. 57–104.

[12] L. HADDAD, P. HELL, and E. MENDELSOHN, *On the complexity of coloring areflexive h-ary relations with given permutation group*, ***Ars Combinatoria***, vol. 48 (1998), pp. 111–128.

[13] L. HADDAD and V. RÖDL, *On the complexity of coloring areflexive relations*, ***Ars Combinatoria***, vol. 33 (1992), pp. 217–225.

[14] L. HADDAD and I. G. ROSENBERG, *Critère général de complétude pour les algèbres partielles finies*, ***Comptes Rendus des Séances de l'Académie des Sciences. Série I. Mathématique***, vol. 304 (1987), no. 17, pp. 507–509.

[15] ———, *Maximal partial clones determined by the areflexive relations*, ***Discrete Applied Mathematics***, vol. 24 (1989), no. 1-3, pp. 133–143.

[16] ———, *Completeness theory for finite partial algebras*, ***Algebra Universalis***, vol. 29 (1992), no. 3, pp. 378–401.

[17] M. HALL, ***The Theory of Groups***, 2nd ed., Chelsea, New York, 1976.

[18] P. HELL and J. NEŠETŘIL, ***Graphs and Homomorphisms***, Oxford University Press, Oxford, 2004.

[19] D. HOBBY and R. MCKENZIE, ***The Structure of Finite Algebras***, Contemporary Mathematics, vol. 76, American Mathematical Society, Providence, RI, 1988.

[20] N JACOBSON, ***Basic Algebra. II***, 2nd ed., Freeman, New York, 1989.

[21] P. G. JEAVONS, *On the algebraic structure of combinatorial problems*, ***Theoretical Computer Science***, vol. 200 (1998), no. 1-2, pp. 185–204.

[22] P. G. JEAVONS, D. COHEN, and M. COOPER, *Constraints, consistency and closure*, ***Artificial Intelligence***, vol. 101 (1998), no. 1-2, pp. 251–265.

[23] P. G. JEAVONS, D. COHEN, and M. GYSSENS, *Closure properties of constraints*, ***Journal of the ACM***, vol. 44 (1997), no. 4, pp. 527–548.

[24] K. KEARNES and Á. SZENDREI, *Collapsing permutation groups*, ***Algebra Universalis***, vol. 45 (2001), no. 1, pp. 35–51.

[25] P. G. KOLAITIS and M. Y. VARDI, *Conjunctive-query containment and constraint satisfaction*, ***Proceedings of the Seventeenth ACM SIGACT-SIGMOD-SIGART Symposium on Principles of Database Systems***, Seattle, Washington, 1998, pp. 205–213.

[26] A. KROKHIN and B. LAROSE, *A monoidal interval of isotone clones on a finite chain*, ***Acta Universitatis Szegediensis. Acta Scientiarum Mathematicarum***, vol. 68 (2002), no. 1-2, pp. 37–62.

[27] A. A. KROKHIN, *On clones, transformation monoids, and associative rings*, ***Algebra Universalis***, vol. 37 (1997), no. 4, pp. 527–540.

[28] B. LAROSE, *Taylor operations on finite reflexive structures*, ***International Journal of Mathematics and Computer Science***, vol. 1 (2006), no. 1, pp. 1–21.

[29] B. LAROSE and ZÁDORI, *Bounded width problems and algebras*, ***Algebra Universalis***, to appear, 2007.

[30] B. LAROSE and L. ZÁDORI, *The complexity of the extendibility problem for finite posets*, ***SIAM Journal on Discrete Mathematics***, vol. 17 (2003), no. 1, pp. 114–121.

[31] R. N. MCKENZIE, G. F. MCNULTY, and W. F. TAYLOR, ***Algebras, Lattices, Varieties***, Wadsworth & Brooks/Cole, Monterey, CA, 1987.

[32] P. P. PÁLFY and Á. SZENDREI, *Unary polynomials in algebras. II*, ***Contributions to General Algebra, 2 (Klagenfurt, 1982)***, Hölder-Pichler-Tempsky, Vienna, 1983, pp. 273–290.

[33] Á. SZENDREI, ***Clones in Universal Algebra***, Séminaire de Mathématiques Supérieures, vol. 99, Presses de l'Université de Montréal, Montreal, 1986.

[34] W. TAYLOR, *Varieties obeying homotopy laws*, ***Canadian Journal of Mathematics***, vol. 29 (1977), no. 3, pp. 498–527.

DEPARTMENT OF MATHEMATICS AND STATISTICS
CONCORDIA UNIVERSITY, 1455 DE MAISONNEUVE WEST MONTRÉAL
QC, CANADA, H3G 1M8
E-mail: larose@mathstat.concordia.ca
URL: `http://cicma.mathstat.concordia.ca/faculty/larose/`

DÉPARTEMENT DE MATHÉMATIQUES ET D'INFORMATIQUE
COLLÈGE MILITAIRE ROYAL DU CANADA, BOîTE POSTALE 17000, STATION "FORCES"
KINGSTON, ONTARIO K7K 7B4 CANADA
E-mail: haddad-l@rmc.ca
URL: `http://www.rmc.ca/academic/math-cs/haddad/index-e.html`

TOLERANCE INTERSECTION PROPERTIES AND SUBALGEBRAS OF SQUARES

PAOLO LIPPARINI

Abstract. Tolerance identities can be used [4] in order to provide a fairly simple proof of a classical result by R. Freese and B. Jónsson [7] asserting that every congruence modular variety is in fact Arguesian. The main advantage of the approach using tolerances is that stronger identities can be obtained: the higher Arguesian identities introduced by M. Haiman [12].

Discouragingly enough, however, the approach using tolerances does not appear to give a proof for the local version of Freese and Jónsson's Theorem: *If every subalgebra of* $\boldsymbol{A} \times \boldsymbol{A}$ *is congruence modular then* $\boldsymbol{A}$ *is Arguesian.*

A common generalization of the results mentioned in the above two paragraphs would furnish the following (a conjecture, so far): *If every subalgebra of* $\boldsymbol{A} \times \boldsymbol{A}$ *is congruence modular then* $\boldsymbol{A}$ *satisfies all Haiman's higher Arguesian identities.*

Towards a possible solution of the above conjecture, we introduce a new tolerance identity, called wTIP, and show that it holds in the algebra $\boldsymbol{A}$ provided that every subalgebra of $\boldsymbol{A} \times \boldsymbol{A}$ is congruence modular. Moreover, wTIP is equivalent to the Shifting Principle introduced by H. P. Gumm. The known methods using tolerance identities apply, if we restrict ourselves to a particular class of lattice terms.

§1. From congruence varieties to tolerance identities. In the present section we briefly describe how two seemingly unrelated notions used in universal algebra came in touch in an unexpected way. We first recall some basic notions and terminology. Introductory textbooks to universal algebra are, among others, [1, 9, 26]. For a reader not familiar with universal algebra, a preliminary reading of the survey [16] would probably be of great help.

An *algebraic structure*, *algebra* for short, is a set endowed with operations and constants. Algebras are denoted by $\boldsymbol{A}$, $\boldsymbol{B}$... A *variety* $\mathcal{V}$ is a class of algebras of the same type which is closed under taking products, subalgebras and homomorphic images. Equivalently, a variety is a class of algebras which can be defined by a set of equations.

2000 *Mathematics Subject Classification.* Primary 08A30; Secondary 08B10.

Key words and phrases. Congruence, tolerance identities; congruence modularity; shifting principle.

The author has received support from MPI and GNSAGA. We thank G. Czédli and K. Kearnes (in alphabetical order) for stimulating correspondence and discussions

Logic Colloquium '04
Edited by A. Andretta, K. Kearnes, and D. Zambella
Lecture Notes in Logic, 29

Many interesting results from *universal algebra* (better named the *general theory of algebraic systems*) have the following form:

THEOREM 1.1 (Prototype). *If every algebra in the variety $\mathcal{V}$ satisfies Property P then every algebra in the variety $\mathcal{V}$ satisfies Property Q,*

where P and Q are appropriately chosen properties P and Q. In the most important cases, an implication as above is non-trivial, in the sense that it holds for varieties, not just for single algebras. This means that there exists some algebra $\boldsymbol{A}$ satisfying P, and not satisfying Q.

Particularly intriguing results appear when dealing with congruences: a congruence of an algebra $\boldsymbol{A}$ is the equivalence relation on A determined by some homomorphism ϕ from $\boldsymbol{A}$ into some algebra $\boldsymbol{B}$. For practical purposes, an "internal" description is better suited: a *congruence* α of $\boldsymbol{A}$ is a compatible equivalence relation on A, where *compatible* means that whenever f is an n-ary operation of $\boldsymbol{A}$, and $a_1 \,\alpha\, b_1, \dots, a_n \,\alpha\, b_n$, then also $f(a_1, \dots, a_n) \,\alpha\, f(b_1, \dots, b_n)$. Here, $a_1 \,\alpha\, b_1$ is a shorthand for "a_1 and b_1 being α-related", that is, $(a_1, b_1) \in \alpha$. According to notational convenience, this will also be denoted by $a_1 \equiv b_1 \mod \alpha$.

The set of congruences of an algebra turns out to be endowed with a lattice structure, with a minimal congruence 0 and a maximal congruence 1, and with meet coinciding with intersection. We shall denote the lattice operations by $+$ and $\cdot$ (or simply juxtaposition). A lattice is *distributive* if and only if it satisfies $\alpha(\beta+\gamma) = \alpha\beta + \alpha\gamma$. The *modular* law asserts that $\alpha(\beta+\gamma) = \alpha\beta+\gamma$, provided $\gamma \leq \alpha$.

A detailed analysis of varieties, and of algebras in a variety $\mathcal{V}$ has proved possible under the assumption that the lattices of congruences of *all* algebras in $\mathcal{V}$ satisfy some non trivial lattice identity. Historically, the first structure theorem has been discovered for algebras in congruence distributive varieties [15]; shortly after, a characterization of congruence modularity has been found [6], and eventually a refined theory for congruence modular varieties emerged: see [8, 10, 28]. Briefly, the theory shows that congruence modular varieties share all the good properties which are in common to congruence distributive and congruence permutable varieties. Recall that an algebra is *permutable* if and only if every two congruences permute, that is, $\alpha \circ \beta = \beta \circ \alpha$. A. I. Mal'cev characterization of permutable varieties [25], in 1954, can be safely considered the starting point for modern universal algebra.

See [16] for further reference about the origins of the subject and its earlier history. A brief introduction to the latter developments (commutator theory) is given in [20]. See also [4] for further references.

Present-day research is obtaining results even for varieties satisfying any non-trivial congruence lattice equation. For further information the reader is referred to K. Kearnes and Á Szendrei's fundamental work [18], and to, among others, [13, 19] and [17]. There are also further results announced by

K. Kearnes and E. Kiss. In the present note, however, we shall be dealing only with properties related to congruence modularity.

J. B. Nation [27] discovered the first non trivial result of the form given in Theorem 1.1, where both P and Q are lattice identities (intended to be satisfied by congruence lattices of algebras in $\mathcal{V}$): he proved that there is an identity lattice-theoretically strictly weaker than modularity which is nevertheless equivalent to modularity for congruence lattices of algebras in a given variety. Other results of this kind soon followed: see [16] for further information; see [4, 21] and references there for more recent results.

Our main concern here is a result of the above kind obtained by R. Freese and B. Jónsson [7] in the '70's; they proved that, within a variety, modularity implies a stronger identity, the *Arguesian identity*. The result is particularly significant since the Arguesian identity is an important identity with a clear geometrical meaning (see again [16] for further information).

An alternative proof of R. Freese and B. Jónsson's result has been found in [20] using commutator theory. More intriguingly, [4] provides still another proof, which uses the following tolerance identity, named *Tolerance Intersection Property* in [3]:

$$\text{(TIP)} \qquad \Gamma^* \cap \Theta^* = (\Gamma \cap \Theta)^*$$

where * denotes transitive closure. A *tolerance* Θ on some algebra $\boldsymbol{A}$ is a compatible symmetric and reflexive relation (in other words, a tolerance is a "not necessarily transitive" congruence). A priori, there is no evidence that tolerances should be useful in the study of congruences; however, astonishingly, tolerances have proved useful and sometimes irreplaceable (at least with present-day available techniques). A satisfying explanation of such an important role played by tolerances has still to be found, in our opinion. We should mention that [4], using tolerances, obtained simple and short proofs for results whose previous proofs were quite tricky and complicated, let alone the fact that usually the results given in [4] are even stronger. For example, considering Freese and Jónsson's result, the main advantage of the proof given in [4] is that it furnishes even stronger identities: the *higher Arguesian identities* introduced by M. Haiman [12]. The higher Arguesian identities are interesting because they hold in all lattices of permutable equivalence relations (see [11, 14]).

However, to our disappointment, neither [20] nor [4] entirely encompass [7], since the local form in which Freese and Jónsson [7] proved their theorem is quite strong, and asserts that *If every subalgebra of* $\boldsymbol{A} \times \boldsymbol{A}$ *is congruence modular then* $\boldsymbol{A}$ *is Arguesian.* Of course, it is not really fundamental to get another proof of the local version of [7], since a proof is available. On the other hand, a new proof along the lines of [4] would probably improve [7] to *If every subalgebra of* $\boldsymbol{A} \times \boldsymbol{A}$ *is congruence modular then* $\boldsymbol{A}$ *satisfies all Haiman's higher Arguesian identities.* We have not yet found a proof of the above conjecture, but we suggest here a possible approach towards an affirmative solution (see also Section 6).

Let us say a few words about the proof given in [4]. [4] shows that every algebra satisfying TIP satisfies all the higher Arguesian identities. Since it was previously known that every congruence modular variety satisfies TIP, we immediately get that every algebra in a congruence modular variety satisfies the higher Arguesian identities. The natural way to extend the methods from [4] would be to show that if every subalgebra of $\boldsymbol{A} \times \boldsymbol{A}$ is congruence modular, then $\boldsymbol{A}$ satisfies TIP; however, this is still an open problem. We only get a weaker property wTIP, and it is possible that wTIP implies that congruence lattices satisfy the higher Arguesian identities, but this is open, too.

However, we show here that wTIP has some interest for itself: it implies, congruence modularity, and it is equivalent, for every algebra, to a principle introduced by H. P. Gumm in his deep study [10] of congruence modular varieties. Moreover, wTIP is strong enough for the methods of [4] to be applied, provided we restrict attention to a particular class of terms, the class of *terms without double couples* (Definition 4.1).

The paper is divided as follows. In Section 2 we introduce some further notation. In Section 3 we prove our main positive result, Theorem 3.2, asserting that *If $\boldsymbol{A}$ is an algebra and every subalgebra of $\boldsymbol{A} \times \boldsymbol{A}$ is congruence modular then $\boldsymbol{A}$ satisfies wTIP*. We state a related (somewhat more general) result in Theorem 3.4. In Section 4 we introduce the notion of a term without double couples, and show that, assuming wTIP, the transitive closures of certain lattice terms coincide, no matter whether they are evaluated in the lattice of tolerances or in the lattice of congruences. As a consequence, wTIP implies congruence modularity. Some further consequences are derived in Corollaries 4.6 and 4.7.

In Section 5 we show that wTIP is equivalent to the Shifting Principle, introduced by H. P. Gumm; moreover, we introduce a stronger version of the Shifting Principle (Definition 5.4), and show that it is equivalent to TIP (Proposition 5.5). Finally, in Section 6, we recall the main problems which led to our definition of wTIP, and to the results proved in the present paper; some further problems are stated.

§2. Notation. Here is the notation we use (if not already introduced in the previous section). R, S denote reflexive binary relations; as in the case of congruences, we shall write $a \mathrel{R} b$ to mean that $(a, b) \in R$. It will be notationally convenient to use chains of the above notation: for example, $a \mathrel{\Theta} b \mathrel{\alpha} c \mathrel{R} d$ means $(a, b) \in \Theta$, $(b, c) \in \alpha$ and $(c, d) \in R$.

Juxtaposition denotes intersection; in particular $\alpha\beta$ denotes the meet of the congruences α and β.

R^* denotes the transitive closure of the binary relation R; in particular, Θ^* is the smallest congruence which contains the tolerance Θ. $\overline{R}$ denotes the smallest compatible relation containing R (where R is a binary relation on some algebra which should be clear from the context). In particular, $\overline{\Theta \cup \Gamma}$ is the smallest tolerance which contains the tolerances Θ and Γ.

$R + S$ denotes $\bigcup_{n \in N} \underbrace{R \circ S \circ R \circ S \dots}_{n\ factors}$ Thus, $R + S = (R \cup S)^* = (R \circ S)^*$, since R and S are assumed to be reflexive. In particular, if α, β are congruences, $\alpha + \beta$ is the join of α and β in the lattice of congruences, while, for Θ, Γ tolerances, $\Theta + \Gamma$ is the smallest *congruence* which contains both Θ and Γ. Notice that $\Theta + \Gamma$ is far larger than the join of Θ and Γ *in the lattice of tolerances.*

R^- denotes the *converse* of R, that is, $a \, \mathrm{R}^- \, b$ if and only if $b \, \mathrm{R} \, a$. In particular, if R is a reflexive relation, $R + R^-$ is the smallest equivalence relation containing R.

§3. The weak tolerance intersection property.

DEFINITION 3.1. Let us say that an algebra $\boldsymbol{A}$ satisfies wTIP when

$$\alpha \cap \Theta^* = (\alpha \cap \Theta)^*$$

holds for every congruence α and tolerance Θ of $\boldsymbol{A}$, where * denotes transitive closure.

THEOREM 3.2. *If $\boldsymbol{A}$ is an algebra and every subalgebra of $\boldsymbol{A} \times \boldsymbol{A}$ generated by* 4 *elements is congruence modular then $\boldsymbol{A}$ satisfies wTIP.*

LEMMA 3.3. *If an algebra $\boldsymbol{A}$ satisfies*

$$\alpha \cap (\Theta \circ \Theta) \subseteq (\alpha \cap \Theta)^*$$

for every tolerance Θ and congruence α of $\boldsymbol{A}$, then $\boldsymbol{A}$ satisfies wTIP.

PROOF. The proof is a slight variation on arguments from [2] (cf. also the proof that condition (5) implies condition (3) in [3]).

The inclusion $(\alpha \cap \Theta)^* \subseteq (\alpha \cap \Theta^*)^* = \alpha \cap \Theta^*$ is trivial, since α is a congruence, hence $\alpha \cap \Theta^*$ is a congruence.

For the converse, suppose that $a, b \in A$, and $(a, b) \in \alpha \cap \Theta^*$. We have to show that $(a, b) \in (\alpha \cap \Theta)^*$.

If $(a, b) \in \alpha \cap \Theta^*$, then there is some natural number n, such that $(a, b) \in \alpha \cap \Theta^n$, where we define $\Theta^n = \Theta \circ \Theta \circ \cdots \circ \Theta$ (n occurrences of Θ). Hence, it is enough to show that $\alpha \cap \Theta^n \subseteq (\alpha \cap \Theta)^*$, for every n.

We shall show by induction on $m \geq 1$ that

$$(*) \qquad \alpha \cap \Theta^{2^m} \subseteq (\alpha \cap \Theta)^*$$

and this is enough, since $\alpha \cap \Theta^n \subseteq \alpha \cap \Theta^{2^m}$, provided we choose m in such a way that $n \leq 2^m$.

The basis $m = 1$ of the induction is the hypothesis of the Lemma.

Now for the inductive step. Suppose that (*) is true for some given m: then

$$\alpha \cap \Theta^{2^{m+1}} = \alpha \cap (\Theta^{2^m} \circ \Theta^{2^m}) \subseteq (\alpha \cap \Theta^{2^m})^* \subseteq (\alpha \cap \Theta)^{**} = (\alpha \cap \Theta)^*$$

where the last inclusion follows from the inductive hypothesis, and the first inclusion follows from the hypothesis of the Lemma by taking the tolerance Θ^{2^m} in place of Θ.

Thus, (*) is true for $m+1$, and the induction is complete. ⊣

Proof of Theorem 3.2. In view of Lemma 3.3, it is enough to show that $\alpha \cap (\Theta \circ \Theta) \subseteq (\alpha \cap \Theta)^*$.

So let $a, c \in A$ and $(a, c) \in \alpha \cap (\Theta \circ \Theta)$. Thus $a \,\alpha\, c$, and there is $b \in A$ such that $a \,\Theta\, b \,\Theta\, c$.

Consider the subalgebra $\boldsymbol{B}$ of $\boldsymbol{A} \times \boldsymbol{A}$ generated by the four elements (a, a), (a, b), (c, b), (c, c). First, observe that if $(x, y) \in \boldsymbol{B}$ then $x \,\Theta\, y$, since all the generators of $\boldsymbol{B}$ are in Θ, and Θ is compatible.

We have that $((a, a), (a, b))$ and $((c, b), (c, c))$ belong to $(0 \times 1)_{|\boldsymbol{B}}$, $((a, b), (c, b))$ belongs to $(\alpha \times 0)_{|\boldsymbol{B}}$ and $((a, a), (c, c))$ belongs to $(\alpha \times \alpha)_{|\boldsymbol{B}}$.

The above relations imply that $((a, a), (c, c))$ belongs to

$$(\alpha \times \alpha)_{|\boldsymbol{B}} \cap \left((0 \times 1)_{|\boldsymbol{B}} + (\alpha \times 0)_{|\boldsymbol{B}}\right)$$

Since $(\alpha \times 0)_{|\boldsymbol{B}} \leq (\alpha \times \alpha)_{|\boldsymbol{B}}$, and $\boldsymbol{B}$ is congruence modular,

$$(\alpha \times \alpha)_{|\boldsymbol{B}} \cap \left((0 \times 1)_{|\boldsymbol{B}} + (\alpha \times 0)_{|\boldsymbol{B}}\right) \leq \left((\alpha \times \alpha)_{|\boldsymbol{B}} \cap (0 \times 1)_{|\boldsymbol{B}}\right) + (\alpha \times 0)_{|\boldsymbol{B}} = (0 \times \alpha)_{|\boldsymbol{B}} + (\alpha \times 0)_{|\boldsymbol{B}}$$

In conclusion, $((a, a), (c, c))$ belongs to

$$(0 \times \alpha)_{|\boldsymbol{B}} + (\alpha \times 0)_{|\boldsymbol{B}}$$

This implies that there is some n, and there are pairs $(x_i, y_i) \in \boldsymbol{B}$ $(0 \leq i \leq n)$ such that

$$\begin{array}{lll} (a, a) = (x_0, y_0) & (x_n, y_n) = (c, c) & \\ (x_i, y_i) \equiv (x_{i+1}, y_{i+1}) & \bmod (0 \times \alpha)_{|\boldsymbol{B}} & \text{for } i \text{ even} \\ (x_i, y_i) \equiv (x_{i+1}, y_{i+1}) & \bmod (\alpha \times 0)_{|\boldsymbol{B}} & \text{for } i \text{ odd} \end{array}$$

In other words,

$$\begin{array}{lll} a = x_0 = y_0 & c = x_n = y_n & \\ x_i = x_{i+1}, & y_i \,\alpha\, y_{i+1} & \text{for } i \text{ even} \\ x_i \,\alpha\, x_{i+1}, & y_i = y_{i+1} & \text{for } i \text{ odd} \end{array}$$

In particular, $a = x_0 = x_1 \,\alpha\, x_2 = x_3 \,\alpha\, x_4 \ldots$, and $a = y_0 \,\alpha\, y_1 = y_2 \,\alpha\, y_3 = y_4 \ldots$, hence $x_i \,\alpha\, y_j$ for all i's and j's, since α is a congruence, and both x_i and y_j are congruent to a modulo α.

Moreover, since $(x_i, y_i) \in \boldsymbol{B}$, then $x_i \, \Theta \, y_i$ for all i's, by the remark made after the definition of $\boldsymbol{B}$.

Hence, for all i's, $(x_i, y_i) \in \alpha \cap \Theta$, and $(y_i, x_i) \in \alpha \cap \Theta$, since Θ is symmetric. In conclusion, the sequence

$$a = x_0 = x_1 \quad y_1 = y_2 \quad x_2 = x_3 \quad y_3 = y_4 \quad \dots \quad x_n = y_n = c$$

witnesses that $(a, c) \in (\alpha \cap \Theta)^*$. ⊣

The proof of Theorem 3.2 can be modified to get (see [22] for details):

THEOREM 3.4. *Suppose that* $\boldsymbol{A}$ *is an algebra such that every subalgebra of* $\boldsymbol{A} \times \boldsymbol{A}$ *generated by* 4 *elements satisfies* $\beta \cap (\gamma \circ \delta \circ \gamma) \subseteq \beta\gamma + \delta$, *for all congruences* β, γ, δ *with* $\delta \leq \beta$.

Then $\boldsymbol{A}$ *satisfies wTIP. More generally,* $\boldsymbol{A}$ *satisfies*

$$\alpha(R + S) \subseteq \alpha(\overline{R \cup S^-}) + \alpha(\overline{R^- \cup S}) = \alpha(\overline{R \cup S}) + \alpha(\overline{R^- \cup S^-})$$

and

$$\alpha(R + R^-) \subseteq \alpha(\overline{R} + \overline{R}^-) = \alpha\overline{R} + \alpha\overline{R}^- = \alpha(Cg(R))$$

for all reflexive relations R *and* S *and every congruence* α, *and where* $Cg(R)$ *denotes the congruence generated by* R.

§4. Terms without double couples. The definition of wTIP and Theorem 3.2 allow us to apply the methods of [4] to the class of *terms without double couples* in the following sense:

DEFINITION 4.1. The class of *wdc-terms* (terms without double couples) is the smallest class (of terms in the language of lattices) which

(a) contains all variables
(b) contains all terms of the form

$$x_i(t^1 + t^2 + \cdots + t^n)$$

where $n > 0$, x_i is a variable, and $t^1, t^2, \dots t^n$ are wdc-terms.

Essentially, a wdc-term is a term with no subterm of the form $(t^1 + t^2)(t^3 + t^4)$. Wdc-terms had been called *simple terms* in [23].

DEFINITION 4.2. If t is a term for the language of lattices $\cdot, +$, define by induction the term t_2 for the language $\cap, \circ$ as follows:
If $t = x_i$ is a variable, let $t_2 = t$;
if $t = r \cdot s$, let $t_2 = r \cap s$, and
if $t = r + s$, let $t_2 = r \circ s$.

Thus, t_2 is obtained from t simply by replacing $\cdot$ by $\cap$ and $+$ by $\circ$.

In the present paper, we are mainly concerned with lattices of congruences. Tolerances, too, form a lattice. In both cases, meet is intersection. Congruences are (particular) tolerances but usually the join of two congruences in the lattice of tolerances is far smaller than their join in the lattice of congruences. Since we need consider both cases simultaneously, given a lattice term t, we shall write t_{con}, t_{tol}, respectively, for the case when join is evaluated in the lattice of congruences, tolerances, respectively. $+$ will always denote join in the lattice of congruences. Recall that the join of two tolerances Γ, Θ (in the lattice of tolerances) is $\overline{\Gamma \cup \Theta}$, and that juxtaposition denotes intersection.

PROPOSITION 4.3. *If A is an algebra which satisfies wTIP, then $t_{tol}(\bar{\alpha})^* = t_2(\bar{\alpha})^* = t_{con}(\bar{\alpha})$, for every wdc-term t, and for every m-tuple $\bar{\alpha}$ of congruences of A.*

PROOF. The proof is essentially the same as the proof of [4, Lemma 1]. wTIP allows us to apply the methods of [4], provided we consider only wdc-terms ([4] can be applied to *every* term, but using the stronger assumption TIP).

We shall first prove $t_{con}(\bar{\alpha}) = t_{tol}(\bar{\alpha})^*$ by induction on the complexity of the wdc-term t.

If $t = x_i$ is a variable, then $t_{con}(\bar{\alpha}) = \alpha_i = t_{tol}(\bar{\alpha})^*$, since α is assumed to be a congruence.

Now let $t = x_i(t^1 + t^2 + \cdots + t^n)$, where the t^j's are wdc-terms. Thus,

$$t_{con}(\bar{\alpha}) = \alpha_i \left(t^1_{con}(\bar{\alpha}) + t^2_{con}(\bar{\alpha}) + \cdots + t^n_{con}(\bar{\alpha})\right) =$$

$$\alpha_i \left(t^1_{tol}(\bar{\alpha})^* + t^2_{tol}(\bar{\alpha})^* + \cdots + t^n_{tol}(\bar{\alpha})^*\right) = \alpha_i \left(\overline{t^1_{tol}(\bar{\alpha}) \cup t^2_{tol}(\bar{\alpha}) \cup \cdots \cup t^n_{tol}(\bar{\alpha})}\right)^* =$$

$$\left(\alpha_i \left(\overline{t^1_{tol}(\bar{\alpha}) \cup t^2_{tol}(\bar{\alpha}) \cup \cdots \cup t^n_{tol}(\bar{\alpha})}\right)\right)^* = t_{tol}(\bar{\alpha})^*$$

by the inductive hypothesis applied to $t^1, t^2, \dots, t^n$, and by wTIP applied to the congruence α_i and to the tolerance $\overline{t^1_{tol}(\bar{\alpha}) \cup t^2_{tol}(\bar{\alpha}) \cup \cdots \cup t^n_{tol}(\bar{\alpha})}$. The induction step is thus complete, hence we have proved $t_{con}(\bar{\alpha}) = t_{tol}(\bar{\alpha})^*$.

Since $t_{tol}(\bar{\alpha}) \subseteq t_2(\bar{\alpha}) \subseteq t_{con}(\bar{\alpha})$ trivially, and since $t_{con}(\bar{\alpha})$ is a congruence, hence $t_{con}(\bar{\alpha})^* = t_{con}(\bar{\alpha})$, we get $t_{tol}(\bar{\alpha})^* \subseteq t_2(\bar{\alpha})^* \subseteq t_{con}(\bar{\alpha})$, by the closure properties of *.

Since we have already proved $t_{con}(\bar{\alpha}) = t_{tol}(\bar{\alpha})^*$, we get $t_{tol}(\bar{\alpha})^* = t_2(\bar{\alpha})^* = t_{con}(\bar{\alpha})$. ⊣

COROLLARY 4.4. *If A satisfies wTIP, s, t are lattice terms, and t is wdc, then for every m-tuple $\bar{\alpha}$ of congruences of A the following are equivalent:*

(i) $t_{tol}(\bar{\alpha}) \subseteq s_{con}(\bar{\alpha})$;
(ii) $t_2(\bar{\alpha}) \subseteq s_{con}(\bar{\alpha})$;
(iii) $t_{con}(\bar{\alpha}) \subseteq s_{con}(\bar{\alpha})$.

PROOF. Since $s_{con}(\bar{\alpha})$ is a congruence, $t_{tol}(\bar{\alpha}) \subseteq s_{con}(\bar{\alpha})$ if and only if $t_{tol}(\bar{\alpha})^* \subseteq s_{con}(\bar{\alpha})$. Similarly, $t_2(\bar{\alpha}) \subseteq s_{con}(\bar{\alpha})$ if and only if $t_2(\bar{\alpha})^* \subseteq s_{con}(\bar{\alpha})$.

The conclusion is now immediate from Proposition 4.3. ⊣

COROLLARY 4.5. *If $\boldsymbol{A}$ satisfies wTIP then $\boldsymbol{A}$ is congruence modular.*

PROOF. Apply Corollary 4.4 to the terms $t(x, y, z) = x(y + (zx))$ and $s(x, y, z) = (xy) + (zx)$.

Since $\alpha \cap (\beta \circ (\gamma \cap \alpha)) = (\alpha \cap \beta) \circ (\gamma \cap \alpha) \subseteq \alpha\beta + \gamma\alpha$, we get, by (ii)⇒(iii) in Corollary 4.4, $\alpha(\beta + (\gamma\alpha)) \leq \alpha\beta + \gamma\alpha$, that is, congruence modularity. ⊣

Indeed, Proposition 4.3 gives more: for example, if $\boldsymbol{A}$ satisfies wTIP then $\alpha(\beta + \gamma) = (\alpha \cap (\beta \circ \gamma))^*$, for all congruences α, β, γ of $\boldsymbol{A}$.

COROLLARY 4.6. *Suppose that $\boldsymbol{A}$ is an algebra such that every subalgebra of $\boldsymbol{A} \times \boldsymbol{A}$ generated by 4 elements satisfies $\beta(\gamma \circ \delta \circ \gamma) \subseteq \beta\gamma + \delta$, for all congruences β, γ, δ with $\delta \leq \beta$.*

Then $\boldsymbol{A}$ is congruence modular.

PROOF. By Theorem 3.4 $\boldsymbol{A}$ satisfies wTIP, hence $\boldsymbol{A}$ is congruence modular by Corollary 4.5. ⊣

COROLLARY 4.7. *If $\boldsymbol{A}$ satisfies wTIP then the following are equivalent:*

(a) *$\boldsymbol{A}$ satisfies $\alpha(\overline{\beta \cup \gamma}) \subseteq \alpha\beta + \alpha\gamma$, for all congruences α, β and γ;*
(b) *$\boldsymbol{A}$ satisfies $\alpha(\beta \circ \gamma) \subseteq \alpha\beta + \alpha\gamma$, for all congruences α, β and γ;*
(c) *$\boldsymbol{A}$ is congruence distributive.*

PROOF. Apply Corollary 4.4 to the terms $t(x, y, z) = x(y+z)$ and $s(x, y, z) = (xy) + (zx)$. ⊣

The proofs of Proposition 4.3 and of Corollary 4.4 furnish the following slight strengthening of some results in [4]:

PROPOSITION 4.8. *Suppose that the algebra $\boldsymbol{A}$ satisfies TIP, s, t are lattice terms, and $\bar{\alpha}$ is an m-tuple of congruences of $\boldsymbol{A}$.*

Then $t_{tol}(\bar{\alpha})^* = t_2(\bar{\alpha})^* = t_{con}(\bar{\alpha})$.

Moreover, the following are equivalent:

(i) $t_{tol}(\bar{\alpha}) \subseteq s_{con}(\bar{\alpha})$;
(ii) $t_2(\bar{\alpha}) \subseteq s_{con}(\bar{\alpha})$;
(iii) $t_{con}(\bar{\alpha}) \subseteq s_{con}(\bar{\alpha})$.

$t_{tol}(\bar{\alpha})^* = t_{con}(\bar{\alpha})$ in Proposition 4.8 can be obtained also as a consequence of the known result stating that TIP implies that * is a homomorphism from the lattice of tolerances of $\boldsymbol{A}$ to the lattice of congruences of $\boldsymbol{A}$ (see e.g. [5, Lemma 3.1]).

§5. Gumm's Shifting Principle. The Shifting Principle has been introduced by H. P. Gumm in connection with his geometrical interpretation of commutator theory in the study of congruence modular varieties. We show here that the Shifting Principle is equivalent to wTIP. Moreover, we introduce a stronger version of the Shifting Principle, and we show that this stronger version is equivalent to TIP.

DEFINITION 5.1. [10, p. 14] An algebra $\boldsymbol{A}$ satisfies *Gumm's Shifting Principle* if and only if whenever α, γ are congruences, and Λ is a tolerance of $\boldsymbol{A}$ such that $\alpha\Lambda \leq \gamma \leq \alpha$ then $\alpha(\Lambda \circ \gamma \circ \Lambda) \subseteq \gamma$ (recall that juxtaposition denotes intersection).

See [10] for a picture, and for geometrical interpretations and consequences. Recall that, for tolerances Λ, Ψ, $\Lambda + \Psi = (\overline{\Lambda \cup \Psi})^*$ is the smallest congruence containing both Λ and Ψ.

PROPOSITION 5.2. *For every algebra, the following are equivalent*:

(i) *wTIP*;
(ii) $\alpha(\Lambda + \alpha\Psi) = \alpha\Lambda + \alpha\Psi$, *for all tolerances* Λ *and* Ψ, *and congruence* α;
(iii) $\alpha(\Lambda \circ \alpha\Psi \circ \Lambda) \subseteq \alpha\Lambda + \alpha\Psi$, *for all tolerances* Λ *and* Ψ, *and congruence* α;
(iv) *The Shifting Principle.*

PROOF. (i)$\Rightarrow$(ii). Let Θ be the tolerance $\overline{\Lambda \cup \alpha\Psi}$. By applying wTIP, and since α is a congruence, we get $\alpha(\Lambda+\alpha\Psi) = \alpha\Theta^* = (\alpha\Theta)^* \subseteq (\alpha(\Lambda\circ\alpha\Psi))^* = (\alpha\Lambda \circ \alpha\Psi)^* = \alpha\Lambda + \alpha\Psi$.

The $\supseteq$ inclusion is trivial, since α is a congruence.

(ii)$\Rightarrow$(iii) is trivial, since $\Lambda \circ \alpha\Psi \circ \Lambda \subseteq \Lambda + \alpha\Psi$.

(iii)$\Rightarrow$(iv). Take $\Psi = \gamma$ in (iii), thus getting $\alpha(\Lambda \circ \gamma \circ \Lambda) = \alpha(\Lambda \circ \alpha\gamma \circ \Lambda) \subseteq \alpha\Lambda + \alpha\gamma = \alpha\Lambda + \gamma = \gamma$.

(iv)$\Rightarrow$(i). Suppose that Gumm's Shifting Principle holds, and let α and Θ be, respectively, a congruence and a tolerance of $\boldsymbol{A}$. Consider the congruence $\gamma = (\alpha\Theta)^*$, and apply the Shifting Principle with Θ in place of Λ. We get $\alpha(\Theta \circ \Theta) \subseteq \alpha(\Theta \circ \gamma \circ \Theta) \subseteq \gamma = (\alpha\Theta)^*$.

Hence wTIP holds by Lemma 3.3. ⊣

[2, Theorem 1] showed that congruence modular varieties satisfy a certain tolerance identity (which turned out to be equivalent to TIP, see [3]), and [2, Corollary 1] showed that this tolerance identity implies the Shifting Principle.

COROLLARY 5.3. *Suppose that* $\boldsymbol{A}$ *is an algebra such that every subalgebra of* $\boldsymbol{A} \times \boldsymbol{A}$ *generated by* 4 *elements satisfies* $\beta(\gamma \circ \delta \circ \gamma) \subseteq \beta\gamma + \delta$, *for all congruences* β, γ, δ *with* $\delta \leq \beta$. *Then* $\boldsymbol{A}$ *satisfies Gumm's Shifting Principle.*

In particular, if every subalgebra of $\boldsymbol{A} \times \boldsymbol{A}$ *is congruence modular then* $\boldsymbol{A}$ *satisfies Gumm's Shifting Principle.*

PROOF. Immediate from Theorem 3.4 and Proposition 5.2. ⊣

DEFINITION 5.4. Let us say that an algebra $\boldsymbol{A}$ satisfies the *Strong Shifting Principle* if and only if whenever γ is a congruence, and Γ, Λ are tolerances of $\boldsymbol{A}$ such that $\Gamma\Lambda \leq \gamma \leq \Gamma^*$ then $\Gamma(\Lambda \circ \gamma \circ \Lambda) \subseteq \gamma$.

Gumm's original Shifting Principle is a particular case of the Strong Shifting Principle: just let $\Gamma = \alpha$ be a congruence.

PROPOSITION 5.5. *For every algebra, TIP is equivalent to the Strong Shifting Principle.*

PROOF. Suppose that TIP holds, and that Γ, γ and Λ are as in the hypothesis of the Strong Shifting Principle.

Since γ is a congruence, $\Gamma\Lambda \subseteq \gamma$ implies $(\Gamma\Lambda)^* \subseteq \gamma$. By TIP we get $\Gamma^*\Lambda^* \subseteq \gamma$, hence $\Gamma^*\Lambda \subseteq \gamma$. Since TIP implies wTIP, then by Proposition 5.2 we can apply Gumm's Shifting Principle with the congruence Γ^* in place of α in order to get $\Gamma(\Lambda \circ \gamma \circ \Lambda) \subseteq \Gamma^*(\Lambda \circ \gamma \circ \Lambda) \subseteq \gamma$.

Conversely, suppose that the Strong Shifting Principle holds, and let Γ, Θ be tolerances of $\boldsymbol{A}$. Consider the congruence $\gamma = (\Gamma\Theta)^*$, and apply the Strong Shifting Principle with Θ in place of Λ. We get $\Gamma(\Theta \circ \Theta) \subseteq \Gamma(\Theta \circ \gamma \circ \Theta) \subseteq \gamma = (\Gamma\Theta)^*$, from which we reach the conclusion, since it is known that $\Gamma(\Theta \circ \Theta) \subseteq (\Gamma\Theta)^*$ implies TIP (see e.g. [3]). ⊣

Other conditions equivalent to TIP have been introduced in G. Czédli and E. Horváth's fundamental contribution [2]. See [5, Lemma 3.1] and [3] for the proof that such conditions are actually equivalent to TIP. Notice that the statement of [5, Lemma 3.1] includes the assumption that $\boldsymbol{A}$ is congruence modular, but this assumption turns out to be unnecessary for Conditions (ii), (iii) and (iv) therein since they all imply modularity (see [4, Lemma 1(D)], or Corollary 4.5).

PROBLEM 5.6. Investigate the geometrical meaning of the Strong Shifting Principle.

In [24] we introduced still another variant of TIP.

(wTIP$_2$)
$$\Gamma^* \cap \Theta^* = (\Gamma \cap (\Theta \circ \Theta))^*$$

COROLLARY 5.7. *For every variety $\mathcal{V}$ the following are equivalent*:

(i) *Every algebra in $\mathcal{V}$ is congruence modular*;
(ii) *every algebra in $\mathcal{V}$ satisfies TIP*;
(iii) *every algebra in $\mathcal{V}$ satisfies wTIP$_2$*;
(iv) *every algebra in $\mathcal{V}$ satisfies wTIP*;
(v) *every algebra in $\mathcal{V}$ satisfies Gumm's Shifting Principle.*
(vi) *every algebra in $\mathcal{V}$ satisfies the Strong Shifting Principle.*

PROOF. See e.g. [3] for a proof that (i)⇒(ii).

Since TIP⇒wTIP$_2$ ⇒wTIP in every algebra, (ii)⇒(iii)⇒(iv) are trivial.

(iv)⇒(v) follows from Proposition 5.2.

(v)⇒(i) is proved in [10, Lemma 3.2]. Alternatively, (v)⇒(iv) follows from Proposition 5.2, and (iv)⇒(i) follows from Corollary 4.5.

(vi)⇔(ii) is Proposition 5.5.

Notice that Theorem 3.2 gives another proof for (i)⇒(iv). ⊣

The equivalence of (i) and (v) is due to Gumm [10, Corollary 3.6]. The equivalence of (i) and (ii) is already known, but we are not in the position to give explicit credits.

§6. Some problems. Let us recall the results we are trying to generalize.

THEOREM 6.1. [7] *If every subalgebra of* $\mathbf{A} \times \mathbf{A}$ *is congruence modular then* $\mathbf{A}$ *is Arguesian.*

THEOREM 6.2. [4] *A congruence modular variety satisfies all the higher Arguesian identities introduced by Haiman.*

PROBLEM 6.3. Suppose that every subalgebra of $\mathbf{A} \times \mathbf{A}$ is congruence modular. Does $\mathbf{A}$ satisfy Haiman's higher Arguesian identities?

PROBLEM 6.4. Does wTIP imply the Arguesian identity? Does wTIP imply the higher Arguesian identities?

In view of Theorem 3.2, affirmative answers to Problem 6.4 would furnish another proof of Theorem 6.1 and, respectively, an affirmative answer to Problem 6.3.

PROBLEM 6.5. [24] Suppose that every subalgebra of $\mathbf{A} \times \mathbf{A}$ is congruence modular. Does $\mathbf{A}$ satisfy TIP?

An affirmative answer to Problem 6.5 would give an affirmative answer to Problem 6.3 since [4] shows that TIP implies all the higher Arguesian identities.

Another route to the solution of Problem 6.3 (to another proof of Theorem 6.1, respectively) would be to show that if $\mathbf{A} \times \mathbf{A}$ is congruence modular then $\mathbf{A}$ satisfies some property stronger than wTIP, which is strong enough in order to show that $\mathbf{A}$ satisfies Haiman's higher Arguesian identities (the Arguesian identity, respectively). In this connection, we have found a proof for the following version of Theorem 3.2.

THEOREM 6.6. [24] *If* $\mathbf{A}$ *is an algebra and every subalgebra of* $\mathbf{A}^4$ *is congruence modular then* $\mathbf{A}$ *satisfies*

$$(\mathrm{wTIP}_2) \qquad \Gamma^* \cap \Theta^* = (\Gamma \cap (\Theta \circ \Theta))^*$$

for all tolerances Γ, Θ *of* $\mathbf{A}$*.*

See [24] for a proof of Theorem 6.6, as well as for further problems and comments.

PROBLEM 6.7. Can the Arguesian identity (the higher Arguesian identities, respectively) be written as $p \leq q$, for some wdc-term p?

If Problem 6.7 has an affirmative answer, then it is conceivable that Problem 6.4 has an affirmative answer, in view of Corollary 4.4. Apparently, the methods of Section 4 should work, unless the equality $p \leq q$ in Problem 6.7 has a surprisingly strange and unexpected form.

PROBLEM 6.8. Even in case Problems 6.3 and 6.5 have a negative answer, we can ask whether the conclusions hold under the stronger assumption that every subalgebra of A^n is congruence modular (for some specific $n > 2$).

PROBLEM 6.9. Find (if it exists) an example of an algebra satisfying wTIP but not TIP.

REFERENCES

[1] STANLEY BURRIS and H. P. SANKAPPANAVAR, *A Course in Universal Algebra*, Graduate Texts in Mathematics, vol. 78, Springer-Verlag, New York, 1981.

[2] GÁBOR CZÉDLI and ESZTER K. HORVÁTH, *Congruence distributivity and modularity permit tolerances*, ***Acta Universitatis Palackianae Olomucensis. Facultas Rerum Naturalium. Mathematica***, vol. 41 (2002), pp. 39–42.

[3] ———, *All congruence lattice identities implying modularity have Mal'tsev conditions*, ***Algebra Universalis***, vol. 50 (2003), no. 1, pp. 69–74.

[4] GÁBOR CZÉDLI, ESZTER K. HORVÁTH, and PAOLO LIPPARINI, *Optimal Mal'tsev conditions for congruence modular varieties*, ***Algebra Universalis***, vol. 53 (2005), no. 2-3, pp. 267–279.

[5] GÁBOR CZÉDLI, ESZTER K. HORVÁTH, and S. RADELECZKI, *On tolerance lattices of algebras in congruence modular varieties*, ***Acta Mathematica Hungarica***, vol. 100 (2003), no. 1-2, pp. 9–17.

[6] ALAN DAY, *A characterization of modularity for congruence lattices of algebras*, ***Canadian Mathematical Bulletin***, vol. 12 (1969), pp. 167–173.

[7] RALPH FREESE and BJARNI JÓNSSON, *Congruence modularity implies the Arguesian identity*, ***Algebra Universalis***, vol. 6 (1976), no. 2, pp. 225–228.

[8] RALPH FREESE and RALPH MCKENZIE, ***Commutator Theory for Congruence Modular Varieties***, London Mathematical Society Lecture Note Series, vol. 125, Cambridge University Press, Cambridge, 1987.

[9] GEORGE GRÄTZER, ***Universal Algebra***, second ed., Springer-Verlag, New York, 1979.

[10] H. PETER GUMM, *Geometrical methods in congruence modular algebras*, ***Memoirs of the American Mathematical Society***, vol. 45 (1983), no. 286, pp. viii+79.

[11] MARK HAIMAN, *Proof theory for linear lattices*, ***Advances in Mathematics***, vol. 58 (1985), no. 3, pp. 209–242.

[12] ———, *Arguesian lattices which are not type-*1, ***Algebra Universalis***, vol. 28 (1991), no. 1, pp. 128–137.

[13] DAVID HOBBY and RALPH MCKENZIE, ***The Structure of Finite Algebras***, Contemporary Mathematics, vol. 76, American Mathematical Society, Providence, RI, 1988.

[14] BJARNI JÓNSSON, *Representation of modular lattices and of relation algebras*, ***Transactions of the American Mathematical Society***, vol. 92 (1959), pp. 449–464.

[15] ———, *Algebras whose congruence lattices are distributive*, ***Mathematica Scandinavica***, vol. 21 (1967), pp. 110–121 (1968).

[16] ———, *Congruence varieties*, ***Algebra Universalis***, vol. 10 (1980), no. 3, pp. 355–394.

[17] KEITH A. KEARNES, *Congruence join semidistributivity is equivalent to a congruence identity*, ***Algebra Universalis***, vol. 46 (2001), no. 3, pp. 373–387.

[18] KEITH A. KEARNES and ÁGNES SZENDREI, *The relationship between two commutators*, ***International Journal of Algebra and Computation***, vol. 8 (1998), no. 4, pp. 497–531.

[19] PAOLO LIPPARINI, *Commutator theory without join-distributivity*, ***Transactions of the American Mathematical Society***, vol. 346 (1994), no. 1, pp. 177–202.

[20] ———, *Congruence modularity implies the Arguesian law for single algebras with a difference term*, ***Journal of Algebra***, vol. 219 (1999), no. 2, pp. 658–681.

[21] ———, *A non-trivial congruence implication between identities weaker than modularity*, ***Acta Scientiarum Mathematicarum (Szeged)***, vol. 68 (2002), no. 3-4, pp. 593–609.

[22] ———, *Intersection properties of relations*, 2005, `arXiv:math.CO/0503103`.

[23] ———, *Tolerance intersection properties (abstract)*, ***The Bulletin of Symbolic Logic***, vol. 11 (2005), no. 2, p. 262.

[24] ———, *A local proof for a tolerance intersection property*, ***Algebra Universalis***, in print (2005).

[25] A. I. MAL'CEV, *On the general theory of algebraic systems*, ***Matematicheskiĭ Sbornik***, vol. 35(77) (1954), pp. 3–20.

[26] RALPH N. MCKENZIE, GEORGE F. MCNULTY, and WALTER F. TAYLOR, ***Algebras, Lattices, Varieties. Vol. I***, The Wadsworth & Brooks/Cole Mathematics Series, Wadsworth & Brooks/Cole Advanced Books & Software, Monterey, CA, 1987.

[27] J. B. NATION, *Varieties whose congruences satisfy certain lattice identities*, ***Algebra Universalis***, vol. 4 (1974), pp. 78–88.

[28] JONATHAN D. H. SMITH, ***Mal'cev Varieties***, Lecture Notes in Mathematics, vol. 554, Springer-Verlag, Berlin, 1976.

DIPARTIMENTO DI MATEMATICA
VIALE DELLA RICERCA SCIENTIFICA
II UNIVERSITA DI ROMA (TOR VERGATA)
I-00133 ROME, ITALY
E-mail: lipparin@axp.mat.uniroma2.it
URL: `http://www.mat.uniroma2.it/~lipparin`

UNIVERSES IN TYPE THEORY PART I — INACCESSIBLES AND MAHLO

ANTON SETZER

Abstract. We give an overview over universes in Martin-Löf type theory and consider the following universe constructions: a simple universe, E. Palmgren's super-universe and the Mahlo universe. We then introduce models for these theories in extensions of Kripke-Platek set theory having the same proof theoretic strength. The extensions of Kripke-Platek set theory used formalise the existence of a recursively inaccessible ordinal, a recursively hyper-inaccessible ordinal, and a recursively Mahlo ordinal. Using these models we determine upper bounds for the proof theoretic strength of the theories in questions. In case of simple universes and the Mahlo universe, these bounds have been shown by the author to be sharp. This article is an overview of the main techniques in developing these models. Full details will be presented in a series of future articles.

§1. **Introduction.** This article presents some results of a research program with the goal of formulating extensions of Martin-Löf type theory (MLTT) that are proof-theoretically as strong as possible while remaining predicatively justified, and of determining their precise proof theoretic strength. We see three main reasons for following this research program:

1. The goal is to develop type theoretic analogues of the theories analysed in proof theory. In this way we hope to make the rather abstract and technically difficult results from proof theory more accessible to the general audience, and we hope as well to give some computational meaning to those results. This will be particularly important in case of the Π_3-reflecting universe (to be presented in the follow-up article [37]), which was developed from Rathjen's ordinal notation system [26] for $\mathrm{KP} + (\Pi_3 - \mathrm{refl})$ (Kripke-Platek set theory extended by the principle of Π_3-reflection).

2. One can consider predicatively justified extensions of MLTT as safe theories, more precisely theories for which there is a philosophical argument which justifies the validity of everything shown in those theories. This philosophical argument is given by Martin-Löf's meaning explanations. The proofs of the lower bounds show that these extensions of MLTT show the consistency of

The author was supported by Nuffield Foundation, grant ref. NAL/00303/G and EPSRC grant GR/S30450/01.

Logic Colloquium '04
Edited by A. Andretta, K. Kearnes, and D. Zambella
Lecture Notes in Logic, 29

corresponding extensions of Kripke-Platek set theory, or more precisely of any approximations of those extensions. If one accepts meaning explanations as a philosophical argument for the validity of statements shown in MLTT, one obtains in this way consistency proofs for extensions of Kripke-Platek set theory, which are strong enough to prove a large portion of mathematical theorems (see the results in reverse mathematics, e.g. Simpson's book [38], which show that rarely more than the strength of $(\Pi^1_1 - \mathrm{CA})_0$ is needed; one exception seems to be the graph minor theorem). These results are in line with a revised Hilbert's program. Hilbert's original program was to prove the consistency of axiom systems for formalising mathematical proofs by finitary arguments. By Gödel's incompleteness theorem we know that this program cannot be carried out. MLTT can be considered as one replacement for finitary methods in a revised Hilbert's program, and our research demonstrates that MLTT can indeed be used for this purpose.

We should note however that meaning explanations haven't been worked out yet in the case of the Mahlo universe — the author himself doesn't have a sufficient background in philosophy to fill in this gap.

3. We hope as well that the new sets developed can be used as data structures in general computing. This hope has been fulfilled in case of the Mahlo universe. The data type of inductive-recursive definitions used in the closed formalisation of inductive-recursive definitions developed by P. Dybjer and the author (see [13, 14, 15, 16]) has a similar character to the Mahlo universe. Variants of this data type have been used in [10, 12] in the area of generic or generative programming.

In generative programming one aims at writing programs with a higher degree of polymorphism. Ordinary programs have as input elements of certain fixed data structures and compute from it elements of other fixed data structures. Polymorphic programs allow both the input and output data structures to be extended by certain type parameters, for instance both the input and output data structure can be $\mathrm{List}(X)$ (the type of lists of type X) depending on a type parameter X. Generative programs have as input data structures and elements of these data structures, and *compute* new data structures and elements of these. The computation can make use of the data and the build-up of the data structures. An example is the function, which takes as input an inductive data type and computes from it an inductive data type with one more constructor, and embeds the original data type into the extended one.

We hope that the Π_3-reflecting universe, which will be developed in the follow-up article [37], will give rise to data structures which have similar interesting applications in computer science.

We should note that whether the Mahlo universe and extensions of it presented in the follow-up article [37] are actually predicatively justifiable is still a matter of debate. However, even if there is a debate on this, it seems at least at the moment to be unlikely that a better construction can be found which avoids the problem in this case. For instance Erik Palmgren's higher order universes [23] look at first sight more directly predicatively justifiable — but a closer look (it took the author a long time to actually discover this) reveals that in Palmgren's construction Set has a similar character as a Mahlo universe. The introduction rule allows to introduce new elements of Set assuming an arbitrary function from families of Set into families of Set (the new element will be a universe closed under this function). This is essentially the same as the introduction rule for the Mahlo universe. Despite this criticism the author wants to express his strong admiration for Palmgren's higher order universes — its a fascinating construction which doesn't seem to have been fully understood yet. Especially its precise proof-theoretical strength still remains an open problem (to obtain as upper bound the strength of KPM by using techniques used in this article shouldn't be a big problem — the real difficulty is to show that this is a lower bound as well).

Follow up articles. The original goal when writing this article was to present the Π_3-reflecting universe. It turned out that much more space is needed to present it, and therefore the presentation of the Π_3-reflecting universe will be given in a planned follow-up article [37]. Even so the current article is rather long, we won't have room to present the models in full detail. We plan to write a series of articles, in which the models will be fully worked out. These future articles will then set up a better infrastructure for developing model constructions for extensions of MLTT. So the current article serves as an overview article, presenting the basic techniques for developing models of type theory suitable for determining upper bounds for the proof theoretic strength. It might be that this article is more accessible than the more detailed technical articles to follow.

Content. The structure of this article is as follows: In Section 2 we briefly develop the small and full logical framework and the basic set constructions (i.e. the sets N_n, N, and the sets formed using $+$, Σ, Π, W, Id). We then briefly introduce Kripke-Platek set theory. Then we develop the basic principles for defining models of type theory for the basic set constructions and the small logical framework. In Section 3 we introduce the rules for universes and the theory $\mathrm{ML_W U}$. Furthermore we determine the principles for introducing nested universes. We will see that there are two ways of defining subuniverses, namely recursively and inductively. We then develop the main concepts for introducing models for universes and develop in particular a model of $\mathrm{ML_W U}$. In Section 4 we introduce Erik Palmgren's super-universe and develop a model of it in a corresponding extension of Kripke-Platek set theory. Finally, in Section 5 we introduce the Mahlo universe and determine a model of it. For all three universe constructions mentioned we obtain proof theoretic upper bounds for the strength of the theories in question.

Related results. The first model of MLTT developed for proof theoretic purposes seems to have been the model of MLTT with one universe but no

W-type, developed by P. Aczel [1]. Rathjen and Griffor have analysed the strength of MLTT with finitely many universes, W-type, and of several variants of it in [17] (where by finitely many universes we mean that one has universes U_n indexed over Meta-natural numbers n; so one cannot form for instance $\Sigma n : N.U_n$). Erik Palmgren has introduced the super-universe and later higher order universes (the latter are conjectured to reach the strength of KPM), which are best presented in [23]. There is an extensive literature on PER models of MLTT — an overview of the techniques is given in M. Hofmann's article [18]. Note that the emphasis of this article is in developing models in order to determine upper bounds for the proof-theoretic strength of those theories, which means as well to develop models using minimal strength in the Meta-theory. In this research programme we are following the steps of ordinal theoretic proof theory, where the major recent steps have been carried out by M. Rathjen, who analysed the theories KPM [26], $\mathrm{KP} + (\Pi_3 - \mathrm{refl})$ [27] and $(\Pi^1_2 - \mathrm{CA}) + (\mathrm{BI})$ ([28]; see [29] for a full version for the theory of parameter free Π^1_2-comprehension). Independently, similar steps have been taken by T. Arai ([5, 6, 7, 8]; there are as well preprints of Arai [2, 3, 4] covering Π_n-reflection, Σ_1-stability and Π_1-collection). This research benefitted very much from intensive discussions with T. Arai, U. Berger, P. Martin-Löf, E. Palmgren, and M. Rathjen.

§2. Basic Martin-Löf type theory. In this section we will briefly introduce the version of Martin-Löf type theory (MLTT) used in this article. We will require some knowledge about MLTT. The reader not familiar with it might refer to [20, 21, 22, 25] or Chapter 11 of [39] (the latter deviates in its description of intensional type theory from the standard versions of MLTT).

2.1. The small logical framework.

The role of the logical framework in this article. In this article we will, when formally introducing theories and analysing them, not make use of the full logical framework. (We call the logical framework as it is used usually the *full* logical framework as opposed to the *small* logical framework, which is introduced in this subsection.) There are two reasons for this:

On one hand, the author has at the moment conceptual problems with the full logical framework. The problem manifests itself particularly when considering meaning explanations. Meaning explanations for the full logical framework don't seem to have been worked out fully at present, whereas the concept of meaning explanations for type theory without the full logical framework seems to be well understood. One should note however that Per Martin-Löf seems to have a clear understanding of meaning explanations for the full logical framework, and has given talks on this topic.

On the other hand, a treatment of the full logical framework causes at the moment still problems when determining proof theoretic bounds. Our techniques for modelling type theories at present don't allow to model the full logical framework directly without using more strength than is actually needed. Therefore, with our techniques we won't be able to obtain precise proof theoretic bounds. In order to avoid this, it seems to be necessary to first eliminate the use of the full logical framework, and then to use the techniques used in this article. Martin Hofmann has shown in Section 4.3 of [18] that type theory with the full logical framework is conservative over type theory without it, which would give the desired reduction. However, we haven't had time yet to study this result in detail in order to be sure that it can be applied to our setting.

For these two reasons all theories presented in the following will only make use of the restriction of the logical framework to Set, which we call the small logical framework. However, in order to explain the heuristics of our constructions, it is sometimes useful to make use of the full logical framework. We will do so for heuristic purposes only — when the formal theories are introduced, we will only make use of the small logical framework.

Judgements. The judgements of MLTT will be written as $x_1 : A_1, \dots, x_n : A_n \Rightarrow \theta$, where $x_1 : A_1, \dots, x_n : A_n$ is the context, often abbreviated as Γ or Δ, and θ is of the form A : Set, $A = B$: Set, $a : A$ or $a = b : A$. In case of the empty context $\emptyset$, we write θ instead of $\emptyset \Rightarrow \theta$. In the presence of the full logical framework, A : Set, $A = B$: Set are replaced by A : Type, $A = B$: Type, respectively. (Note that A : Set and $A = B$: Set are in the presence of the full logical framework instances of the judgements $a' : A'$ and $a' = b' : A'$ with $A' :=$ Set, since Set : Type.)

The small logical framework. The small logical framework contains the dependent function set and product as in the full logical framework, but limited to sets. So we have, under the assumptions A : Set, $x : A \Rightarrow B$: Set, the following set constructions:

- The dependent function set $(x : A) \to B$: Set.
  - The formation rule allows to form $(x : A) \to B$: Set from A : Set and $x : A \Rightarrow B$: Set.
  - The introduction rule expresses that we can form $(x : A)t : (x : A) \to B$ provided $x : A \Rightarrow t : B$.
  - The elimination rule expresses that we can apply $f : (x : A) \to B$ to $a : A$ and obtain $f(a) : B[x := a]$.
  - We have as equality rule β-equality, i.e. $((x : A)t)(s) = t[x := s]$.
  - Additionally we have the η-rule (which doesn't fall under the general principle of having formation, introduction, elimination and equality rules): if $f : (x : A) \to B$ then $f = (x : A)f(x)$.
  - Furthermore we have equality versions of the formation, introduction, and elimination rule:
    - ∗ The equality version of the formation rule expresses that if $A = A'$: Set, $x : A \Rightarrow B = B'$: Set, then $(x : A) \to B = (x : A') \to B'$: Set.

  * The equality version of the introduction rule expresses that if $x : A \Rightarrow t = t' : B$ then $(x : A)t = (x : A)t' : (x : A) \to B$ (that's the ξ-rule).
  * The equality version of the elimination rule expresses that if $s = s' : (x : A) \to B$ and $t = t' : A$ then $s(t) = s'(t') : B[x := t]$.
  * The principle of forming the equality versions of the formation, introduction and elimination rules from the standard (non-equality) rules is a straightforward principle. Therefore the convention is that whenever we introduce in the following formation, introduction and elimination rules, we silently introduce as well their equality versions. Note that there is no equality version of an equality rule.
  * α-equivalent terms are considered to be identical, therefore there is no explicit rule for α-equality. This extends to judgements as well, i.e. $x : A, y : B \Rightarrow C : \text{Set}$ and $u : A, v : B[x := u] \Rightarrow C[x := u, y := v] : \text{Set}$ are considered as the same judgement.
- We mention the following abbreviations:
  * We write $f(a, b)$ instead of $f(a)(b)$, similarly for longer applications.
  * We usually omit the type A in $(x : A)t$ and write $(x)t$ instead.
  * $(x, y)t := (x)(y)t$, similarly for longer expressions.
  * We write $A \to B$ for $(x : A) \to B$ for a fresh variable x, $(x, y : A) \to B$ for $(x : A) \to (y : A) \to B$, $(x : A, y : B) \to C$ for $(x : A) \to (y : B) \to C$. Similar abbreviations are to be understood in the same way.
  * If $f : A \to B$ and $g : B \to C$, then $g \circ f := (x)g(f(x))$.

• The dependent product $(x : A) \times B : \text{Set}$.
  - The introduction rule allows to form $\langle a, b \rangle$ for $a : A$ and $b : B[x := a]$.
  - The elimination rule forms the projections of an element $c : (x : A) \times B : \text{Set}$, written as $\pi_0(c) : A$ and $\pi_1(c) : B[x := \pi_0(c)]$.
  - The equality rule expresses that $\pi_0(\langle a, b \rangle) = a$ and $\pi_1(\langle a, b \rangle) = b$.
  - We have as well the η-rule associated with it, so if $c : (x : A) \times B$ then $c = \langle \pi_0(c), \pi_1(c) \rangle$.
  - We have similar abbreviations for the product as for the function set, especially $A \times B := (x : A) \times B$ for some fresh variable x.

Definition 2.1. By the rules of the small logical framework we mean the structural rules of type theory and the rules for the dependent function set and the dependent product restricted to Set.

2.2. The full logical framework. The full logical framework (which is not part of any of the type theories in this article unless explicitly stated), is obtained in the following way: In the small logical framework we were introducing sets, and elements of such sets. These sets are small types. In the full logical framework we have as well large types. There is a judgement A : Type for A is a large type. We have Set : Type, and if A : Set then A : Type. So we have the rules

$$\text{Set} : \text{Type} \qquad \frac{A : \text{Set}}{A : \text{Type}} \qquad \frac{A = B : \text{Set}}{A = B : \text{Type}}$$

If A : Type, then A can occur on the right hand side of the colon, especially we can conclude from $\Gamma \Rightarrow A : \text{Type}$ that $\Gamma, x : A$ is a context and $\Gamma, x : A \Rightarrow x : A$. Apart from Set, Type will as well be closed under the dependent function type and the dependent product. So if A : Type and $x : A \Rightarrow B : \text{Type}$, then

- $(x : A) \to B : \text{Type}$,
- $(x : A) \times B : \text{Type}$.

These types have essentially the same rules as the corresponding sets of the small logical framework, except that one refers to Type instead of Set.

2.3. The basic set constructions. The basic set constructions are the following sets (or more precisely principles for forming sets) and their corresponding rules:

- The finite sets N_n : Set, where $n \in \{0, 1, \dots\}$. Note that here n is not an internal natural number inside type theory, but exists on the Meta-level. The introduction rules are $\mathrm{A}_i^n : \mathrm{N}_n$ for $i = 0, \dots, n-1$ (where i are again natural numbers on the Meta-level).
- The set N of natural numbers. The introduction rules are $0 : \mathrm{N}$ and $\mathrm{S} : \mathrm{N} \to \mathrm{N}$.
- The disjoint union $A + B$ of two sets A, B, with introduction rules $\mathrm{inl} : A \to (A + B)$ and $\mathrm{inr} : B \to (A + B)$.
- For A : Set and $x : A \Rightarrow B$: Set we have the following sets:
  - The Π-set $\Pi x : A.B$. The introduction rule is $\lambda : ((x : A) \to B) \to \Pi x : A.B$.
    One writes $\lambda x : A.t$ for $\lambda((x : A)t)$.
  - The Σ-set $\Sigma x : A.B$. The introduction rule is $\mathrm{p} : (a : A) \to (b : B[x := a]) \to \Sigma x : A.B$.
    (The differences between $\Pi x : A.B$, $\Sigma x : A.B$, and $(x : A) \to B$, $(x : A) \times B$, respectively, will be explained below).
  - The W set $\mathrm{W}x : A.B$, which is the set of well-founded trees with branching degrees $B[x := a]$ for $a : A$. The introduction rule is $\sup : (a : A, b : B[x := a] \to \mathrm{W}x : A.B) \to \mathrm{W}x : A.B$.
- The intensional identity set $\mathrm{Id}(A, a, b)$ for A : Set, $a : A$, $b : A$. The introduction rule is $\mathrm{refl}_A : (a : A) \to \mathrm{Id}(A, a, a)$.

- For all of the above set constructions, the elimination rules express that the above sets are the least sets introduced by these constructors. This will be for instance in case of N primitive recursion into arbitrary sets (which may depend on the element n : N we are eliminating), in case of W induction over those trees, and in case of $A + B$ case distinction on whether $ab : A + B$ is of the form $\mathrm{inl}(a)$ or $\mathrm{inr}(b)$.
- Furthermore, we have the standard equality rules for the above mentioned sets.

The main difference between the dependent function set and product and the sets $\Pi x : A.B$, $\Sigma x : A.B$ is that we have the η-rule for the constructions of the logical framework, but not for Π and Σ. The conceptual reason for this becomes clear when considering inductive-recursive definitions. All the basic set constructions are instances of inductive-recursive definitions (in case of $\mathrm{Id}(A, a, b)$ general indexed inductive-recursive definitions) as introduced originally by P. Dybjer with a formalisation using finitely many rules by P. Dybjer and the author; see [11, 12, 13, 14, 15, 16]. Indexed inductive-recursive definitions allow to introduce all sets in MLTT by determining their introduction rules (there are of course restrictions on which introduction rules are allowed). The elimination and equality rules are then derived automatically. There is no η-rule involved in this schema, which would be unnatural in general. For instance, it does not make much sense to consider an η-rule for N. Therefore it is natural to exclude the η-rule from Π and Σ, and to have separate logical framework versions which contain the η-rule. In this article, the difference between the constructions of the small logical framework and the Π- and Σ-set won't play a big rôle.

Our models will admit as well the addition of an extensional equality set to the type theory. Since lower bounds will be obtained using intensional equality only, it will follow that the proof theoretic strength of the type theories under consideration with intensional equality and with extensional equality coincides.

DEFINITION 2.2. (a) By the basic set constructions in type theory we mean the above mentioned constructions N_n, N, $+$, Σ, Π, W, Id for forming sets, and the corresponding formation/introduction/elimination/equality rules.
(b) Basic MLTT is the type theory consisting of the rules of the small logical framework and of the basic set constructions.

2.4. Kripke-Platek set theory. In this article, we will develop models of the type theories considered in extensions of Kripke-Platek set theory. This will allow us to determine upper bounds for the proof theoretic strength of the theories in question.

Kripke-Platek set theory (KP) is a weak version of set theory, based on classical logic. The "bible" of KP is the book by Barwise [9]. KP can be used in order to develop most concepts in generalised recursion theory, as demonstrated in that book. KP has been pioneered by Jäger [19] as a reference theory for proof theoretic studies (see as well the handbook article by Pohlers [24], which contains an overview over proof theoretic results up to the strength of KPI in this area). For many theories, there exists a variant of KP of equal strength, and often one can determine upper bounds for theories by first determining upper bounds for the variant for KP of equal strength, and then by modelling the original theory in that variant. This is as well the approach taken in this article.

We don't have room to introduce KP in full (we highly recommend the reader not familiar with it to study the first chapters of [9]). We briefly repeat here its axioms (see p. 10 of [9]; the theory presented in [9] adds to KP as well urelements, and is therefore called KPU for KP plus urelements; the variants of KP used in this article don't have urelements; Δ_0-formulae are formulae containing only bounded quantifiers, $\mathrm{FV}(\varphi)$ are the free variables of φ):

(*Extensionality*) $\forall x, y.(\forall z.z \in x \leftrightarrow z \in y) \to x = y\,.$

(*Foundation*) $\forall \vec{z}.(\forall x.(\forall y \in x.\varphi(y, \vec{z})) \to \varphi(x, \vec{z})) \to \forall x.\varphi(x, \vec{z})\,,$
where $\mathrm{FV}(\varphi(x, \vec{z})) \subseteq \{x, \vec{z}\}$.

(*Pair*) $\forall x, y.\exists z.x \in z \wedge y \in z\,.$

(*Union*) $\forall x.\exists y.\forall z \in x.\forall u \in z.u \in y\,.$

(Δ_0-*Separation*) $\forall \vec{u}, x.\exists y.\forall z.(z \in y \leftrightarrow (z \in x \wedge \varphi(z, \vec{u})))\,,$
where $\varphi(z, \vec{u})$ is Δ_0, $\mathrm{FV}(\varphi(z, \vec{u})) \subseteq \{z, \vec{u}\}$.

(Δ_0-*Collection*) $\forall \vec{u}, x.(\forall y \in x.\exists z.\varphi(y, z, \vec{u})) \to$
$\exists v.\forall y \in x.\exists z \in v.\varphi(y, z, \vec{u})\,,$
where $\varphi(y, z, \vec{u})$ is Δ_0, $\mathrm{FV}(\varphi(y, z, \vec{u})) \subseteq \{y, z, \vec{u}\}$.

The variants of KP used in this article will always be augmented by the axiom of infinity

$$(\textit{Infinity}) \quad \exists x.\emptyset \in x \wedge \forall y \in x.y \cup \{y\} \in x\,,$$

where $\emptyset \in x$ and $y \cup \{y\} \in x$ are to be understood in the usual way as Δ_0-formulae. Let $\mathrm{KP}\omega$ be the theory KP plus the axiom of infinity.

When forming extensions of $\mathrm{KP}\omega$, one adds a predicate $\mathrm{Ad}(x)$ for "x is an admissible containing ω", following the approach taken by Jäger (e.g. [19]; see as well Pohlers' survey article [24]). Here an admissible is a transitive inner model of KP, and by an admissible $> \omega$ we mean an admissible containing ω, i.e. a transitive inner model of $\mathrm{KP}\omega$. So one has the following additional axioms (trans(x) expresses that x transitive):

(Ad.1) $\forall x.\mathrm{Ad}(x) \to \mathrm{trans}(x)$.
(Ad.2) $\forall x, y.\mathrm{Ad}(x) \wedge \mathrm{Ad}(y) \to (x \in y \vee x = y \vee y \in x)$.
(Ad.3) $\forall x.\mathrm{Ad}(x) \to \psi^x$,
for every instance ψ of (*Pair*), (*Union*), (Δ_0-*Separation*), (Δ_0-*Collection*), (*Infinity*).

ψ^x is obtained by replacing all unbounded quantifiers $\forall y$, $\exists y$ occurring in ψ by $\forall y \in x$, $\exists y \in x$ respectively, but leaving bounded quantifiers as they are (where a bounded quantifier is a quantifier of the form $\forall y \in z$, $\exists y \in z$). The extensions of KPω are then obtained by adding to KPω the above axioms for Ad and axioms expressing that there exists an admissible with sufficiently strong closure properties. The convention in this article is that whenever we introduce an extension of KP, we always add as well the infinity axiom and, if Ad is involved in any of the additional axioms, the axioms *(Ad.1-3)*.

We introduce as well the theory KPl^{r} which expresses that there are finitely many admissibles (i.e. for Meta-each n we have a chain of n admissibles $a_0 \in a_1 \in \cdots \in a_{n-1}$) but that the set theoretic universe itself is not necessarily a model of KP. The standard model of KPl^{r} is $\mathrm{a} := \bigcup_{n\in\omega} \mathrm{b}_n$, where b_n is the nth admissible ($n \in \omega$). Note that a is not an admissible. Formally, KPl^{r} is obtained by taking KPω without (Δ_0-*Collection*), with *(Foundation)* restricted to Δ_0-formulae $\varphi(x, \vec{z})$. and by adding the axioms *(Ad.1-3)* and the following axiom *(Lim)*:

$$(Lim) \quad \forall x.\exists y.\mathrm{Ad}(y) \wedge x \in y$$

The theory KPl is defined as KPl^{r}, but with *(Foundation)* for all formulae $\varphi(x, \vec{z})$.

When working in extensions of KPω, we will frequently make use of the constructible hierarchy, and refer to the definition of L_α in [9]. We define as well the following:

Definition 2.3. (a) $\mathrm{L}_{<\alpha} := \bigcup_{\beta<\alpha} \mathrm{L}_\beta$,
(b) $\mathrm{L}'_\alpha := \mathrm{L}_{\omega\cdot\alpha}$,
(c) $\mathrm{L}'_{<\alpha} := \bigcup_{\beta<\alpha} \mathrm{L}'_\beta$.
(d) An admissible ordinal is an ordinal α such that L_α is admissible.

We will often make use of the following essentially trivial remark:

Remark 2.4. (a) If $c = \{x \in a \mid \varphi(x, \vec{b})\}$, where $\varphi(x, \vec{y})$ is Δ_0, $\mathrm{FV}(\varphi(x, \vec{y})) \subseteq \{x, \vec{y}\}$, and $a, \vec{b} \in \mathrm{L}'_\alpha \cup \{\mathrm{L}'_\alpha\}$, then $c \in \mathrm{L}'_{\alpha+1}$.
(b) $\alpha \in \mathrm{L}'_{\alpha+1}$.
(c) $\omega \in \mathrm{L}'_2$.
(d) $\mathrm{L}'_\beta = \mathrm{L}_\beta$, if β is a multiplicative principal number, e.g. an admissible ordinal $> \omega$.

Proof. Let Ord be the class of ordinals. (a): Assumption II.5.2 and Theorem II.6.4 of [9]. (b): One easily can show that $\mathrm{L}_\alpha \cap \mathrm{Ord} \in \mathrm{Ord}$. We show the assertion by induction on α. The cases $\alpha = 0$ and $\alpha \to \alpha + 1$ follow easily or by induction hypothesis (IH). In case α being a limit ordinal we have by IH $\alpha \subseteq \mathrm{L}'_{<\alpha} \cap \mathrm{Ord} =: \beta \in \mathrm{L}'_{\alpha+1}$, and therefore $\alpha \in \mathrm{L}'_{\alpha+1}$. (c): $\omega \subseteq \mathrm{L}_{<\omega}$, $\mathrm{L}_{<\omega} \cap \mathrm{Ord} \subseteq \omega$, therefore $\omega := \{x \in \mathrm{L}_{<\omega} \mid \mathrm{Ord}(x)\} \in \mathrm{L}_{\omega+\omega} = \mathrm{L}'_2$. (d): $\omega \cdot \beta = \beta$. ⊣

2.5. Models of basic MLTT. As said before, we will develop models of the type theories in question in extensions of KPω. These models will extend the model of MLTT with one universe and the W-type, as worked out in the author's PhD thesis [30]. (See as well [32] for a draft version of an article on it, and [31] for an article containing a basic model construction for MLTT with one Mahlo universe.) We will first start to explain how to model basic MLTT.

Basic principles of the model construction. The basic idea of all our models is that we define a PER model, i.e. a model in which sets are interpreted as partial equivalence relations on a set of basic objects, where in this article the set of basic objects will be a set of terms, which is a slightly simplified version of the set of terms occurring in the type theory. Let ML$'$ denote the type theory for which we want to determine an upper bound for its proof theoretic strength. We will model ML$'$ in an extension KP$'$ of KPω. KP$'$ will be defined in such a way that it has the same expected proof theoretic strength. Once we have modelled ML$'$ in KP$'$ one will be able to conclude (this step requires some techniques which can be found in [30]) that the proof theoretic strength $|\mathrm{ML}'|$ of ML$'$ is less than or equal to the proof theoretic strength $|\mathrm{KP}'|$ of KP$'$, i.e. $|\mathrm{ML}'| \leq |\mathrm{KP}'|$.

The soundness theorem for the model of type theory will be a Meta-theorem (otherwise we would obtain $|\mathrm{ML}'| < |\mathrm{KP}'|$): For Meta-every judgement $\Gamma \Rightarrow \theta$ such that $\mathrm{ML}' \vdash \Gamma \Rightarrow \theta$ we have that KP$'$ proves the correctness statement associated with $\Gamma \Rightarrow \theta$. But ML$'$ doesn't prove that for all $\Gamma \Rightarrow \theta$ the correctness statement holds. In fact, ML$'$ cannot even *express* such a general correctness statement.

Let Term be the underlying set of terms occurring in ML$'$. We will introduce in the model a set of closed terms $[\![\mathrm{Term}]\!]$ using essentially the same term and type constructors and destructors as in the type theory. α-equivalent terms (as well sets, context, judgements) will always be identified. However, we will, when introducing the models of the universes below occasionally throw away arguments needed for type checking purposes only (this is not really necessary, but simplifies matters).

One introduces as well a *deterministic* reduction relation $\longrightarrow$ on $[\![\mathrm{Term}]\!]$ corresponding to weak head reductions following the equality rules of the type theory. $\longrightarrow^*$ is the reflexive and transitive closure of $\longrightarrow$.

DEFINITION 2.5 (Environments).

(a) An environment relative to a sequence of variables $x_1, \dots, x_n$ is an expression $[x_1 = s_1, \dots, x_n = s_n]$, where x_i are distinct variables and $s_i \in [\![\,\mathrm{Term}\,]\!]$.
(b) $\emptyset$ stands for the empty environment $[]$.
(c) $[x_1 = s_1, \dots, x_n = s_n], y = t := [x_1 = s_1, \dots, x_n = s_n, y = t]$, where we assume that after some renaming of variables corresponding to α-conversion in the situation used, y is different from x_i.
(d) $\mathrm{FV}([x_1 = s_1, \dots, x_n = s_n]) := \{x_1, \dots, x_n\}$.
(e) $[x_1 = s_1, \dots, x_n = s_n](x_i) := s_i$.
(f) $\mathrm{Env}(x_1, \dots, x_n) := \{[x_1 = s_1, \dots, x_n = s_n] \mid s_1, \dots, s_n \in [\![\,\mathrm{Term}\,]\!]\}$.

The interpretation $[\![\,t\,]\!]_\rho \in [\![\,\mathrm{Term}\,]\!]$ for environments ρ and terms $t \in \mathrm{Term}$ such that $\mathrm{FV}(t) \subseteq \mathrm{FV}(\rho)$ is the result of omitting in t the arguments of constructors/destructors to be thrown away (as indicated when introducing the constants) and substituting free variables x in t by $\rho(x)$.

We write $[\![\,t\,]\!]$ for $[\![\,t\,]\!]_\emptyset$.

Then one defines the interpretation $[\![\,A\,]\!]_\rho$ of sets A relative to environments ρ such that $\mathrm{FV}(A) \subseteq \mathrm{FV}(\rho)$ as a set of pairs $\langle s, t\rangle$ of closed terms. The intended meaning for $\langle s, t\rangle \in [\![\,A\,]\!]_\rho$ is that s and t are equal elements of the interpretation of A. We write $[\![\,A\,]\!]$ for $[\![\,A\,]\!]_\emptyset$.

We define some basic operations on sets $X \subseteq [\![\,\mathrm{Term}\,]\!]^2$:

DEFINITION 2.6. Let $X \subseteq [\![\,\mathrm{Term}\,]\!]^2$.

(a) The closure of X under reductions is defined as

$$\mathrm{Clos}_{\longrightarrow}(X) := \{\langle s, t\rangle \mid \exists \langle s', t'\rangle \in X \mid s \longrightarrow^* s' \wedge t \longrightarrow^* t'\}.$$

(b) $\mathrm{Flat}(X) := \{a \in [\![\,\mathrm{Term}\,]\!] \mid \exists b \in [\![\,\mathrm{Term}\,]\!].\langle a, b\rangle \in X \vee \langle b, a\rangle \in X\}$.
(c) $\mathrm{PER}(X)$ iff X is a partial equivalence relation (i.e. symmetric and transitive) and X is closed upwards under $\longrightarrow$, i.e. $\mathrm{Clos}_{\longrightarrow}(X) \subseteq X$.

$\mathrm{Flat}([\![\,A\,]\!]_\rho)$ is the set of elements of the interpretation of A, and $\langle a, b\rangle \in [\![\,A\,]\!]_\rho$ means that a and b are equal elements of $[\![\,A\,]\!]_\rho$. This is meaningful if $\mathrm{PER}([\![\,A\,]\!]_\rho)$ holds, which we will show for all derivable sets A as expressed by the correctness statement below.

In the models for the type theories considered in this article, we can as well assign to every set A with $\mathrm{FV}(A) \subseteq \{x_1, \dots, x_n\}$ an ordinal $\mathrm{o}(A) \geq 3$ such that for all $\rho \in \mathrm{Env}(x_1, \dots, x_n)$ we have $[\![\,A\,]\!]_\rho \in \mathrm{L}'_{\mathrm{o}(A)}$, independent of ρ. Note that $[\![\,\mathrm{Term}\,]\!]$ is a Δ_0-definable subset of ω and therefore $[\![\,\mathrm{Term}\,]\!] \in \mathrm{L}'_3$.

The definition of $[\![\,A\,]\!]_\rho$ and $\mathrm{o}(A)$ for A formed by one of the basic set constructions or constructions of the small logical framework is as follows:

- $[\![\,\mathrm{N}\,]\!]_\rho := [\![\,\mathrm{N}\,]\!] := \mathrm{Clos}_{\longrightarrow}(\{\langle \mathrm{S}^n(0), \mathrm{S}^n(0)\rangle \mid n \in \omega\})$.
  $\mathrm{o}(\mathrm{N}) := 3$ (this is probably not the minimal ordinal one can associate

with it, similarly for the later ordinals, but this bound suffices for our proofs).

- $[\![\Sigma x : A.B]\!]_\rho := [\![\Sigma]\!]([\![A]\!]_\rho, \lambda s \in \mathrm{Flat}([\![A]\!]_\rho).[\![B]\!]_{\rho,x=s})$,
where, if $A \subseteq [\![\mathrm{Term}]\!]^2$, and for $a \in \mathrm{Flat}(A)$ we have $B(a) \subseteq [\![\mathrm{Term}]\!]^2$, then

$$[\![\Sigma]\!](A, B) := \mathrm{Clos}_{\rightarrow}(\{\langle \mathrm{p}(a, b), \mathrm{p}(a', b')\rangle \mid \langle a, a'\rangle \in A \wedge (B(a) = B(a')) \wedge \langle b, b'\rangle \in B(a)\}) .$$

(Here $\lambda x \in A.s(x) := \{\langle x, s(x)\rangle \mid x \in A\}$; note that p is the constructor of Σ as defined in Subsection 2.3, and that according to our convention $f(a, b)$ is an abbreviation for $f(a)(b)$, therefore $\mathrm{p}(a, b)$ stands for $\mathrm{p}(a)(b)$).

$$\mathrm{o}(\Sigma x : A.B) := \max\{\mathrm{o}(A), \mathrm{o}(B)\} + 2.$$

- Similarly one defines the semantics and the ordinals of N_n, $\Pi x : A.B$, $\mathrm{W}x : A.B$, $A + B$, $\mathrm{Id}(A, a, b)$, $(x : A) \rightarrow B$, $(x : A) \times B$, making use of suitable definitions of $[\![\mathrm{N}_n]\!]$, $[\![\Pi]\!]$, $[\![\mathrm{W}]\!]$, $[\![+]\!]$, $[\![\mathrm{Id}]\!]$, $[\![\rightarrow]\!]$ and $[\![\times]\!]$, and incrementing the ordinals as for Σ by 2 (except for W, see below; $\mathrm{o}(\mathrm{N}_n) := 2$).
  We introduce here as well the following notations to be used later:
  - $(x \in A)\, [\![\rightarrow]\!]\, B := [\![\rightarrow]\!](A, \lambda x \in \mathrm{Flat}(A).B)$,
  - $A\, [\![\rightarrow]\!]\, B := (x \in A)\, [\![\rightarrow]\!]\, B$ where x is fresh,
  - $(x \in A)\, [\![\times]\!]\, B := [\![\times]\!](A, \lambda x \in \mathrm{Flat}(A).B)$,
  - $A\, [\![\times]\!]\, B := (x \in A)\, [\![\times]\!]\, B$ where x is fresh.
  - We can omit brackets in the above, using for $[\![\rightarrow]\!]$, $[\![\times]\!]$ the same conventions as we have for $\rightarrow$, $\times$.

  In the case of W only, one defines $\mathrm{o}(\mathrm{W}x : A.B) := (\max\{\mathrm{o}(A), \mathrm{o}(B)\})^+ + 1$, where α^+ is the next admissible ordinal above α. The reason for using $\alpha^+ + 1$ in case of $[\![\mathrm{W}]\!]$ is that $[\![\mathrm{W}]\!](A, B)$ is defined by iterating α^+-times an operator Φ. Here $\Phi(A)$ is the set of trees with immediate subtrees in A.

We interpret contexts as sets of environments as follows:

$$\begin{aligned}
&[\![x_1 : A_1, \ldots, x_n : A_n]\!] := \\
&\{\langle [x_1 = s_1, \ldots, x_n = s_n], [x_1 = t_1, \ldots, x_n = t_n]\rangle \mid \\
&\qquad \langle s_1, t_1\rangle \in [\![A_1]\!]_\emptyset \wedge \\
&\qquad \langle s_2, t_2\rangle \in [\![A_2]\!]_{[x_1=s_1]} \wedge \cdots \wedge \\
&\qquad \langle s_n, t_n\rangle \in [\![A_n]\!]_{[x_1=s_1,\ldots,x_{n-1}=s_{n-1}]}\} .
\end{aligned}$$

DEFINITION 2.7. The correctness of a judgement is defined as follows by induction on the length:

(a) $\mathrm{Correct}(\emptyset \Rightarrow \mathrm{Context})$ (where $\emptyset$ is the empty context) is defined as a true formula, e.g. $\forall x.x = x$.

(b)

$$\begin{aligned}&\mathrm{Correct}(\Gamma \Rightarrow A = B : \mathrm{Set})\\ &:= \mathrm{Correct}(\Gamma \Rightarrow \mathrm{Context}) \wedge \forall \langle \rho, \rho' \rangle \in [\![\Gamma]\!].\mathrm{PER}([\![A]\!]_\rho) \wedge [\![A]\!]_\rho = [\![B]\!]_{\rho'} .\end{aligned}$$

(c)

$$\begin{aligned}&\mathrm{Correct}(\Gamma, y : A \Rightarrow \mathrm{Context}) := \mathrm{Correct}(\Gamma \Rightarrow A : \mathrm{Set})\\ &:= \mathrm{Correct}(\Gamma \Rightarrow A = A : \mathrm{Set}),\end{aligned}$$

as just defined.

(d)

$$\begin{aligned}&\mathrm{Correct}(\Gamma \Rightarrow a = b : A)\\ &:= \mathrm{Correct}(\Gamma \Rightarrow A : \mathrm{Set}) \wedge \forall \langle \rho, \rho' \rangle \in [\![\Gamma]\!].\langle [\![a]\!]_\rho, [\![b]\!]_{\rho'} \rangle \in [\![A]\!]_\rho .\end{aligned}$$

(e) $\mathrm{Correct}(\Gamma \Rightarrow a : A) := \mathrm{Correct}(\Gamma \Rightarrow a = a : A)$ as just defined.

The correctness statement to be shown by induction on the derivations in ML′ is as follows: For Meta-every judgement $\Gamma \Rightarrow \theta$ we have

$$(\mathrm{ML}' \vdash \Gamma \Rightarrow \theta) \rightarrow \mathrm{KP}' \vdash \mathrm{Correct}(\Gamma \Rightarrow \theta) .$$

§3. Universes and recursively inaccessibles.

3.1. Rules for universes. A universe (à la Tarski) in MLTT (Martin-Löf type theory) is given by a set U : Set of codes for sets together with a decoding function T : U → Set, which without the full logical framework is given by the following rule:

$$\frac{u : \mathrm{U}}{\mathrm{T}(u) : \mathrm{Set}}$$

All universes in the following will be closed under the basic set constructions (i.e. N_n, N, Σ, Π, W, +, Id). This means that we have in case of N and Σ the rules:

$$\widehat{\mathrm{N}} : \mathrm{U} \qquad\qquad \mathrm{T}(\widehat{\mathrm{N}}) = \mathrm{N}$$

$$\frac{a : \mathrm{U} \qquad b : \mathrm{T}(a) \rightarrow \mathrm{U}}{\widehat{\Sigma}(a, b) : \mathrm{U}} \qquad \mathrm{T}(\widehat{\Sigma}(a, b)) = \Sigma x : \mathrm{T}(a).\mathrm{T}(b(x))$$

(The codes for N_n, Π, W, +, Id will be called $\widehat{\mathrm{N}}_n$, $\widehat{\Pi}$, $\widehat{\mathrm{W}}$, $\widehat{+}$, $\widehat{\mathrm{Id}}$ respectively.)

In particular we consider the theory $\mathrm{ML_WU}$ consisting of the small logical framework, the basic set constructions and one universe (which is by our convention closed under the standard set constructors including the W-set, but not under U itself; the subscript W stands for the closure under the W-set).

We will not add any elimination rules for universes. In case of the theory $\mathrm{ML_WU}$ and of the super-universe (as introduced in Section 4), they could be added without any problems, but won't add in the presence of the W-set any strength.[1] In case of the Mahlo universe (Section 5) and stronger universes, such elimination rules result in inconsistencies (see Theorem 6.1 in [23]).

Type theories with several universes. We will already here look at how to extend this approach to type theories having several universes. In this case, the codes $\widehat{\mathrm{N}}, \widehat{\Sigma}$ etc. have special names and additional parameters so that we can distinguish them from each other. These parameters will always be omitted when moving to $[\![\,\mathrm{Term}\,]\!]$. E.g. if we have universes V and $\mathrm{U}_{a,b}$, we have in Term codes $\widehat{\mathrm{N}}_{\mathrm{V}}$ and $\widehat{\mathrm{N}}_{\mathrm{U},a,b}$, whereas in $[\![\,\mathrm{Term}\,]\!]$ we have only one code $\widehat{\mathrm{N}}$ and define $[\![\,\widehat{\mathrm{N}}_{\mathrm{V}}\,]\!]_\rho := [\![\,\widehat{\mathrm{N}}_{\mathrm{U},a,b}\,]\!]_\rho := \widehat{\mathrm{N}}$.

By a *recursive subuniverse* $(\mathrm{U}, \widehat{\mathrm{T}}_{\mathrm{U}})$ of a universe $(\mathrm{V}, \mathrm{T}_{\mathrm{V}})$ we mean that we have $\mathrm{U} : \mathrm{Set}$, but, instead of having $\mathrm{T}_{\mathrm{U}} : \mathrm{U} \to \mathrm{Set}$ (or, if one avoids the full logical framework, the rule stating that if $a : \mathrm{U}$ then $\mathrm{T}_{\mathrm{U}}(a) : \mathrm{Set}$), we have the rule formulating the existence of a function $\widehat{\mathrm{T}}_{\mathrm{U}} : \mathrm{U} \to \mathrm{V}$, which recursively decodes codes by their corresponding codes in the other set. If we for $a : \mathrm{U}$ define $\mathrm{T}_{\mathrm{U}}(a) := \mathrm{T}_{\mathrm{V}}(\widehat{\mathrm{T}}_{\mathrm{U}}(a)) : \mathrm{Set}$, we have therefore the rules

$$\widehat{\mathrm{N}}_{\mathrm{U}} : \mathrm{U} \qquad\qquad \widehat{\mathrm{T}}_{\mathrm{U}}(\widehat{\mathrm{N}}_{\mathrm{U}}) = \widehat{\mathrm{N}}_{\mathrm{V}}$$

$$\frac{a : \mathrm{U} \qquad b : \mathrm{T}_{\mathrm{U}}(a) \to \mathrm{U}}{\widehat{\Sigma}_{\mathrm{U}}(a, b) : \mathrm{U}} \qquad \widehat{\mathrm{T}}_{\mathrm{U}}(\widehat{\Sigma}_{\mathrm{U}}(a, b)) = \widehat{\Sigma}_{\mathrm{V}}(\widehat{\mathrm{T}}_{\mathrm{U}}(a), \widehat{\mathrm{T}}_{\mathrm{U}} \circ b)$$

similarly for the other basic set constructions.

Sometimes it will not be possible to develop the subuniverses introduced in this article as recursive subuniverses. The problem occurs as soon as we have infinite chains of universes. Consider for instance a sequence of universes $(\mathrm{U}_\alpha)_{\alpha \le \beta}$ such that U_α is a subuniverse of $\mathrm{U}_{\alpha'}$ for $\alpha < \alpha'$. There is no problem of using recursive subuniverses, if we have a finite sequence of such universes (i.e. $\beta < \omega$): Then we can consider for $\alpha < \beta$ the universe U_α as a recursive subuniverse of $\mathrm{U}_{\alpha+1}$, whereas the universe U_β itself is defined as a standard universe. The set corresponding to $a : \mathrm{U}_\alpha$ is then $\mathrm{T}_\beta(\widehat{\mathrm{T}}_{\beta-1}(\cdots(\widehat{\mathrm{T}}_\alpha(a))\cdots))$. If $\beta = \omega$, we would however with this approach end up with a nonterminating chain

$$\mathrm{U}_0 \xrightarrow{\widehat{\mathrm{T}}_0} \mathrm{U}_1 \xrightarrow{\widehat{\mathrm{T}}_1} \mathrm{U}_2 \xrightarrow{\widehat{\mathrm{T}}_2} \cdots$$

and we cannot determine a set corresponding to $a : \mathrm{U}_0$.

[1] One can show this by extending the model given in the following to include the elimination rules for the universe while still using the same Meta theory for modelling it. Therefore the upper bound for the proof theoretic strength of the type theory with elimination rules coincides with the upper bound for the theory without elimination rules which in turn coincides with the lower bound for the theory without elimination rules obtained using a well-ordering proof. The idea for a more direct proof — we haven't carried out the details yet — is to use the W-set in order to select out of the ordinary universe a subuniverse which is the least universe closed under the basic set constructions. That would show that a type theory with elimination rules for the universe could be interpreted in the type theory without those rules. Both techniques wouldn't work for constructions like the Mahlo universe.

The way around this is to follow Per Martin-Löf's approach, by forming *inductive subuniverses*. An inductive subuniverse U of a universe $(\mathrm{V}, \mathrm{T_V})$ is given by

- a set U,
- a recursively defined decoding function $\mathrm{T_U} : \mathrm{U} \to \mathrm{Set}$, and
- a *constructor* $\widehat{\mathrm{T}}_\mathrm{U} : \mathrm{U} \to \mathrm{V}$.

So $\widehat{\mathrm{T}}_\mathrm{U} : \mathrm{U} \to \mathrm{V}$ forms an additional introduction rule for V.

If we avoid the full logical framework, we get the following rules for an inductive subuniverse $(\mathrm{U}, \mathrm{T_U})$ of $(\mathrm{V}, \mathrm{T_V})$:

$$\mathrm{U} : \mathrm{Set} \qquad \frac{a : \mathrm{U}}{\mathrm{T_U}(a) : \mathrm{Set}}$$

$$\frac{a : \mathrm{U}}{\widehat{\mathrm{T}}_\mathrm{U}(a) : \mathrm{V}} \qquad \mathrm{T_V}(\widehat{\mathrm{T}}_\mathrm{U}(a)) = \mathrm{T_U}(a)$$

$$\widehat{\mathrm{N}}_\mathrm{U} : \mathrm{U} \qquad \mathrm{T_U}(\widehat{\mathrm{N}}_\mathrm{U}) = \mathrm{N}$$

$$\frac{a : \mathrm{U} \qquad b : \mathrm{T_U}(a) \to \mathrm{U}}{\widehat{\Sigma}_\mathrm{U}(a, b) : \mathrm{U}} \qquad \mathrm{T_U}(\widehat{\Sigma}_\mathrm{U}(a, b)) = \Sigma x : \mathrm{T_U}(a).\mathrm{T_U}(b(x))$$

Similarly for the other basic set constructions.

The reason why we prefer recursive subuniverses to inductive ones is that we obtain in inductive subuniverses many copies of the same element. In the simple situation, we have for instance two codes for N in V, namely $\widehat{\mathrm{N}}_\mathrm{V}$ and $\widehat{\mathrm{T}}_\mathrm{V}(\widehat{\mathrm{N}}_\mathrm{U})$. This is not very aesthetic (it doesn't cause many problems when actually making use of such universes). However, one should be aware that we anyway get often doubling of codes. For instance, if in case of the super-universe to be introduced in Section 4 we form a recursive subuniverse $(\mathrm{U}, \widehat{\mathrm{T}}_\mathrm{U})$ of $(\mathrm{V}, \mathrm{T_V})$ containing the family $\langle \widehat{\mathrm{N}}_\mathrm{V}, (x)\widehat{\mathrm{N}}_\mathrm{V} \rangle$, then U will have infinitely many codes for N, namely $\widehat{\mathrm{N}}_\mathrm{U}$, $\widehat{\mathrm{a}}$ and $\widehat{\mathrm{b}}(n)$ for $n : \mathrm{N}$.

When modelling subuniverses, we will treat them as mere subsets of the interpretation of the original universe. $\widehat{\mathrm{T}}_\mathrm{U}$ will for both the recursive and inductive subuniverses be essentially the identity function, expressed by the reduction rule $\widehat{\mathrm{T}}_\mathrm{U}(a) \longrightarrow a$. This means that in the model we will identify codes coming from the same element in U, so for instance all the codes for N in the above example will be equal in the model.

3.2. Modelling universes. For modelling universes closed under the W-set in extensions of KPω one needs the notion of a recursively inaccessible ordinal. A recursively inaccessible ordinal I is an admissible ordinal I such that for every $\beta \in \mathrm{I}$ there exists a $\gamma \in \mathrm{I}$ such that $\beta \in \gamma$ and γ is an admissible ordinal. If one refers to admissible sets rather than admissible ordinals, then a recursively inaccessible set is a set a such that

$$\mathrm{Ad}(a) \wedge \forall y \in a.\exists z \in a.y \in z \wedge \mathrm{Ad}(z)\,.$$

Let KPI^{+} be the theory of $\mathrm{KP}\omega$ extended by the existence of one recursively inaccessible set and finitely many admissibles (i.e. for Meta-each n we have n admissibles) above it. So KPI^{+} has constants $\mathrm{a}_{\mathrm{I},k}$ for Meta $k \in \omega$, axioms expressing that $\mathrm{a}_{\mathrm{I},0}$ is a recursively inaccessible set, and for Meta $k \in \omega$ the axioms

$$\mathrm{Ad}(\mathrm{a}_{\mathrm{I},k}) \wedge \mathrm{a}_{\mathrm{I},k} \in \mathrm{a}_{\mathrm{I},k+1}\,.$$

Proof theoretically equivalent is the theory $\mathrm{KPI}^{\mathrm{r}}$ plus the existence of one recursively inaccessible set.

A universe is a family of sets, and therefore it will be modelled by a family of terms, i.e. an element of $[\![\,\mathrm{Fam}\,]\!](\mathrm{Set})$, as given by the following definition ($\mathrm{dom}(f)$ is the domain of the function f):

Definition 3.1 (in the language of Kripke-Platek set theory). (a) By Set we mean the class of all sets.

(b) $$\begin{aligned}[\![\,\mathrm{Fam}\,]\!](\mathrm{Set}) := \{(A, B)\in \mathrm{Set}^2 \mid\ & A \subseteq [\![\,\mathrm{Term}\,]\!]^2\\ & \wedge B \text{ is a function}\\ & \wedge \mathrm{dom}(B) = \mathrm{Flat}(A)\\ & \wedge \forall x \in \mathrm{Flat}(A).B(x) \subseteq [\![\,\mathrm{Term}\,]\!]^2\\ & \wedge \forall \langle x, x'\rangle \in A.B(x) = B(x')\}\,.\end{aligned}$$

(c) Assume $(A, B) \in [\![\,\mathrm{Fam}\,]\!](\mathrm{Set})$. Then the set $[\![\,\mathrm{IFam}\,]\!](A, B)$ of families of sets internal to (A, B) ($[\![\,\mathrm{IFam}\,]\!]$ stands for internal family) is defined as

$$[\![\,\mathrm{IFam}\,]\!](A, B) := (x \in A)\,[\![\,\times\,]\!]\,(B(x)\,[\![\rightarrow]\!]\,A)\,,$$

(which is

$$\begin{aligned}\mathrm{Clos}_{\rightarrow}(\{\langle c, c'\rangle \in [\![\,\mathrm{Term}\,]\!]^2 \mid\ & \langle \pi_0(c), \pi_0(c')\rangle \in A\\ & \wedge\ (B(\pi_0(c))\,[\![\rightarrow]\!]\,A) = (B(\pi_0(c'))\,[\![\rightarrow]\!]\,A)\\ & \wedge\ \forall \langle x, x'\rangle \in B(\pi_0(c)).\langle \pi_1(c)(x), \pi_1(c')(x')\rangle\\ & \quad \in A\}))\,.\end{aligned}$$

(Note that π_i are the projections of pairs $\langle a, b\rangle$, which are part of the small logical framework, see Subsection 2.1.)

(d) Assume $(A, B), (A', B') \in [\![\,\mathrm{Fam}\,]\!](\mathrm{Set})$. Then

$$(A, B) \leq (A', B') :\Leftrightarrow A \subseteq A' \wedge B \subseteq B'\,.$$

Note that by $(A, B), (A', B') \in [\![\,\mathrm{Fam}\,]\!](\mathrm{Set})$, $B \subseteq B'$ is equivalent to

$$B \restriction \mathrm{Flat}(A) = B' \restriction \mathrm{Flat}(A)\,.$$

(e) Assume $(U^{\alpha}, T^{\alpha}) \in [\![\,\mathrm{Fam}\,]\!](\mathrm{Set})$ for $\alpha < \beta$ such that

$$\forall \alpha < \alpha' < \beta.(U^{\alpha}, T^{\alpha}) \leq (U^{\alpha'}, T^{\alpha'})\,.$$

Then we define

$$(U^{<\beta}, T^{<\beta}) := (\bigcup_{\alpha<\beta} U^\alpha, \bigcup_{\alpha<\beta} T^\alpha) \in [\![\,\mathrm{Fam}\,]\!](\mathrm{Set})\,.$$

Note that, if $x \in \mathrm{Flat}(U^\alpha)$, $\alpha < \beta$, then

$$T^{<\beta}(x) = T^\alpha(x)\,.$$

A model of a universe (U, T) can be obtained by defining sets $[\![\,\mathrm{U}\,]\!]$ and $[\![\,\mathrm{T}\,]\!]$ such that

$$([\![\,\mathrm{U}\,]\!], [\![\,\mathrm{T}\,]\!]) \in [\![\,\mathrm{Fam}\,]\!](\mathrm{Set})\,.$$

Then

$$[\![\,\mathrm{T}(s)\,]\!]_\rho := [\![\,\mathrm{T}\,]\!]([\![\,s\,]\!]_\rho)\,.$$

We will introduce first by induction on α sets

$$(\mathrm{U}^\alpha, \mathrm{T}^\alpha) \in [\![\,\mathrm{Fam}\,]\!](\mathrm{Set})$$

and define (assuming the existence of a recursively inaccessible ordinal I)

$$[\![\,\mathrm{U}\,]\!] = \mathrm{U}^{<\mathrm{I}}\,,\ [\![\,\mathrm{T}\,]\!] = \mathrm{T}^{<\mathrm{I}}$$

for the least recursively inaccessible ordinal I. (When we later in this article consider universes closed under additional constructions, the ordinal I will be replaced by a recursively inaccessible ordinal, which has some additional closure properties.)

We will guarantee that the following assumptions are fulfilled:

Assumption 3.2. $(\mathrm{U}^\alpha, \mathrm{T}^\alpha) \in [\![\,\mathrm{Fam}\,]\!](\mathrm{Set})$ *are defined in such a way that the following holds*:

(a) $\alpha < \beta \to (\mathrm{U}^\alpha, \mathrm{T}^\alpha) \leq (\mathrm{U}^\beta, \mathrm{T}^\beta)$.
(b) $\mathrm{PER}(\mathrm{U}^\alpha) \wedge \forall x \in \mathrm{Flat}(\mathrm{U}^\alpha).\mathrm{PER}(\mathrm{T}^\alpha(x))$.
(c) $\mathrm{U}^\alpha \in \mathrm{L}'_{\alpha+1} \wedge \forall x \in \mathrm{Flat}(\mathrm{U}^\alpha).\mathrm{T}^\alpha(x) \in \mathrm{L}'_\alpha$.
(d) $(\lambda\beta < \alpha.\mathrm{U}^\beta) \in \mathrm{L}'_{\alpha+1} \wedge (\lambda\beta < \alpha.\lambda x \in \mathrm{U}^\beta.\mathrm{T}^\beta(x)) \in \mathrm{L}'_{\alpha+1}$.
(e) "$x \in \mathrm{U}^\beta$" *and* "$x \in \mathrm{T}^\beta(y)$" *are* Σ_1*-formulae, and the unbounded existential quantifiers in those* Σ_1*-formulae can be, if we restrict* β *to ordinals less than* α *(where* $\alpha \geq 2$*), restricted to* L'_α.

Here (e) follows from the fact that the unbounded existential quantifiers in the definition of U^β and $\mathrm{T}^\beta(x)$, restricted to $\beta < \alpha$, are of the form $\exists f$, where f is $\lambda\delta < \beta.\mathrm{U}^\delta$, $\lambda\delta < \beta.\lambda x \in \mathrm{U}^\delta.\mathrm{T}^\delta(x)$, $\lambda\delta < \beta.\mathrm{L}'_\delta$, ω or a primitive recursive function on ω, and therefore f is an element of $\mathrm{L}'_{\beta+1} \subseteq \mathrm{L}'_\alpha$. (Here $\lambda\delta < \beta.\mathrm{U}^\delta$ stands for the set theoretic function with domain β mapping δ to U^δ.) More precisely, when showing that this assumption is fulfilled one shows simultaneously (c) - (e) by induction on α.

Because of $\mathrm{PER}(\mathrm{U}^\alpha)$, U^α needs to be closed under reductions. This is more formally obtained as follows: When defining U^α below, what we actually define are sets U'^{α} and functions T'^{α} (while still referring to $\mathrm{U}^{<\alpha}$ and $\mathrm{T}^{<\alpha}$). Then one defines $(\mathrm{U}^{\alpha}, \mathrm{T}^{\alpha})$ itself as the closure of $(\mathrm{U}'^{\alpha}, \mathrm{T}'^{\alpha})$ under reductions (note that $\longrightarrow$ is a deterministic reduction relation):

$$\mathrm{U}^{\alpha} := \mathrm{Clos}_{\rightarrow}(\mathrm{U}'^{\alpha}),$$
$$\mathrm{T}^{\alpha}(a) := \mathrm{T}'^{\alpha}(a') \text{ if } a \longrightarrow^{*} a' \in \mathrm{U}'^{\alpha}.$$

We will however in the following not explicitly mention U'^{α}, T'^{α} in this article, assuming that the reader can adapt the model correspondingly.

U^{α} is the closure of $\mathrm{U}^{<\alpha}$ under one application of each basic set construction of type theory, provided the results are in L'_{α}, as expressed by the following:

- $[\![\mathrm{N}]\!] \in \mathrm{L}'_{\alpha} \rightarrow \langle \widehat{\mathrm{N}}, \widehat{\mathrm{N}} \rangle \in \mathrm{U}^{\alpha}$, $\mathrm{T}^{\alpha}(\widehat{\mathrm{N}}) = [\![\mathrm{N}]\!]$.
- Assume $\langle \langle a, b \rangle, \langle a'b' \rangle \rangle \in [\![\mathrm{IFam}]\!](\mathrm{U}^{<\alpha}, \mathrm{T}^{<\alpha})$, and

$$A := [\![\Sigma]\!](\mathrm{T}^{<\alpha}(a), \lambda x \in \mathrm{T}^{<\alpha}(a).\mathrm{T}^{<\alpha}(b(x))) \in \mathrm{L}'_{\alpha}.$$

  Then $\langle \widehat{\Sigma}(a, b), \widehat{\Sigma}(a', b') \rangle \in \mathrm{U}^{\alpha}$ and $\mathrm{T}^{\alpha}(\widehat{\Sigma}(a, b)) = A$.
- Similarly for the other basic set constructions.

In order to obtain a model of a universe closed under some other principles, we will in later sections of this article add to U^{α} additional elements. In this article, these closure principles will always be the closure under certain universe constructions.

We show that $(\mathrm{U}^{<\mathrm{I}}, \mathrm{T}^{<\mathrm{I}})$ is a model of the standard universe. In fact we show a more general theorem:

THEOREM 3.3. *Assume* $(\mathrm{U}^{\alpha}, \mathrm{T}^{\alpha})$ *is defined as above, except that one adds possibly additional elements to* U^{α} *and defines* $\mathrm{T}^{\alpha}(s)$ *for each such additional element* s*, and that this is done in such a way that Assumption* 3.2 *is fulfilled. Let* κ *be a recursively inaccessible ordinal. Then* $(\mathrm{U}^{<\kappa}, \mathrm{T}^{<\kappa})$ *is closed under the basic set constructions.*

PROOF. First one observes that if $(A, B) \in [\![\mathrm{Fam}]\!](\mathrm{Set})$ and $A, B \in \mathrm{L}'_{\alpha}$, then $[\![\Sigma]\!](A, B)$, $[\![\Pi]\!](A, B) \in \mathrm{L}'_{\alpha+2}$ and $[\![\mathrm{W}]\!](A, B) \in \mathrm{L}'_{\alpha^{+}+1}$. Similar properties hold for $[\![+]\!]$ and $[\![\mathrm{Id}]\!]$ (making use of $\alpha + 2$).

Now, if I is a recursively inaccessible ordinal, then $(\mathrm{U}^{<\mathrm{I}}, \mathrm{T}^{<\mathrm{I}})$ is closed under $\widehat{\mathrm{N}}_n$, $\widehat{\mathrm{N}}$, $\widehat{\Sigma}$, $\widehat{\Pi}$, $\widehat{\mathrm{W}}$, $\widehat{+}$, $\widehat{\mathrm{Id}}$: We consider only the most difficult case $\widehat{\mathrm{W}}$. First we show a small Lemma:

LEMMA 3.4. *Assume Assumption* 3.2 *is fulfilled. Let* κ *be an admissible ordinal and*

$$\langle \langle a, b \rangle, \langle a', b' \rangle \rangle \in [\![\mathrm{IFam}]\!](\mathrm{U}^{<\kappa}, \mathrm{T}^{<\kappa}).$$

Then there exists an $\alpha < \kappa$ *such that*

$$\langle \langle a, b \rangle, \langle a', b' \rangle \rangle \in [\![\mathrm{IFam}]\!](\mathrm{U}^{<\alpha}, \mathrm{T}^{<\alpha}).$$

Proof of the Lemma. We have $\langle a, a'\rangle \in \mathrm{U}^{<\kappa}$, and since κ is a limit ordinal, it follows $\langle a, a'\rangle \in \mathrm{U}^{<\alpha_0}$ for some $\alpha_0 < \kappa$. Furthermore, $\mathrm{T}^{<\alpha_0}(a) \in \mathrm{L}'_{\alpha_0}$, and for all $\langle x, x'\rangle \in \mathrm{T}^{<\alpha_0}(a)$ there exists $\alpha_1 < \kappa$ such that $\langle b(x), b'(x')\rangle \in \mathrm{U}^{<\alpha_1}$. Since "$u \in \mathrm{U}^{\beta}$" is Σ_1, the unbounded quantifier in this Σ_1-formula can be relativised to L_κ for $\beta < \kappa$ (by Assumption 3.2 (e)), $\mathrm{T}^{<\alpha_0}(a) \in \mathrm{L}'_{\alpha_0}$, and κ is admissible, we can find a uniform α_1 for this. Now let $\alpha := \max\{\alpha_0, \alpha_1\}$. ⊣

We continue with the main proof. Assume

$$\langle\langle a, b\rangle, \langle a', b'\rangle\rangle \in [\![\, \mathrm{IFam} \,]\!](\mathrm{U}^{<\mathrm{I}}, \mathrm{T}^{<\mathrm{I}}).$$

Then by Lemma 3.4 we find a β such that

$$\langle\langle a, b\rangle, \langle a', b'\rangle\rangle \in [\![\, \mathrm{IFam} \,]\!](\mathrm{U}^{<\beta}, \mathrm{T}^{<\beta}).$$

We have $\mathrm{T}^{<\mathrm{I}}(a) = \mathrm{T}^{<\beta}(a) \in \mathrm{L}'_\beta$, $\mathrm{T}^{<\mathrm{I}}(b(x)) = \mathrm{T}^{<\beta}(b(x))$, therefore $(\lambda x \in \mathrm{T}^{<\mathrm{I}}(a).\mathrm{T}^{<\mathrm{I}}(b(x))) \in \mathrm{L}'_{\beta+1}$, therefore $[\![\, \mathrm{W} \,]\!](\mathrm{T}^{<\mathrm{I}}(a), \lambda x \in \mathrm{T}^{<\mathrm{I}}(a).\mathrm{T}^{<\mathrm{I}}(b(x))) \in \mathrm{L}'_{\beta^+ +1}$. Therefore $\langle \widehat{\mathrm{W}}(a, b), \widehat{\mathrm{W}}(a', b')\rangle \in \mathrm{U}^{\beta^+ +2} \subseteq \mathrm{U}^{<\mathrm{I}}$. ⊣

One obtains a model of the type theory $\mathbf{ML_WU}$ in $\mathbf{KPI}^+$ as follows. We have modelled all the ingredients already, and need only to make sure that the constructions can be carried out in $\mathbf{KPI}^+$. Let κ_n be the nth admissible above I. First, one observes, that $[\![\, \mathrm{U} \,]\!] \in \mathrm{L}'_{\mathrm{I}+1}$, for $x \in \mathrm{Flat}([\![\, \mathrm{U} \,]\!])$ we have $[\![\, \mathrm{T_U} \,]\!](x) \in \mathrm{L}'_{\mathrm{I}}$, and for $\langle x, x'\rangle \in [\![\, \mathrm{U} \,]\!]$ we have $[\![\, \mathrm{T_U} \,]\!](x) = [\![\, \mathrm{T_U} \,]\!](x')$. Therefore one defines $\mathrm{o}(\mathrm{U}) := \mathrm{I} + 1$, $\mathrm{o}(\mathrm{T_U}(s)) := \mathrm{I}$. For all other set constructions we have seen that if the maximum of the ordinals of the subterms is α, then the ordinal of the term itself is $\alpha + 2$ or $\alpha^+ + 1$. Therefore we obtain that for any set term A there exists an n such that $\mathrm{o}(A) \le \kappa_n$ and therefore for all ρ $[\![\, A \,]\!]_\rho \in \mathrm{L}'_{\kappa_n}$. One sees now that we can interpret all sets in $\mathbf{KPI}^+$, and can show that this model is sound with respect to the above mentioned correctness conditions. So we have given a sketch of the following theorem of which a detailed proof can be found in [30]; see as well [32]; Griffor and Rathjen have shown related results in [17]:

Theorem 3.5. (a) $\mathbf{ML_WU}$ *can be modelled in* $\mathbf{KPI}^+$.

(b) $|\mathbf{ML_WU}| \le |\mathbf{KPI}^+|$.

(c) *The previous statements hold as well if we replace intensional by extensional equality.*

In the other direction, we have developed in [34] (see as well our PhD thesis [30], and the overview articles [33, 36]; Rathjen and Griffor have obtained related results in [17]) a well-ordering proof which shows the other direction, namely:

Theorem 3.6. (a) $|\mathbf{ML_WU}| = |\mathbf{KPI}^+|$.

(b) *The previous statements hold as well if we replace intensional by extensional equality.*

§4. Palmgren's super-universe and recursively hyperinaccessibles.

4.1. Definition of the super-universe. The first substantial step beyond the type theory $\mathrm{ML_WU}$ was the introduction of a hierarchy of finitely many universes by Per Martin-Löf. We won't cover this extension in this article, and will instead move to the next step, namely the super-universe, as introduced by Erik Palmgren. He introduced a super-universe operator which takes as arguments a family of sets, and forms a universe containing this family of sets. (More precisely, since we are using universes à la Tarski, the resulting universe contains a code for the index set and codes for each element of the family. The decoding function maps these codes to the corresponding sets.) The super-universe is defined as a universe closed under this operator.

In order to introduce this in detail, let us first introduce some notions:

DEFINITION 4.1 (Refers to the full logical framework).

(a) The type of families of sets is defined (using the full logical framework) as

$$\mathrm{Fam}(\mathrm{Set}) := (X : \mathrm{Set}) \times (X \to \mathrm{Set}) .$$

(b) Let $\langle \mathrm{U}, \mathrm{T} \rangle \in \mathrm{Fam}(\mathrm{Set})$. The family of sets internal to (U, T) (or internal family of sets in (U, T)) is the set of families of sets in U, indexed over elements of U:

$$\mathrm{IFam}(\mathrm{U}, \mathrm{T}) := (x : \mathrm{U}) \times (\mathrm{T}(x) \to \mathrm{U}) .$$

(c) Let $\langle \mathrm{U}, \mathrm{T} \rangle \in \mathrm{Fam}(\mathrm{Set})$. The lifting of an element $a : \mathrm{IFam}(\mathrm{U}, \mathrm{T})$ to an element of $\mathrm{Fam}(\mathrm{Set})$ is defined as

$$\mathrm{T}[a] := \langle \mathrm{T}(\pi_0(a)), (x)\mathrm{T}(\pi_1(a)(x)) \rangle : \mathrm{Fam}(\mathrm{Set}) .$$

Palmgren first introduced the super-universe operator

$$\mathrm{SU} : \mathrm{Fam}(\mathrm{Set}) \to \mathrm{Fam}(\mathrm{Set}) .$$

$\mathrm{SU}(C)$ is a universe (U', T') closed under the standard set constructions and containing C. This means that if $C = \langle A, B \rangle$, then U' contains

- a code $\widehat{\mathrm{a}}$ for A, i.e. $T'(\widehat{\mathrm{a}}) = A$,
- and for $a : A$ a code $\widehat{\mathrm{b}}(a)$ for $B(a)$, i.e. $T'(\widehat{\mathrm{b}}(a)) = B(a)$.

The next step was to introduce the super-universe as a universe $(\mathrm{V}, \mathrm{T_V})$, which is as usual closed under the standard set constructions and which reflects the super-universe operator: if $a : \mathrm{IFam}(\mathrm{V}, \mathrm{T_V})$, then V contains codes $\langle \widehat{\mathrm{U}}_a, (x)\widehat{\mathrm{T}}_a(x) \rangle$ for $\mathrm{SU}(\mathrm{T_V}[a])$, which means

$$\langle \widehat{\mathrm{U}}_a, (x)\widehat{\mathrm{T}}_a(x) \rangle : \mathrm{IFam}(\mathrm{V}, \mathrm{T_V}) , \quad \mathrm{T_V}\big[\langle \widehat{\mathrm{U}}_a, (x)\widehat{\mathrm{T}}_a(x) \rangle\big] = \mathrm{SU}(\mathrm{T_V}[a]) .$$

We will in the following present a version, in which $(\mathrm{T}(\widehat{\mathrm{U}}_a), \widehat{\mathrm{T}}_a)$ is a recursive subuniverse of $(\mathrm{V}, \mathrm{T_V})$. Furthermore, we will define the universe operator only restricted to elements of $\mathrm{IFam}(\mathrm{V}, \mathrm{T_V})$ (which will be written as $\mathrm{U}_{a,b}$ for $\langle a, b\rangle : \mathrm{IFam}(\mathrm{V}, \mathrm{T_V})$). We uncurry SU and obtain the following rules:

$$\mathrm{V} : \mathrm{Set} \qquad \frac{a : \mathrm{V}}{\mathrm{T_V}(a) : \mathrm{Set}}$$

$$\frac{a : \mathrm{V} \qquad b : \mathrm{T_V}(a) \to \mathrm{V}}{\mathrm{U}_{a,b} : \mathrm{Set}}$$

$$\frac{a : \mathrm{V} \qquad b : \mathrm{T_V}(a) \to \mathrm{V}}{\widehat{\mathrm{U}}_{a,b} : \mathrm{V}} \qquad \mathrm{T_V}(\widehat{\mathrm{U}}_{a,b}) = \mathrm{U}_{a,b}$$

$$\frac{a : \mathrm{V} \qquad b : \mathrm{T_V}(a) \to \mathrm{V} \qquad c : \mathrm{U}_{a,b}}{\widehat{\mathrm{T}}_{\mathrm{U},a,b}(c) : \mathrm{V}}$$

Let $\mathrm{T}_{\mathrm{U},a,b}(x) := \mathrm{T_V}(\widehat{\mathrm{T}}_{\mathrm{U},a,b}(x)) : \mathrm{Set}$.

$(\mathrm{V}, \mathrm{T_V})$ is a universe, which means it is closed under the basic set constructions. The codes are written as $\widehat{\mathrm{N}}_{\mathrm{V}}$, $\widehat{\Sigma}_{\mathrm{V}}$, etc.

One demands for $a : \mathrm{V}$, $b : \mathrm{T_V}(a) \to \mathrm{V}$ that $(\mathrm{U}_{a,b}, \widehat{\mathrm{T}}_{\mathrm{U},a,b})$ is a recursive subuniverse of $(\mathrm{V}, \mathrm{T_V})$, which means as indicated in Section 3 that $\mathrm{U}_{a,b}$ is closed under constructors for forming codes for the standard set constructions like $\widehat{\mathrm{N}}_{\mathrm{U},a,b}$, $\widehat{\Sigma}_{\mathrm{U},a,b}$, etc., and that $\widehat{\mathrm{T}}_{\mathrm{U},a,b}$ decodes them as the corresponding codes in V, e.g. $\widehat{\mathrm{T}}_{\mathrm{U},a,b}(\widehat{\mathrm{N}}_{\mathrm{U},a,b}) = \widehat{\mathrm{N}}_{\mathrm{V}}$.

Furthermore, $\mathrm{U}_{a,b}$ contains codes for a, b, which means that one has the following rules:

$$\widehat{\mathrm{a}}_{a,b} : \mathrm{U}_{a,b} \qquad \widehat{\mathrm{T}}_{\mathrm{U},a,b}(\widehat{\mathrm{a}}_{a,b}) = a$$

$$\widehat{\mathrm{b}}_{a,b} : \mathrm{T_V}(a) \to \mathrm{U}_{a,b} \qquad \widehat{\mathrm{T}}_{\mathrm{U},a,b}(\widehat{\mathrm{b}}_{a,b}(c)) = b(c)$$

Let $\mathrm{ML_W}$ + (Superuniv) be the type theory consisting of the rules of the small logical framework, the basic set constructions, and the rules for V, $\mathrm{T_V}$, $\mathrm{U}_{a,b}$, $\widehat{\mathrm{U}}_{a,b}$, $\widehat{\mathrm{T}}_{\mathrm{U},a,b}$ as above.

In this type theory, one can define using the elimination rules for N a hierarchy of universes $\widehat{\mathrm{U}}'_n : \mathrm{V}$ by

$$\widehat{\mathrm{U}}'_0 = \widehat{\mathrm{U}}_{\widehat{\mathrm{N}}_0,(x)\widehat{\mathrm{N}}_0} \quad \widehat{\mathrm{U}}'_{\mathrm{S}(n)} = \widehat{\mathrm{U}}_{\widehat{\mathrm{N}}_1,(n)\widehat{\mathrm{U}}'_n}.$$

In a next step one can define a universe containing this hierarchy of universes, namely $\mathrm{U}_{\widehat{\mathrm{N}},(n)\widehat{\mathrm{U}}'_n}$.

This is the first step towards forming a transfinite hierarchy of universes: Assume any well-founded relation $(R, \prec)$ such that $(R, \prec)$ can be represented in $(\mathrm{V}, \mathrm{T_V})$, i.e. such that there exist $\widehat{\mathrm{R}} : \mathrm{V}$ and $\widehat{\prec} : R \to R \to \mathrm{U}$ such that $\mathrm{T_V}(\widehat{\mathrm{R}}) = R$ and $\mathrm{T_V}(\widehat{\prec}(r, r')) = \prec (r, r')$. Assume the well-foundedness of $(R, \prec)$ can be shown in this type theory. Then we can define a hierarchy of universes $\widehat{\mathrm{U}}''_\alpha$ for $\alpha \in R$ such that $\widehat{\mathrm{U}}''_\alpha$ contains codes for $\widehat{\mathrm{U}}''_\beta$ for $\beta \prec \alpha$.

4.2. A model of the super-universe. $\mathrm{ML_W}$ + (Superuniv) has strength $(\mathrm{KP} + (\mathrm{hyper} - \mathrm{inacc}))^+$, where $(\mathrm{KP} + (\mathrm{hyper} - \mathrm{inacc}))^+$ is Kripke-Platek set theory plus the existence of a recursively hyper-inaccessible set plus finitely many admissibles (i.e. for Meta-each n we have n admissibles) above it, which is proof theoretically equivalent to $\mathrm{KPI^r}$ plus the existence of a recursively hyper-inaccessible set.

Here a recursively hyper-inaccessible set is an admissible set a such that for all $x \in a$ there exists a $y \in a$ such that $x \in y$ and y is recursively inaccessible. A recursively hyper-inaccessible ordinal is an admissible ordinal HI such that

$$\forall \alpha \in \mathrm{HI}.\exists \beta \in \mathrm{HI}.\alpha \in \beta \wedge \mathrm{Inacc}(\beta),$$

where $\mathrm{Inacc}(\beta)$ means that β is a recursively inaccessible ordinal.

The precise formulation of $(\mathrm{KP} + (\mathrm{hyper} - \mathrm{inacc}))^+$ is similar to KPI^+, except that one replaces "inaccessible" by "hyper-inaccessible".

An upper bound for the strength of $\mathrm{ML_W}$ + (Superuniv) can be obtained by modelling it in $(\mathrm{KP} + (\mathrm{hyper} - \mathrm{inacc}))^+$. We will give in the following a sketch of the model, full details will be presented in a future paper.

The basic construction is as in Section 2. As pointed out there, in $[\![\mathrm{Term}]\!]$ we omit arguments of the constructors of V, $\mathrm{U}_{a,b}$ used for type checking only, except in $\widehat{\mathrm{a}}$, $\widehat{\mathrm{b}}$. (By "used for type checking only" we mean that they have no computational content — in an equality rule of the form $\widetilde{\mathrm{T}}(c) = d$, where $\widetilde{\mathrm{T}} = \mathrm{T_V}$ or $\widetilde{\mathrm{T}} = \widehat{\mathrm{T}}_{\mathrm{U},a,b}$, d will not depend on those parameters.) For instance, we have a constant $\widehat{\mathrm{N}} \in [\![\mathrm{Term}]\!]$ and interpret

$$[\![\widehat{\mathrm{N}}_{\mathrm{V}}]\!]_\rho = [\![\widehat{\mathrm{N}}_{\mathrm{U},a,b}]\!]_\rho = \widehat{\mathrm{N}},$$

similarly for $\widehat{\mathrm{N}}_n$, $\widehat{\Sigma}$, $\widehat{\Pi}$, $\widehat{+}$, $\widehat{\mathrm{Id}}$, $\widehat{\mathrm{W}}$. $\widehat{\mathrm{U}}$, $\widehat{\mathrm{T}}_{\mathrm{U}}$, $\widehat{\mathrm{a}}$ and $\widehat{\mathrm{b}}$ will depend on a, b. We will in the model identify $\widehat{\mathrm{a}}_{a,b}$ with a, $\widehat{\mathrm{b}}_{a,b}$ with b, and $\widehat{\mathrm{T}}_{\mathrm{U},a,b}(c)$ with c as expressed by the reduction rules

$$\begin{array}{rcl}\widehat{\mathrm{a}}_{a,b} & \longrightarrow & a \\ \widehat{\mathrm{b}}_{a,b}(c) & \longrightarrow & b(c) \\ \widehat{\mathrm{T}}_{\mathrm{U},a,b}(c) & \longrightarrow & c\end{array}$$

As for standard universes, the super-universe is interpreted, by defining first by induction on α sets

$$(\mathrm{V}^\alpha, \mathrm{T}^\alpha) \in [\![\mathrm{Fam}]\!](\mathrm{Set})$$

such that

$$\alpha < \beta \rightarrow (\mathrm{V}^\alpha, \mathrm{T}^\alpha) \leq (\mathrm{V}^\beta, \mathrm{T}^\beta).$$

As for all models of universes in this article, we add to V^α the codes for standard set constructors applied to arguments, if the elements of the universe they refer to are already in $\mathrm{V}^{<\alpha}$ and the sets added are in L'_α. Additionally, whenever there exists a β such that $\beta + 1 < \alpha$ and $\mathrm{V}^{<\beta}$ is closed under all universe constructions, and such that

$$\langle\langle a, b\rangle, \langle a', b'\rangle\rangle \in [\![\,\mathrm{IFam}\,]\!](\mathrm{V}^{<\beta}, \mathrm{T}^{<\beta}),$$

then

$$\langle \widehat{\mathrm{U}}_{a,b}, \widehat{\mathrm{U}}_{a',b'}\rangle \in \mathrm{V}^{\alpha}$$

and we define

$$\mathrm{T}^{\alpha}(\widehat{\mathrm{U}}_{a,b}) := \mathrm{V}^{<\beta}$$

for the minimal β, which has the mentioned closure properties. The construction fulfils Assumption 3.2 with U replaced by V.

Let HI be the least recursively hyper-inaccessible ordinal. Then one interprets

$$\begin{array}{rcl} [\![\,\mathrm{V}\,]\!] & := & \mathrm{V}^{<\mathrm{HI}}, \\ [\![\,\mathrm{T_V}\,]\!] & := & \mathrm{T}^{<\mathrm{HI}}, \\ [\![\,\mathrm{U}_{a,b}\,]\!]_\rho & := & \mathrm{T}^{<\mathrm{HI}}([\![\,\widehat{\mathrm{U}}_{a,b}\,]\!]_\rho). \end{array}$$

Note that this model construction won't admit elimination rules for $\mathrm{U}_{a,b}$, since it is not interpreted as the least set closed under a, b and the basic set constructions — the $\mathrm{V}^{\beta'}$ chosen above might contain elements of the form $\widehat{\mathrm{U}}_{a',b'}$, which have been added, since the model of V needs to be closed under $\widehat{\mathrm{U}}$, but which are not elements of the least subuniverse closed under a, b and the standard set constructions. One could modify the above model in order to admit elimination rules: Then one would try to form at each stage α the least subset of $\mathrm{V}^{<\alpha}$ closed under the universe constructions and containing a and $b(x)$ for $x \in \mathrm{Flat}(\mathrm{T}^{<\alpha}(a))$. If this definition succeeds and the least such set is A, one would add $\widehat{\mathrm{U}}_{a,b}$ to V^{α} and define $\mathrm{T}^{\alpha}(\widehat{\mathrm{U}}_{a,b}) := A$.

We have to verify that $\mathrm{V}^{<\mathrm{HI}}$ has the necessary closure properties of the super-universe. Since HI is a recursively inaccessible, $\mathrm{V}^{<\mathrm{HI}}$ will be closed under universe constructions as before. $\mathrm{V}^{<\mathrm{HI}}$ is as well closed under $(a, b)\widehat{\mathrm{U}}_{a,b}$: Assume

$$\langle\langle a, b\rangle, \langle a', b'\rangle\rangle \in [\![\,\mathrm{IFam}\,]\!](\mathrm{V}^{<\mathrm{HI}}, \mathrm{T}^{<\mathrm{HI}}).$$

By the admissibility of HI it follows using Lemma 3.4 (note that we observed above that Assumption 3.2 is fulfilled)

$$\langle\langle a, b\rangle, \langle a', b'\rangle\rangle \in [\![\,\mathrm{IFam}\,]\!](\mathrm{V}^{<\alpha}, \mathrm{T}^{<\alpha})$$

for some $\alpha < \mathrm{HI}$. Since HI is a recursively hyper inaccessible, there exists a recursively inaccessible ordinal $\alpha < \kappa < \mathrm{HI}$. $\mathrm{V}^{<\kappa}$ will be closed under the universe constructions and

$$\langle\langle a, b\rangle, \langle a', b'\rangle\rangle \in [\![\,\mathrm{IFam}\,]\!](\mathrm{V}^{<\kappa}, \mathrm{T}^{<\kappa}),$$

therefore

$$\langle \widehat{\mathrm{U}}_{a,b}, \widehat{\mathrm{U}}_{a',b'} \rangle \in \mathrm{V}^{\kappa+2} \subseteq \mathrm{V}^{<\mathrm{HI}}\,.$$

That $[\![\mathrm{U}_{a,b}]\!]_\rho$ has the closure properties needed in order to interpret $\mathrm{U}_{a,b}$ follows by the construction.

The remaining construction is as for any standard model of type theory, i.e. we interpret the constructions of the small logical framework and the basic set constructions on top of U. As for $\mathrm{ML_W U}$, we need $< \omega$ many admissibles in order to interpret W-types built on top of $\mathrm{V}^{<\mathrm{HI}}$, therefore finite approximations of the type theory can be interpreted into Kripke Platek set theory plus one recursively hyper-inaccessible and finitely many admissibles (i.e. for Meta-each n we have n admissibles) above it. So we have given the essence of a proof of the following theorem:

THEOREM 4.2. (a) *We can model* $\mathrm{ML_W}$ + (Superuniv) *in*

$$(\mathrm{KP} + (\mathrm{hyper} - \mathrm{inacc}))^{+}.$$

(b) $|\mathrm{ML_W} + (\mathrm{Superuniv})| \leq |(\mathrm{KP} + (\mathrm{hyper} - \mathrm{inacc}))^{+}|$

(c) *The previous statements hold as well if we replace intensional by extensional equality.*

In order to obtain the precise proof theoretic strength, one has to carry out a well-ordering proof in order to obtain

$$|\mathrm{ML_W} + (\mathrm{Superuniv})| \geq |(\mathrm{KP} + (\mathrm{hyper} - \mathrm{inacc}))^{+}|\,.$$

We note that if one adds as in Palmgren's original type theory the full universe operator $\mathrm{SU} : \mathrm{Fam(Set)} \to \mathrm{Fam(Set)}$, one obtains a slightly stronger theory. Formulated without the full logical framework, in this type theory one has depending on A : Set and $x : A \Rightarrow B$: Set the set $\mathrm{U}(A, (x)B)$: Set and decoding function

$$u : \mathrm{U}(A, (x)B) \Rightarrow \mathrm{T}(A, (x)B, u) : \mathrm{Set}\,.$$

$(\mathrm{U}(A, (x)B), \mathrm{T}(A, (x)B))$ is a universe containing codes for A and $B[x := a]$. Now one can replace in the type theory for the super universe the rules for $\mathrm{U}_{a,b}$, $\mathrm{T}_{\mathrm{U},a,b}$ and defines

$$\begin{aligned} \mathrm{U}_{a,b} &:= \mathrm{U}(\mathrm{T_V}(a), (x)\mathrm{T_V}(b(x)))\,, \\ \mathrm{T}_{\mathrm{U},a,b}(c) &:= \mathrm{T}(\mathrm{T_V}(a), (x)\mathrm{T_V}(b(x)), c)\,. \end{aligned}$$

$\mathrm{U}_{a,b}$ is now an inductive subuniverse of V, so $\widehat{\mathrm{T}}_{\mathrm{U},a,b}$ is a constructor of V with equality rule $\mathrm{T_V}(\widehat{\mathrm{T}}_{\mathrm{U},a,b}(c)) = \mathrm{T}_{\mathrm{U},a,b}(c)$. In this theory one can apply the super-universe operator as well to V itself, and to elements formed from it. However, only finite nestings of the super-universe operator are possible. The modelling of one application of the super-universe to arguments (A, B) requires one recursively inaccessible above $\max\{\mathrm{o}(A), \mathrm{o}(B)\}$. The modelling of V requires one recursively hyper-inaccessible, so in total one needs for

Palmgren's original type theory one recursively hyper-inaccessible ordinal and finitely many recursively inaccessible ordinals (i.e. for Meta-each n we have n recursively inaccessible ordinals) above it.

§5. The Mahlo universe.

5.1. Definition of the Mahlo universe. Once one has defined the super-universe, one can move to a hyper-universe by forming a universe U^2 with decoding function T^2, such that for any family of sets (a, b) internal to $(\mathrm{U}^2, \mathrm{T}^2)$ there exists a super-universe $\mathrm{U}^1_{a,b}$ and an element $\widehat{\mathrm{U}}^1_{a,b}$ in V, such that $\mathrm{T}_\mathrm{V}(\widehat{\mathrm{U}}^1_{a,b}) = \mathrm{U}^1_{a,b}$, where $\mathrm{U}^1_{a,b}$ contains codes which decode as a, b as before. So we have a decoding function $\widehat{\mathrm{T}}^1_{a,b} : \mathrm{U}^1_{a,b} \to \mathrm{U}^2$, and define $\mathrm{T}^1_{a,b} := \mathrm{T}^2 \circ \widehat{\mathrm{T}}^1_{a,b}$. $\mathrm{U}^1_{a,b}$ is closed under the standard universe constructions. That $\mathrm{U}^1_{a,b}$ is a super-universe means, that for any family of sets (c, d) internal to $(\mathrm{U}^1_{a,b}, \mathrm{T}^1_{a,b})$ there exists a subuniverse $\mathrm{U}^0_{a,b,c,d}$ of $\mathrm{U}^1_{a,b}$ which contains codes decoding as c, d, is closed under the standard universe constructions, and which is represented as a code in $\mathrm{U}^1_{a,b}$.

One could define now hyper$^\alpha$-super-universes, but we won't spell this out in this article. The construction is similar to the step from a Mahlo-universe to a hyper$^\alpha$-Mahlo universe, which will be discussed in the follow-up article [37].

We note that E. Palmgren [23] has introduced higher order universes. In those universes, the lowest level is a universe, the next level is a universe of universe operators, the next level is a universe of operators on universe operators, etc. E. Palmgren added rules expressing the closure of higher order universes under the application of a universe operator from the next level. It is conjectured that E. Palmgren's universe reaches the strength of KPM, which is slightly below the strength of the Mahlo universe. We will not investigate this construction any further (see the remark in the introduction that Palmgren's universe contains a Mahlo universe in disguise).

We will investigate in which sense the hyper$^\alpha$-super-universes are universes which for specific operators f on families of sets internal to it contains a universe internal to it, which is closed under f, and in which sense the Mahlo universe to be introduced generalises this.

Definition 5.1 (Assuming the full logical framework). (a) An operator on families of sets is a function $f : \mathrm{Fam}(\mathrm{Set}) \to \mathrm{Fam}(\mathrm{Set})$.

(b) Let $\langle \mathrm{U}, \mathrm{T} \rangle : \mathrm{Fam}(\mathrm{Set})$. An operator on families of sets internal to (U, T) is a function $f : \mathrm{IFam}(\mathrm{U}, \mathrm{T}) \to \mathrm{IFam}(\mathrm{U}, \mathrm{T})$.

(c) Let $\langle A, B \rangle, \langle A', B' \rangle : \mathrm{Fam}(\mathrm{Set})$. Assume elimination of $+$ into Set. Then

$$\langle A, B \rangle \cup_{\mathrm{Fam}(\mathrm{Set})} \langle A', B' \rangle := \langle C, D \rangle,$$

where

$$C := (A + \mathrm{N}_1) + (A' + \mathrm{N}_1)$$

and D is such that (remember that A_0^1 is the canonical element of N_1)

$$D(\mathrm{inl}(\mathrm{inl}(a))) = B(a)\,, \qquad D(\mathrm{inl}(\mathrm{inr}(\mathrm{A}_0^1))) = A\,,$$
$$D(\mathrm{inr}(\mathrm{inl}(a'))) = B'(a')\,, \quad D(\mathrm{inr}(\mathrm{inr}(\mathrm{A}_0^1))) = A'\,.$$

The super-universe operator SU takes A : Fam(Set) and forms a universe closed under the operator $(X)A$: Fam(Set) $\to$ Fam(Set). The super-universe is a universe which is closed under the super-universe operator SU. One can define now a hyper-universe operator HU, which takes A : Fam(Set) and forms a super-universe closed $(X)A$. One obtains essentially the same set by taking a universe closed under $(X)\mathrm{SU}(X) \cup_{\mathrm{Fam(Set)}} A$ (assuming this is definable). Then one can form a hyper^2-universe operator, a universe closed under it etc.

The step towards the Mahlo universe is now to form a universe $(\mathrm{V}, \mathrm{T_V})$ such that for *any* operator on families of sets internal to $(\mathrm{V}, \mathrm{T_V})$ there exists a universe, which is closed under this operator and represented in $(\mathrm{V}, \mathrm{T_V})$. So for any f : $\mathrm{IFam}(\mathrm{V}, \mathrm{T_V}) \to \mathrm{IFam}(\mathrm{V}, \mathrm{T_V})$ there exists a subuniverse $(\mathrm{U}_f, \mathrm{T}_{\mathrm{U},f})$ of V, closed under f and represented as an element $\widehat{\mathrm{U}}_f$ in V.

Spelled out without using the full logical framework, we obtain the following rules:

- We have the rules of the small logical framework and of the basic set constructions.
- We have rules expressing that $(\mathrm{V}, \mathrm{T_V})$ is a universe closed under the basic set constructions. The constructors for the codes for these set constructions will be denoted by $\widehat{\mathrm{N}}_\mathrm{V}$, $\widehat{\Sigma}_\mathrm{V}$, etc.
- We have that for every operator f on families of sets internal to $(\mathrm{V}, \mathrm{T_V})$ there is a subuniverse U_f closed under it. We split f into two components, uncurry it, and obtain the following rules:

$$\frac{f : (a : \mathrm{V}, b : \mathrm{T_V}(a) \to \mathrm{V}) \to \mathrm{V} \quad g : (a : \mathrm{V}, b : \mathrm{T_V}(a) \to \mathrm{V}, \mathrm{T_V}(f(a, b)) \to \mathrm{V}}{\mathrm{U}_{f,g} : \mathrm{Set}}$$

- We have rules expressing that $\mathrm{U}_{f,g}$ is a recursive subuniverse closed under the standard set constructions. So we have the decoding function

$$\widehat{\mathrm{T}}_{\mathrm{U},f,g} : \mathrm{U}_{f,g} \to \mathrm{V}$$

and define

$$\mathrm{T}_{\mathrm{U},f,g}(a) := \mathrm{T_V}(\widehat{\mathrm{T}}_{\mathrm{U},f,g}(a)) : \mathrm{Set}\,.$$

The codes for the basic set constructions will be denoted by $\widehat{\mathrm{N}}_{\mathrm{U},f,g}$, $\widehat{\Sigma}_{\mathrm{U},f,g}$, etc.

- We have rules expressing that $(\mathrm{U}_{f,g}, \mathrm{T}_{\mathrm{U},f,g})$ is closed under f and g:

$$\begin{array}{rcl}
\widehat{\mathrm{f}}_{f,g} & : & (a:\mathrm{U}_{f,g}, b:\mathrm{T}_{\mathrm{U},f,g}(a)\to \mathrm{U}_{f,g})\to \mathrm{U}_{f,g} \\
\widehat{\mathrm{T}}_{\mathrm{U},f,g}(\widehat{\mathrm{f}}_{f,g}(a,b)) & = & f(\widehat{\mathrm{T}}_{\mathrm{U},f,g}(a), \widehat{\mathrm{T}}_{\mathrm{U},f,g}\circ b). \\
\widehat{\mathrm{g}}_{f,g} & : & (a:\mathrm{U}_{f,g}, \\
 & & \;b:\mathrm{T}_{\mathrm{U},f,g}(a)\to \mathrm{U}_{f,g}, \\
 & & \;\mathrm{T}_{\mathrm{V}}(f(\widehat{\mathrm{T}}_{\mathrm{U},f,g}(a), \widehat{\mathrm{T}}_{\mathrm{U},f,g}\circ b))) \\
 & & \;\to \mathrm{U}_{f,g} \\
\widehat{\mathrm{T}}_{\mathrm{U},f,g}(\widehat{\mathrm{g}}_{f,g}(a,b,c)) & = & g(\widehat{\mathrm{T}}_{\mathrm{U},f,g}(a), \widehat{\mathrm{T}}_{\mathrm{U},f,g}\circ b, c).
\end{array}$$

- And finally, we have the rule that V contains a code for $\mathrm{U}_{f,g}$. As P. Martin-Löf observed this is the rule which makes the Mahlo universe so strong — having just subuniverses closed under f, g doesn't add any strength to $\mathrm{ML_W U}$.

$$\frac{f:(a:\mathrm{V}, b:\mathrm{T}_{\mathrm{V}}(a)\to \mathrm{V})\to \mathrm{V} \qquad g:(a:\mathrm{V}, b:\mathrm{T}_{\mathrm{V}}(a)\to\mathrm{V}, \mathrm{T}_{\mathrm{V}}(f(a,b))\to \mathrm{V}}{\widehat{\mathrm{U}}_{f,g}:\mathrm{V}}$$

$$\mathrm{T}_{\mathrm{V}}(\widehat{\mathrm{U}}_{f,g}) = \mathrm{U}_{f,g}$$

We call the resulting type theory $\mathrm{ML_W} + (\mathrm{Mahlo})$.

5.2. A model of the Mahlo universe. $\mathrm{ML_W} + (\mathrm{Mahlo})$ has strength $(\mathrm{KPM})^+$, where $(\mathrm{KPM})^+$ is $\mathrm{KP}\omega$ plus the existence of one recursively Mahlo set plus finitely many admissibles above it (i.e. for Meta-each n we have a chain of n admissibles above it), which is equivalent to $\mathrm{KPl^r}$ plus the existence of one recursively Mahlo set.

Here a *recursively Mahlo set* is an admissible set $\mathrm{a_M}$ such that

$$\begin{array}{ll}
\textit{(Mahlo)} & \forall \vec{z}\in \mathrm{a_M}.((\forall x\in \mathrm{a_M}.\exists y\in \mathrm{a_M}.\varphi(x,y,\vec{z})) \\
 & \quad\to \exists u\in \mathrm{a_M}.\mathrm{Ad}(u)\wedge \vec{z}\in u \wedge (\forall x\in u.\exists y\in u.\varphi(x,y,\vec{z}))) \\
 & \text{for any } \Delta_0\text{-formula } \varphi(x,y,\vec{z}) \text{ such that } \mathrm{FV}(\varphi(x,y,\vec{z}))\subseteq\{x,y,\vec{z}\}
\end{array}$$

A *recursively Mahlo ordinal* is an ordinal M such that $\mathrm{L_M}$ is a recursively Mahlo set.

The basic construction of a model of the Mahlo universe is as for the super-universe: We define by recursion over α approximations $(\mathrm{V}^\alpha, \mathrm{T}^\alpha)$ of $([\![\mathrm{V}]\!], [\![\mathrm{T}]\!])$ by closing it systematically under the basic set constructions, and by adding suitable codes $\widehat{\mathrm{U}}_{f,g}$ to it.

The question is when to add codes $\widehat{\mathrm{U}}_{f,g}$ to the Mahlo universe. In the introduction rule for $\widehat{\mathrm{U}}_{f,g}$, f, g need to be total functions from families of sets internal to (V,T) to the same kind of families. But we only know this type of total functions, once the definition of $[\![\mathrm{V}]\!]$ is complete — so in the definition of V^α we cannot refer to these total functions. The trick to get around this problem is to add more elements $\widehat{\mathrm{U}}_{f,g}$ to the universe than are actually justified by the introduction rules. We only demand that f, g are total on the subuniverse $\mathrm{U}_{f,g}$ itself. So we take f, g to be arbitrary terms. Then we try to form a subuniverse of the current approximation V^{α} of $[\![\,\mathrm{V}\,]\!]$, which is closed under f and g. If we succeed, we add $\widehat{\mathrm{U}}_{f,g}$ to $[\![\,\mathrm{V}\,]\!]$ and define $[\![\,\mathrm{T_V}\,]\!](\widehat{\mathrm{U}}_{f,g})$ as the least such subuniverse.

If we assume now total functions f, g from families of sets internal to $(\mathrm{V}^{<\mathrm{M}}, \mathrm{T}^{<\mathrm{M}})$ into itself, then there is at least one subuniverse of $(\mathrm{V}^{<\mathrm{M}}, \mathrm{T}^{<\mathrm{M}})$ closed under f, g, namely $\mathrm{V}^{<\mathrm{M}}$ itself. Then we will use the fact that M is recursively Mahlo in order to find such a subuniverse inside some $(\mathrm{V}^{<\alpha}, \mathrm{T}^{<\alpha})$ for some $\alpha < \mathrm{M}$, and that therefore $\widehat{\mathrm{U}}_{f,g}$ is in $\mathrm{V}^{<\alpha+1} \subseteq \mathrm{V}^{<\mathrm{M}}$.

As for the super-universe, we simplify this construction slightly and don't search for the least subuniverse closed under f and g, but only for the least β such that V^{β} is closed under f and g.

More precisely we proceed as follows: As in the model for the super-universe, we omit in the terms of the model the arguments needed in $\widehat{\mathrm{N}}$, $\widehat{\Sigma}$, etc. (but not in $\widehat{\mathrm{f}}$, $\widehat{\mathrm{g}}$) for type-checking purposes only, so for instance

$$[\![\,\widehat{\mathrm{N}}_{\mathrm{V}}\,]\!]_{\rho} := [\![\,\widehat{\mathrm{N}}_{\mathrm{U},f,g}\,]\!]_{\rho} := \widehat{\mathrm{N}}\,.$$

We obtain essentially the same reduction rules for $\widehat{\mathrm{T}}_{\mathrm{U},f,g}$ as the reduction rules for $\widehat{\mathrm{T}}_{\mathrm{U},a,b}$ in case of the super-universe. Furthermore, we have the reductions

$$\begin{array}{rcl} \widehat{\mathrm{f}}_{f,g}(a,b) & \longrightarrow & f(a,b)\,, \\ \widehat{\mathrm{g}}_{f,g}(a,b,c) & \longrightarrow & g(a,b,c)\,. \end{array}$$

As for the super-universe, we form $(\mathrm{V}^{\alpha}, \mathrm{T}^{\alpha}) \in [\![\,\mathrm{Fam}\,]\!](\mathrm{Set})$ by closing it under the standard set constructions and by adding codes for suitable subuniverses as defined below. We will then interpret $[\![\,\mathrm{V}\,]\!] = \mathrm{V}^{<\mathrm{M}}$, $[\![\,\mathrm{T_V}\,]\!] = \mathrm{T}^{<\mathrm{M}}$.

Whereas the super-universe was closed under $\widehat{\mathrm{U}}_{a,b}$ for

$$\langle a, b\rangle \in [\![\,\mathrm{IFam}\,]\!](\mathrm{V}^{<\alpha}, \mathrm{T}^{<\alpha}),$$

we close the approximation V^{α} of the Mahlo universe under $\widehat{\mathrm{U}}_{f,g}$, where f, g form the two components of a function

$$[\![\,\mathrm{IFam}\,]\!](\mathrm{V}^{<\alpha}, \mathrm{T}^{<\alpha})\;[\![\rightarrow]\!]\;[\![\,\mathrm{IFam}\,]\!](\mathrm{V}^{<\alpha}, \mathrm{T}^{<\alpha})\,.$$

We call such a pair $\langle f, g\rangle$ an operator on families of sets internal to $(\mathrm{V}^{<\alpha}, \mathrm{T}^{<\alpha})$. More precisely, we introduce the following definition, where we first define a notion for the two components of an operator on families of sets internal to (A, B), and then define what it means to be an operator on families of such sets as the conjunction of the two:

DEFINITION 5.2. Let $\langle A, B\rangle \in [\![\,\mathrm{Fam}\,]\!](\mathrm{Set})$.

(a) $$[\![\,\mathrm{IFamOper}\,]\!]_0(A, B) := (x \in A)\;[\![\rightarrow]\!](B(x)\;[\![\rightarrow]\!]\;A)\;[\![\rightarrow]\!]A\,.$$

(b) Assume $f \in \mathrm{Flat}([\![\,\mathrm{IFamOper}\,]\!]_0(A, B))$. Then

$$[\![\,\mathrm{IFamOper}\,]\!]_1(A, B, f) := (x \in A) \\ [\![\rightarrow]\!](y \in (B(x)\,[\![\rightarrow]\!]\,A)) \\ [\![\rightarrow]\!]B(f(x, y)) \\ [\![\rightarrow]\!]A\,.$$

(c) The set of operators on families of sets internal to $\langle A, B\rangle$ is given as

$$[\![\,\mathrm{IFamOper}\,]\!](A, B) := \\ \{\langle\langle f, g\rangle, \langle f', g'\rangle\rangle \in ([\![\,\mathrm{Term}\,]\!]^2)^2 \mid \langle f, f'\rangle \in [\![\,\mathrm{IFamOper}\,]\!]_0(A, B) \wedge \\ \langle g, g'\rangle \in [\![\,\mathrm{IFamOper}\,]\!]_1(A, B, f)\}.$$

With this notation, the addition of $\widehat{\mathrm{U}}_{f,g}$ to V^α is defined as follows:

Assume $f, f', g, g' \in [\![\,\mathrm{Term}\,]\!]$. Assume $\alpha < \mathrm{M}$ and β such that $\beta + 1 < \alpha$ and such that $\mathrm{V}^{<\beta}$ is closed under all universe constructions. Assume that

$$\langle\langle f, g\rangle, \langle f', g'\rangle\rangle \in [\![\,\mathrm{IFamOper}\,]\!](\mathrm{V}^{<\beta}, \mathrm{T}^{<\beta}).$$

Let β be minimal that this property holds. Then

$$\langle\widehat{\mathrm{U}}_{f,g}, \widehat{\mathrm{U}}_{f',g'}\rangle \in \mathrm{V}^\alpha\,,$$

and we define

$$\mathrm{T}^\alpha(\widehat{\mathrm{U}}_{f,g}) := \mathrm{V}^{<\beta}\,.$$

The construction fulfils Assumption 3.2 with U replaced by V.

Now define

$$\begin{aligned} [\![\,\mathrm{U}\,]\!]_\rho &:= \mathrm{V}^{<\mathrm{M}}\,, \\ [\![\,\mathrm{T}_\mathrm{V}(c)\,]\!]_\rho &:= \mathrm{T}^{<\mathrm{M}}([\![\,c\,]\!]_\rho)\,, \\ [\![\,\mathrm{U}_{f,g}\,]\!]_\rho &:= \mathrm{T}^{<\mathrm{M}}(\widehat{\mathrm{U}}_{[\![\,f\,]\!]_\rho, [\![\,g\,]\!]_\rho}). \end{aligned}$$

where M is the least recursively Mahlo ordinal.

We note that $\mathrm{T}^\alpha(\widehat{\mathrm{U}}_{f,g})$ is not defined as a minimal set closed under universe constructions and under f and g. We could have done so by modifying the above definition and trying at stage α to define the least subset of V^α closed under f and g and the standard universe constructions. If this succeeds, this subset is independent of α (provided α is big enough so that this subset lies inside V^α) and we could have interpreted $\mathrm{T}^\alpha(\widehat{\mathrm{U}}_{f,g})$ as this set. This would interpret elimination rules for $\mathrm{U}_{f,g}$ as well. But since elimination rules for $\mathrm{U}_{f,g}$ don't add any proof theoretic strength, and such a model construction would be more complicated, we use the original approach.

We have to verify that $V^{<M}$ has the necessary closure properties of the Mahlo universe. Every recursively Mahlo ordinal is recursively inaccessible, therefore $V^{<M}$ will be closed under universe constructions as before. $V^{<M}$ is as well closed under $(f,g)\widehat{U}_{f,g}$: Assume $f, f', g, g' \in [\![\,\mathrm{Term}\,]\!]$. Assume

$$\begin{aligned}\langle f, f'\rangle &\in [\![\,(x : \mathrm{V}, y : \mathrm{T_V}(x) \to \mathrm{V}) \to \mathrm{V}\,]\!]\,,\\ \langle g, g'\rangle &\in [\![\,(x : \mathrm{V}, y : \mathrm{T_V}(x) \to \mathrm{V}, \mathrm{T_V}(z(x,y)) \to \mathrm{V}\,]\!]_{[z=f]}\,.\end{aligned}$$

This means

$$\langle\langle f, g\rangle, \langle f', g'\rangle\rangle \in [\![\,\mathrm{IFamOper}\,]\!]\big(\mathrm{V}^{<\mathrm{M}}, \mathrm{T}^{<\mathrm{M}}\big).$$

We need to find a $\kappa < \mathrm{M}$ such that $\mathrm{V}^{<\kappa}$ is closed under universe constructions and that the previous conditions hold with M replaced by κ. For being closed under universe constructions it suffices that κ is inaccessible.

We show the following statement $(*)$: For every $\beta < \mathrm{M}$ there exists a $\rho < \mathrm{M}$ such that ρ is admissible, $\beta < \rho$ and such that

$$\begin{aligned}&\langle\langle a, b\rangle, \langle a', b'\rangle\rangle \in [\![\,\mathrm{IFam}\,]\!]\big(\mathrm{V}^{<\beta}, \mathrm{T}^{<\beta}\big)\\ &\quad \to \langle\langle f(a,b), g(a,b)\rangle, \langle f'(a',b'), g'(a',b')\rangle\rangle \in [\![\,\mathrm{IFam}\,]\!](\mathrm{V}^{<\rho}, \mathrm{T}^{<\rho})\,.\end{aligned}$$

So assume $\beta < \mathrm{M}$ and

$$\langle\langle a, b\rangle, \langle a', b'\rangle\rangle \in [\![\,\mathrm{IFam}\,]\!]\big(\mathrm{V}^{<\beta}, \mathrm{T}^{<\beta}\big).$$

Then

$$\langle\langle f(a,b), g(a,b)\rangle, \langle f'(a,b), g'(a,b)\rangle\rangle \in [\![\,\mathrm{IFam}\,]\!]\big(\mathrm{V}^{<\mathrm{M}}, \mathrm{T}^{<\mathrm{M}}\big).$$

By Lemma 3.4 there exists a $\rho < \mathrm{M}$ such that

$$\langle\langle f(a,b), g(a,b)\rangle, \langle f'(a,b), g'(a,b)\rangle\rangle \in [\![\,\mathrm{IFam}\,]\!](\mathrm{V}^{<\rho}, \mathrm{T}^{<\rho})\,.$$

Furthermore, using the fact that every recursively Mahlo ordinal is recursively inaccessible, we can achieve $\beta < \rho$ admissible. This shows $(*)$.

The statement $(*)$ can be expressed as a Π_2-formula, i.e. a formula of the form

$$\forall\beta < \mathrm{M}.\exists\rho < \mathrm{M}.\varphi(\beta, \rho)\,,$$

where φ is Δ_0.

By M being recursively Mahlo we obtain the existence of an admissible $\kappa < \mathrm{M}$ such that

$$\forall\beta < \kappa.\exists\rho < \kappa.\varphi(\beta, \rho)\,.$$

Since ρ is an admissible $> \beta$, it follows that κ is recursively inaccessible, therefore $(\mathrm{V}^{<\kappa}, \mathrm{T}^{<\kappa})$ is closed under the basic set constructions. Furthermore if

$$\langle\langle a, b\rangle, \langle a', b'\rangle\rangle \in [\![\,\mathrm{IFam}\,]\!](\mathrm{V}^{<\kappa}, \mathrm{T}^{<\kappa})\,,$$

then by Lemma 3.4 there exists a $\beta < \kappa$ such that

$$\langle\langle a,b\rangle,\langle a',b'\rangle\rangle \in [\![\,\mathrm{IFam}\,]\!](\mathrm{V}^{<\beta},\mathrm{T}^{<\beta})\,.$$

Therefore there exists a $\rho < \kappa$ such that

$$\begin{aligned}\langle\langle f(a,b),g(a,b)\rangle,\langle f'(a',b'),g'(a',b')\rangle\rangle &\in [\![\,\mathrm{IFam}\,]\!](\mathrm{V}^{<\rho},\mathrm{T}^{<\rho})\\ &\subseteq [\![\,\mathrm{IFam}\,]\!](\mathrm{V}^{<\kappa},\mathrm{T}^{<\kappa})\,.\end{aligned}$$

This means that $\mathrm{V}^{<\kappa}$ is a candidate for the interpretation of $\langle\widehat{\mathrm{U}}_{f,g},\widehat{\mathrm{U}}_{f',g'}\rangle$, and therefore

$$\langle\widehat{\mathrm{U}}_{f,g},\widehat{\mathrm{U}}_{f',g'}\rangle \in \mathrm{V}^{\kappa+2} \subseteq \mathrm{V}^{<\mathrm{M}}\,.$$

That $[\![\,\mathrm{U}_{f,g}\,]\!]_\rho$ has the closure properties needed in order to interpret $\mathrm{U}_{f,g}$ follows by the construction.

The remaining steps are as for the model of simple $\mathrm{ML_WU}$, i.e. we need $< \omega$ many admissibles in order to interpret the basic set constructions on top of $\mathrm{V},\mathrm{T_V}$. Therefore the type theory can be interpreted in Kripke Platek set theory plus one recursively Mahlo ordinal and finitely many admissibles (i.e. for Meta-each n we have n admissibles) above it. So we have given the essence of a proof of the following theorem:

THEOREM 5.3. (a) *We can model* $\mathrm{ML_W}$ + (Mahlo) *in* $(\mathrm{KPM})^+$.

(b) $|\mathrm{ML_W} + (\mathrm{Mahlo})| \leq |(\mathrm{KPM})^+|$.

(c) *The previous statements hold as well if we replace intensional by extensional equality.*

By [35] we obtain the other direction and obtain therefore

THEOREM 5.4. (a) $|\mathrm{ML_W} + (\mathrm{Mahlo})| = |(\mathrm{KPM})^+|$.

(b) *The previous statements hold as well if we replace intensional by extensional equality.*

REFERENCES

[1] P. Aczel, *The strength of Martin-Löf's intuitionistic type theory with one universe*, ***Proceedings of the Symposium on Mathematical Logic (Oulu 1974)*** (S. Miettinen and J. Väänänen, editors), Report No. 2, Department of Philosophy, University of Helsinki, 1977.

[2] T. Arai, *Proof theory of theories of ordinals I: Reflecting ordinals*, preprint, 1996.

[3] ———, *Proof theory of theories of ordinals II:* Σ_1 *stability*, preprint, 1997.

[4] ———, *Proof theory of theories of ordinals III:* Π_1 *collection*, preprint, 1997.

[5] ———, *Ordinal diagrams for* Π_3*-reflection*, ***The Journal of Symbolic Logic***, vol. 65 (2000), no. 3, pp. 1375–1394.

[6] ———, *Ordinal diagrams for recursively Mahlo universes*, ***Archive for Mathematical Logic***, vol. 39 (2000), no. 5, pp. 353–391.

[7] ———, *Proof theory for theories of ordinals. I. Recursively Mahlo ordinals*, ***Annals of Pure and Applied Logic***, vol. 122 (2003), no. 1-3, pp. 1–85.

[8] ———, *Proof theory for theories of ordinals. II.* Π_3*-reflection*, ***Annals of Pure and Applied Logic***, vol. 129 (2004), no. 1-3, pp. 39–92.

[9] J. BARWISE, ***Admissible Sets and Structures. An Approach to Definability Theory***, Omega-series, Springer, 1975.

[10] M. BENKE, P. DYBJER, and P. JANSSON, *Universes for generic programs and proofs in dependent type theory*, ***Nordic Journal of Computing***, vol. 10 (2003), no. 4, pp. 265–289.

[11] P. DYBJER, *A general formulation of simultaneous inductive-recursive definitions in type theory*, ***The Journal of Symbolic Logic***, vol. 65 (2000), no. 2, pp. 525–549.

[12] P. DYBJER and A. SETZER, *Finite axiomatizations of inductive and inductive-recursive definitions*, ***Workshop on Generic Programming, Marstrand, Sweden, 18 June 1998***, `http://www.cs.ruu.nl/people/johanj/programme_wgp98.html`, 1998.

[13] ———, *A finite axiomatization of inductive-recursive definitions*, ***Typed Lambda Calculi and Applications*** (J.-Y. Girard, editor), Lecture Notes in Computer Science, vol. 1581, Springer, April 1999, pp. 129–146.

[14] ———, *Indexed induction-recursion*, ***Proof Theory in Computer Science*** (R. Kahle, P. Schroeder-Heister, and R. Stärk, editors), Lecture Notes in Computer Science, vol. 2183, Springer, 2001, pp. 93–113.

[15] ———, *Induction-recursion and initial algebras*, ***Annals of Pure and Applied Logic***, vol. 124 (2003), pp. 1–47.

[16] ———, *Indexed induction-recursion*, ***Journal of Logic and Algebraic Programming***, vol. 66 (2006), no. 1, pp. 1–49.

[17] E. GRIFFOR and M. RATHJEN, *The strength of some Martin-Löf type theories*, ***Archive for Mathematical Logic***, vol. 33 (1994), pp. 347–385.

[18] M. HOFMANN, *Syntax and semantics of dependent types*, ***Semantics and Logics of Computation*** (A. M. Pitts and P. Dybjer, editors), Cambridge University Press, Cambridge, 1997, pp. 79–130.

[19] G. JÄGER, ***Theories for Admissible Sets: A Unifying Approach to Proof Theory***, Bibliopolis, Naples, 1986.

[20] P. MARTIN-LÖF, ***Intuitionistic Type Theory***, Bibliopolis, Naples, 1984.

[21] B. NORDSTRÖM, K. PETERSSON, and J. SMITH, ***Programming in Martin-Löf's Type Theory. an Introduction***, Oxford University-Press, Oxford, 1990.

[22] ———, *Martin-löf's type theory*, ***Handbook of Logic in Computer Science, Vol. 5*** (S. Abramsky, D. M. Gabbay, and T. S. E. Maibaum, editors), Oxford University Press, 2000, pp. 1–37.

[23] E. PALMGREN, *On universes in type theory*, ***Twenty Five Years of Constructive Type Theory*** (G. Sambin and J. Smith, editors), Oxford University Press, Oxford, 1998, pp. 191–204.

[24] W. POHLERS, *Subsystems of set theory and second order number theory*, ***Handbook of Proof Theory*** (S. R. Buss, editor), Elsevier, 1998, pp. 209–335.

[25] A. RANTA, ***Type-Theoretical Grammar***, Clarendon Press, Oxford, 1995.

[26] M. RATHJEN, *Proof-theoretical analysis of KPM*, ***Archive for Mathematical Logic***, vol. 30 (1991), pp. 377–403.

[27] ———, *Proof theory of reflection*, ***Annals of Pure and Applied Logic***, vol. 68 (1994), pp. 181–224.

[28] ———, *Recent advances in ordinal analysis: Π^1_2-CA and related systems*, ***The Bulletin of Symbolic Logic***, vol. 1 (1995), pp. 468–485.

[29] ———, *An ordinal analysis of parameter free Π^1_2-comprehension*, ***Archive for Mathematical Logic***, vol. 44 (2005), no. 3, pp. 263–362.

[30] A. SETZER, ***Proof Theoretical Strength of Martin-Löf Type Theory with W-Type and One Universe***, Ph.D. thesis, Universität München, 1993, Available at `http://www.cs.swan.ac.uk/~csetzer`.

[31] ———, *A model for a type theory with Mahlo Universe*, 10pp. Draft, available at `http://www.cs.swan.ac.uk/~/articles/uppermahlo.pdf`, 1996.

[32] ———, *An upper bound for the proof theoretical strength of Martin-Löf Type Theory with W-type and one Universe*, 33pp. Draft, available at `http://www.cs.swan.ac.uk/~/articles/1papdiss.dvi` or `.ps` or `.pdf`, 1996.

[33] ———, *An introduction to well-ordering proofs in Martin-Löf's type theory*, ***Twenty-Five Years of Constructive Type Theory*** (G. Sambin and J. Smith, editors), Clarendon Press, Oxford, 1998, pp. 245–263.

[34] ———, *Well-ordering proofs for Martin-Löf type theory*, ***Annals of Pure and Applied Logic***, vol. 92 (1998), pp. 113–159.

[35] ———, *Extending Martin-Löf type theory by one Mahlo-universe*, ***Archive for Mathematical Logic***, vol. 39 (2000), pp. 155–181.

[36] ———, *Proof theory of Martin-Löf type theory — an overview*, ***Mathematiques et Sciences Humaines***, vol. 42 année, n°165 (2004), pp. 59–99.

[37] ———, *Universes in type theory part II – Π_3-reflection*, submitted, 2006.

[38] S. G. Simpson, ***Subsystems of Second-Order Arithmetic***, Springer, 1999.

[39] A. Troelstra and D. van Dalen, ***Constructivism in Mathematics. An Introduction, Vol. II***, North-Holland, Amsterdam, 1988.

DEPARTMENT OF COMPUTER SCIENCE
UNIVERSITY OF WALES SWANSEA
SINGLETON PARK, SWANSEA SA2 8PP, UK
E-mail: a.g.setzer@swan.ac.uk
URL: http://www.cs.swan.ac.uk/~csetzer/

HAUSDORFF-DIMENSION AND WEAK TRUTH-TABLE REDUCIBILITY

FRANK STEPHAN

Abstract. It is shown that in some relativized world B there is a set A such that the effective Hausdorff-dimension of the wtt-degree and of the lower wtt-cone of A are both strictly between 0 and 1.

Hausdorff [2] introduced a general theory of dimension which also covered the case of intermediate fractal dimensions. Starting from Hausdorff's original definition, one introduced an effective variant of the dimension [1, 5, 12, 14]. Reimann [9] investigated the important question which type of degrees can have intermediate Hausdorff-dimension properly between 0 and 1. Reimann and Terwijn [9, Theorem 3.10] found positive results for many-one and bounded truth-table reducibilities. Later Reimann and Slaman [11] announced an extension of the result to truth-table reducibility. In the present work, an oracle B defining a relativized world is constructed together with a set A such that the wtt-degree of A in this relativized world has an intermediate Hausdorff-dimension between $\frac{1}{4}$ and $\frac{1}{2}$. The same applies for the lower wtt-cone of A. Before proving the main result, some more details are provided and basic notation is introduced.

Notation. The reader should refer to the books of Li and Vitányi [4], Odifreddi [8] and Soare [13] for unexplained facts and notations from recursion theory and Kolmogorov complexity. The following conventions are a bit unusual but simplify the presentation of the proofs.

It is convenient to identify the string x with the number n such that $n+1$ has the binary representation $1x$, so the string 11 stands for 6 (since $6 + 1$ has the binary representation 111) and 000 stands for 7 (since 1000 represents $7 + 1$). This permits to extend the notion of Kolmogorov complexity to numbers. Given a number x, $|x|$ denotes the length of the string identified with x. That is, $|6| = 2$, $|7| = 3$ and, in general, $|x|$ is the down-rounded value of the dual logarithm of $x + 1$.

F. Stephan is supported in part by NUS grant number R252–000–212–112.

Logic Colloquium '04
Edited by A. Andretta, K. Kearnes, and D. Zambella
Lecture Notes in Logic, 29

Also, given a set A and a finite set $\{a_1, a_2, \dots, a_n\}$ of natural numbers with $a_1 < a_2 < \cdots < a_n$, let the notion $A(\{a_1, a_2, \dots, a_n\})$ stand for the string $A(a_1)A(a_2)\dots A(a_n)$ where $A(x) = 1$ if $x \in A$ and $A(x) = 0$ if $x \notin A$.

Hausdorff [2] formulated a measure-theoretic approach to dimension. His notion of dimension turned out to be invariant under bi-Lipschitz transformations. This approach permits to study classes of noninteger dimension. It is a natural question how such classes look like.

A further characteristic of Hausdorff-dimension is countable stability: the dimension of a countable union of sets is the supremum of the dimensions of each member of the union. When effectivizing the dimension, it does no longer measure the dimension of the class but the effective dimension of its most unapproximable members. That is, a class is as difficult as its most difficult members and mathematically this means that effective Hausdorff-dimension has uncountable stability. So every class S of sets of natural numbers satisfies $\dim_H(S) = \sup\{\dim_H(\{A\}) : A \in S\}$. Since the classical Hausdorff-dimension of a class is a lower bound for its effective Hausdorff-dimension, there are be singletons having positive effective Hausdorff-dimension in every class of positive classical Hausdorff-dimension. There are several characterizations of the effective Hausdorff-dimension of a singleton $\{A\}$ which, for convenience, is from now on just written as $\dim_H(A)$. An elegant characterization is the one in terms of the Kolmogorov complexity C [1, 5, 12, 14] which is here taken as a definition.

DEFINITION 1. The effective Hausdorff-dimension of a set A is given as

$$\dim_H(A) = \liminf_{n\to\infty} \frac{1}{n} \cdot C(A(0)A(1)\dots A(n-1)).$$

Since the present work deals only with the effective version and so there is no danger of confusing the terms, this notion is just called "Hausdorff-dimension" from now on.

Recall that the Kolmogorov complexity $C(x)$ of a string x is the length of the shortest program generating x. These programs are measured with respect of a universal machine and the actual value of the Kolmogorov complexity depends only up to a constant on this machine. It is convenient to define the machine such that it is universal for any oracle E. So one chooses U to be a unary universal partial-recursive function accessing an oracle with the property that for every partial-recursive unary function V with oracle access there is a constant c such that

$$\forall p\, \exists q\, (|q| \le |p| + c \wedge U^E(q) = V^E(p))$$

where the equality "$U^E(q) = V^E(p)$" means that either both sides are undefined or both sides are defined and equal. Now one defines the relativized

Kolmogorov complexity as

$$C^E(x) = \min\{|p| : p \in \{0,1\}^* \wedge U^E(p)\downarrow = x\}$$

for any oracle E where one writes simply $C(x)$ and $U(p)$ instead of $C^{\emptyset}(x)$ and $U^{\emptyset}(p)$ for the case where the oracle is the empty set. Kolmogorov complexity can be approximated from above and one assumes that $U_s^E(p)$ does not query the oracle at places $a \geq s$. Therefore, one can write $U_s^{\sigma}(p)$ for partial functions σ with domain $\{t : t < s\}$.

The notion of relativized Kolmogorov complexity also permits to introduce a relativized version of Hausdorff-dimension which follows the straightforward definition

$$\dim_H^E(A) = \liminf_{n\to\infty} \frac{1}{n} \cdot C^E(A(0)A(1)\ldots A(n-1))$$

and is extended to classes using stability. These relativized notions will be the major tool for the main result of the paper that in some relativized world there is a wtt-degree of intermediate Hausdorff-dimension.

Recursively enumerable sets have Hausdorff-dimension 0. The reason is that one can compute $A(0)A(1)\ldots A(n-1)$ just from n and from the number m of elements which A has below n: Having the numbers m and n, one can enumerate A until a stage s where A_s has m elements below n is reached and then output $A_s(0)A_s(1)\ldots A_s(n-1)$. This output is then correct and the Kolmogorov complexity of this string is bounded by $c + 2\log(n)$ for some constant c.

On the other hand, Martin-Löf gave a measure-theoretic criterion for a set R to be algorithmically random [6]. The class of Martin-Löf random sets has measure 1 and every Martin-Loef random set has Hausdorff dimension 1. Then, for every Martin-Löf random set R, $\dim_H(R \oplus \emptyset) = \frac{1}{2}$, so there are sets of intermediate Hausdorff-dimension. Reimann [9, Theorem 3.10] showed that there is a whole many-one degree where all members have intermediate dimension. In the next example it is shown how this can be done.

Example 2. [9] There is a set A such that $\dim_H(A) > 0$ and $\dim_H(B) < 1$ for all $B \leq_m A$.

Proof. Let R be a Martin-Löf random set and define

$$A = \{x \in R : C(x) \geq |x|\}.$$

Now it is shown that every $B \leq_m A$ has intermediate Hausdorff-dimension. First this is done for A itself. There is a constant c such that for almost all n, $\{x \in \{0,1\}^n : C(x) < n\}$ has more than 2^{n-c} elements and more than 2^{n-c} nonelements. Thus, given any n and the number m of $x < n$ with $C(x) < |x|$, one can enumerate these m numbers and then knows that $A(x) = 0$ for all these $x < n$ with $C(x) < |x|$. This gives

$$C(A(0)A(1)\ldots A(n-1)) \leq n - m + 3\log(n)$$

for all sufficiently large n where m depends on n as defined above. In particular, $\dim_H(A) \leq 1-2^{-c-1}$. On the other hand, one can use that the $n-m$ numbers $x < n$ not enumerated satisfy $R(x) = A(x)$ in order to get that

$$C(R(0)R(1)\dots R(n-1)) < C(A(0)A(1)\dots A(n-1)) + m + 3\log(n)$$

for all sufficiently large n. One can conclude that

$$C(A(0)A(1)\dots A(n-1)) \geq n - m - 5\log(n)$$

for sufficiently large n and using that $n - m \geq n \cdot 2^{-1-c}$, one gets a lower bound for the Hausdorff-dimension of A: $\dim_H(A) \geq 2^{-1-c}$.

Of course the construction of a single set A of intermediate dimension could be done more easily by considering sets like $R \oplus \emptyset$, but the special thing about this construction is that no $B \leq_m A$ has Hausdorff-dimension 1. So assume that $B \leq_m A$ via f. Either one has for infinitely many n that $f(x) \in \{f(0), f(1), \dots, f(x-1)\}$ for at least a fourth of all $x < n$ resulting in $\dim_H(B) \leq \frac{3}{4}$. Or one has for almost all n that $|f(x)| \geq |x| - 1$ for at least one half of the $x < n$. The latter then implies that there is a constant c' with $C(f(x)) < |f(x)|$ for at least $2^{-c'} \cdot m$ of the m numbers $x < n$ with $|f(x)| \geq |x| - 1$. From this condition one can conclude that at least $2^{-c'-1} \cdot n$ of the numbers $x < n$ satisfy $C(f(x)) < |f(x)|$ and therefore $B(x) = A(f(x)) = 0$. From this, one can then conclude in the same way as above that $\dim_H(B) < 1$. ⊣

Actually, Reimann and Terwijn [9, Theorem 3.10] obtained a better result via concepts not introduced in this paper: first, the whole many-one degree has intermediate Hausdorff dimension; second, this Hausdorff dimension coincides with a prechosen rational number. Reimann and Terwijn also generalized their result to bounded truth-table degrees. Reimann and Slaman [11] announced that there is even a truth-table degree of Hausdorff-dimension $\frac{1}{2}$. Reimann asked whether the same result holds for weak truth-table reducibility and Turing reducibility. So he posed the following open problem.

QUESTION 3. [9] Is there a set A such that $0 < \dim_H(\{B : B \leq_T A\}) < 1$?

Reimann and Slaman [9, 10] obtained the following result on the way to attack this question: There is a set D such that no Martin-Löf random set is Turing reducible to D but there is a nontrivial lower bound on the Kolmogorov complexity of D in the sense that there is a computable function f with $C(D(0)D(1)\dots D(f(n)-1)) > n$ for all n. Kjos-Hanssen, Merkle and Stephan [3] obtained the same result independently. At the time of submitting this paper, Question 3 was still unsolved for wtt-reducibility and the present work obtains a partial result is obtained by settling it for a relativized world by proving that there is a lower wtt-cone of intermediate Hausdorff-dimension in some relativized world. Indeed, when revising his paper, the author learned that Nies and Reimann [7] obtained an unrelativized positive solution to this question for wtt-reducibility.

THEOREM 4. *There is a set A and an oracle B such that*

$$\tfrac{1}{4} \leq \dim_H^B \left(\{D : D \leq_{wtt}^B A\}\right) \leq \tfrac{1}{2}.$$

PROOF. Proposition 5 below states that there are a $\{0,1\}$-valued partial-recursive function and a partition of the complement of its domain into intervals $I_0, I_1, \ldots$ satisfying four properties (a), (b), (c) and (d) listed in the proposition. Proposition 6 below states that one can choose sets A, B such that B extends ψ and $\frac{1}{4} \leq \dim_H^B(\{D : D \leq_{wtt}^B A\}) \leq \frac{1}{2}$. ⊣

Recall that B extends ψ iff the characteristic function of B coincides with ψ on its domain: $\forall x\, (\psi(x)\downarrow \Rightarrow \psi(x) = B(x))$. Furthermore, let use$(\psi, p)$ denote the largest supremum of all numbers s such that there is some B extending ψ for which $U^B(p)$ converges in exactly s steps of computation and therefore queries B only below s. Note that use(ψ, p) is only infinite iff there is for every t an extension B of ψ such that $U^B(p)$ is defined but needs more than t steps. Furthermore, one assumes without loss of generality for every oracle B and every $p \in \{0,1\}^*$ that $U^B(p)$ needs always at least $|p|$ computation steps before termination, that for every $l \geq 2$ there is $q \in \{0,1\}^l$ such that $U^B(q)$ terminates and that $U^B(\lambda)$, $U^B(0)$, $U^B(1)$ are undefined for all oracles B.

PROPOSITION 5. *There are a partial-recursive $\{0,1\}$-valued function ψ and a partition of its complement in intervals $I_0, I_1, \ldots$ such that the following conditions hold*:

(a) *each I_n is an interval with* $\max(I_n) < \min(I_{n+1})$ *and* $2^{|I_n|} < |I_{n+1}|$;
(b) $\min(I_n) = \sup\{use(\psi, p) : |p| < |I_n|/2\}$;
(c) *if B extends ψ then* $\dim_H^B(A) \geq \frac{1}{4} \cdot \dim_H(A)$ *for any set A*;
(d) *if B extends ψ and $D(\{1, \ldots, |I_n|\}) = B(I_n)$ for infinitely many n then* $\dim_H^B(D) \leq \frac{1}{2}$.

PROOF. The partial recursive $\{0,1\}$-valued function ψ is defined in stages so that (a), (b), (c) and (d) are satisfied. For each stage s there is a number n such that $I_{0,s}, I_{1,s}, \ldots, I_{n,s}$ are defined and each element of $\{0, \ldots, s\}$ is either in the domain of ψ_s or in one interval $I_{m,s}$ with $m \leq n$ but not in both. In stage s, an interval might be shifted or even be made undefined but in most stages it just remains where it is.

At stage 0, ψ_0 is everywhere undefined. Let $I_{m,s}$ is the interval I_m at stage s whenever it exists at that stage, in particular $I_{0,0} = \{0, 1, 2, 3\}$ and $I_{m,0}$ does not exist for $m > 0$.

At stage $s + 1$, take the first of the following cases to apply. Let n be the maximal number such that the interval I_n exists after stage s.

- There are an extension τ of ψ_s with domain $\{0, 1, \ldots, s\}$ and a $p \in \{0, 1\}^*$ with $|p| < |I_n|/2$ at stage s such that $U^{\tau}_{s+1}(p)$ but not $U^{\tau}_{s}(p)$ is defined.
If there are several such (p, τ) then choose the pair with the shortest p; otherwise just take the unique (p, τ). Let m be the minimal number such that $|p| < |I_{m,s}|$. Now update
$$\psi_{s+1}(x) = \begin{cases} \psi_s(x) & \text{if defined;} \\ \tau(x) & \text{if } \min(I_{m,s}) \leq x \leq s; \\ \uparrow & \text{otherwise.} \end{cases}$$
The intervals I_k with $k < m$ remain unchanged ($I_{k,s+1} = I_{k,s}$), the interval I_m is moved beyond s ($I_{m,s+1} = \{s+1, s+2, \ldots, s+|I_{m,s}|\}$) and the intervals I_k with $k > m$ are undefined after stage $s + 1$.
- $s + 1 \in I_n$ after stage s.
Then everything remains unchanged including ψ and the intervals.
- Otherwise, that is, no interval is shifted and $s = \max(I_n)$ after stage s. Then I_{n+1} is newly defined as $\{s+1, s+2, \ldots, s+2^{s+2}\}$ at stage $s + 1$. $I_{m,s+1} = I_{m,s}$ for all $m \leq n$ and $\psi_{s+1} = \psi_s$.

It is easy to see that every stage of the construction is finite and recursive. Now it will be shown by induction that every interval becomes only finitely often undefined and moves only finitely often. As I_n becomes only undefined at stage s if some I_m with $m < n$ moves at stage s, it is sufficient to show that each I_n moves only finitely often and is almost always defined.

The interval I_0 never moves as $|I_0| = 4$ and $U^B(p)$ is undefined for all p with $|p| < 2$. So the assumption is true for $n = 0$. Now assume that it is proven for n and that one has to prove it for $n + 1$. Let $s = \max(I_n) + 1$ where I_n denotes in this phrase the final position of this interval. Then no interval with $m \leq n$ does move at stage s or later since otherwise $s \leq \min(I_n)$ in contradiction to the choice of s. So I_{n+1} is newly defined at stage $s + 1$ and $I_{n+1,s+1} = \{s+1, s+2, \ldots, s+2^{s+2}\}$. From then on, I_{n+1} will never become undefined again but it will only be moved. In particular, I_{n+1} will never change its length again. Now for every $\eta \in \{0, 1\}^{\max(I_n)+1}$ and every p with $|p| < |I_{n+1}|$ it happens at most once that some $t > s$ that I_{n+1} will be moved because of an extension τ of both, η and ψ_t, with $U^{\tau}(p)$ halting exactly at t steps. Thus one can conclude that I_{n+1} moves at most $2^{\max(I_n)+|I_{n+1}|/2}$ times.

This concludes the proof of the inductive step and now the properties (a), (b), (c) and (d) are verified.

(a): $\max(I_n) < \min(I_{n+1})$ and $2^{|I_n|} < |I_{n+1}|$.

The first condition is easy to see. The second one follows from the above verification. If $s = \max(I_n)$ then $I_{n+1,s+1} = \{s+1, s+2, \ldots, s+2^{s+2}\}$ and $|I_{n+1}| = 2^{s+2}$. Using that $|I_n| \leq s + 1$ completes this part of the proof.

(b): $\min(I_n) = \sup\{\text{use}(\psi, p) : |p| < |I_n|/2\}$ for almost all n.

Let I_n be an interval that is defined, for the last time, at some stage s and has length 2^{s+2}. If n is sufficiently large, so is s and there is a program p which computes $B(s)$. This program needs time $O(s)$ and thus the interval I_n is moved at least once. There is a stage t such that I_n is moved at stage t for the last time, that is, $I_{n,t} \neq I_{n,t+1} = I_{n,t'}$ for all $t' > t$. Thus there is a τ extending ψ_t and a p such that $|p| < 2^{s+2}/2$ and $U^B(p)$ halts in exactly $t+1$ steps. As the intervals I_m with $m < n$ have all reached their final positions at s, the string τ is not only consistent with ψ_t and ψ_{t+1} but also with the final function ψ. Thus some extension B of τ is also an extension of ψ and use$(\psi, p) = t+1$. Furthermore, there is no $q \in \{0,1\}^*$ and no extension B of ψ such that $|q| < |I_n|/2$ and $U^B(q)$ halts after exactly $t'+1$ steps with $t' > t$: otherwise the interval I_n would be moved again because the program q together with the prefix of length $t'+1$ of B would qualify in the search of stage $t'+1$ in contrary to the assumption that t is the last stage where this interval I_n moves.

Although this completes the proof of (b), a short note should be added. This note says that for some constant c and all n, $C(\min(I_n)) \leq |I_n|/2 + 2\log(|I_n|) + c$. To see this, consider the s such that $|I_n| = 2^{s+2}$ and the p of length less than $|I_n|/2$ such that there is an extension B of ψ with $U^B(p)$ halting in exactly $\min(I_n)$ steps. If a place queried by $U^B(p)$ is larger than s then ψ is defined at that place and the simulation of ψ provides the answer. If it is smaller, one might look up the answer in a tabled string containing $s+1$ bits. Thus one can compute $\min(I_n)$ from $0^s 1 B(0)B(1)\ldots B(s)p$. As $|I_n| = 2^{s+2}$, this additional information is logarithmic in s and one could use some additional bits in order to make the code prefix-free and having the length $2^{s+1} + 2s + 3$, so one would use the code $0^s 1 B(0)B(1)\ldots B(s)0^{2^{s+1}-|p|}1p$. So there is a constant c such that both, the prefix-free and the plain Kolmogorov complexity of $\min(I_n)$ is bounded by $|I_n|/2 + 2\log(|I_n|) + c$ for all n where this constant c is so large that it also covers the case $n = 0$ where $\min(I_n)$ where the interval is not moved and therefore the p with $|p| < |I_n|/2$ and $\min(I_n) = \text{use}(\psi, p)$ does not exist.

(c): If B extends ψ then $\dim_H^B(A) \geq \frac{1}{4} \cdot \dim_H(A)$.

Let n be any number and p be the shortest program with $U^B(p) = A(\{0,1,\ldots,n\})$. Let m be the largest number such that $U^B(p)$ queries B at some element of I_m. Due to (b), $|p| \geq |I_m|/2$. Let s be the number with $|I_m| = 2^{s+2}$, that is, $s+1$ is the stage where I_m is newly defined for the last time; from then on I_m is only moved and never becomes undefined again. Now one can compute $U^B(p)$ from the inputs $p, s, B(\{0,1,\ldots,s\}), B(I_m)$ and $\min(I_m)$. Let q be the prefix-free coding for $\min(I_m)$ and s from the end of part (b) of length $2^{s+1}+2s+3$. Then $qB(\{0,1,\ldots,s\})B(I_m)p$ is a description for $U^B(p)$ in some machine (not necessarily U) of length $3 \cdot 2^{s+1} + 3s + 4 + |p|$. Using that $2^{s+1} \leq |p|$, one gets that $C(A(\{0,1,\ldots,n\})) \leq 4|p| + 4\log(|p|) + c' = 4(C^B(A(\{0,1,\ldots,n\})) + \log(C^B(A(\{0,1,\ldots,n\})))) + c'$ for some constant c'

independent of n, m, p. It follows that $\dim_H(A) \leq 4 \dim_H^B(A)$ and $\dim_H^B(A) \geq \frac{1}{4} \cdot \dim_H(A)$.

(d): If B extends ψ and $D(\{1, \dots, |I_n|\}) = B(I_n)$ for infinitely many n then $\dim_H^B(D) \leq \frac{1}{2}$.

Recall that $D(\{1, \dots, |I_n|\}) = B(I_n)$ means that $D(1) = B(\min(I_n))$, $D(2) = B(\min(I_n) + 1)$, …, $D(|I_n|) = B(\max(I_n))$. By (b) there is a constant c such that $C(\min(I_n)) \leq |I_n|/2 + 2\log(|I_n|) + c$ for all n. Then $D(\{1, \dots, |I_n|\})$ can be computed relative to B by reading out the oracle B on the interval I_n. Recall that the description for $\min(I_n)$ also revealed s and therefore $|I_n| = 2^{s+2}$. Thus, up to an additive constant, the relation $C^B(D(\{1, \dots, |I_n|\})) \leq |I_n|/2 + 2\log(|I_n|)$ holds for infinitely many n and so the Hausdorff-dimension of D relative to B is at most $\frac{1}{2}$. ⊣

PROPOSITION 6. *Assume that a partial-recursive $\{0, 1\}$-valued function ψ and a partition of its complement into intervals $I_0, I_1, \dots$ satisfy the following four conditions from Proposition 5:*

(a) *each I_n is an interval with $\max(I_n) < \min(I_{n+1})$ and $2^{|I_n|} < |I_{n+1}|$;*
(b) $\min(I_n) = \sup\{use(\psi, p) : |p| < |I_n|/2\}$;
(c) *if $\tilde{B}$ extends ψ then $\dim_H^{\tilde{B}}(\tilde{A}) \geq \frac{1}{4} \cdot \dim_H(\tilde{A})$ for any set $\tilde{A}$;*
(d) *if $\tilde{B}$ extends ψ and $D(\{1, \dots, |I_n|\}) = \tilde{B}(I_n)$ for infinitely many n then $\dim_H^{\tilde{B}}(D) \leq \frac{1}{2}$.*

Then there are sets A, B such that B extends ψ and the lower wtt-cone of A relative to B has nonintegral dimension: $\frac{1}{4} \leq \dim_H^B(\{D : D \leq_{wtt}^B A\}) \leq \frac{1}{2}$.

PROOF. The sets A and B are constructed by a finite extension method. The set A will be an infinite branch of a recursive tree T where T is chosen such that every infinite branch of T has Hausdorff-dimension 1. The tree T is constructed as some type of complement of the class O of a universal Martin-Löf test; O has Lebesgue measure less than one half, O is effectively open and O contains all sets of Hausdorff-dimension less than 1. Indeed $O = \cup_s \vartheta_s \cdot \{0, 1\}^\infty$ for a recursive enumeration $\vartheta_0, \vartheta_1, \dots$ of binary strings with $\sum_s 2^{-|\vartheta_s|} < \frac{1}{2}$ and $\tau \in T$ iff there is no $s \leq |\tau|$ such that τ extends ϑ_s.

A itself will be build by a finite extension method inside T. B will be chosen by inductively by constructing inductively extensions θ_e of ψ which are defined on the domain of ψ plus finitely many intervals I_n. These extensions θ_e are still partial-recursive, but the choice of the extensions is not. At the beginning, one starts with $\sigma_0 = \lambda$ as the empty string and θ_0 being the partial-recursive function ψ. Then all wtt-reductions are addressed inductively, note that every wtt-reduction has infinitely many indices.

In stage e, the e-th wtt-reduction is considered. The construction follows the first case of the three below which applies.

Case 1: There is an extension $\tilde{A}$ of σ_e inside T and $\tilde{B}$ of θ_e and an x such that $\varphi_e^{\tilde{B}}(x; \tilde{A})$ is undefined because the computation of the query bound fails.

Then one chooses a program p for the universal Turing machine U such that $U^E(p)$ computes the bound of the e-th wtt-reduction on input x relative to an oracle E whenever this bound is defined. By condition (b), for all intervals I_n with $|I_n| > 2 \cdot |p|$ the convergence of the computation $U^E(p)$ does not depend on the choice of the values of E on I_n since the computation either converges within $\min(I_n)$ computation steps or diverges.

Let θ_{e+1} be the extension of θ_e which takes the value $\tilde{B}(z)$ whenever z is in the domain of θ_e or z is in an interval I_n with $|I_n| \leq 2 \cdot |p|$. θ_{e+1} is undefined on all other inputs z. The string σ_{e+1} is just σ_e as the oracle $\tilde{A}$ was never queried.

Case 2: There is an extension $\tilde{A}$ of σ_e inside T and $\tilde{B}$ of θ_e and an x such that $\varphi_e^{\tilde{B}}(x; \tilde{A})$ is undefined although the computation of the query bound terminates.

Let r be the query bound. Then there is a program p that codes r, e and the oracle answers of $\tilde{A}$ up to r such that $U^E(p) = \varphi_e^E(x; \tilde{A}(0)\tilde{A}(1)\ldots\tilde{A}(r))$ whenever E is an oracle extending θ_e. Again one can argue using U as above and justify the following extension.

Let θ_{e+1} be the extension of θ_e which takes the value $\tilde{B}(z)$ whenever z is in the domain of θ_e or z is in an interval I_n with $|I_n| \leq 2 \cdot |p|$. θ_{e+1} is undefined on all other inputs z. The string σ_{e+1} is the prefix of $\tilde{A}$ of length $|\sigma_e| + r$.

Case 3: For every extension $\tilde{A}$ of σ_e inside T and $\tilde{B}$ of θ_e and every x the wtt-reduction $\varphi_e^{\tilde{B}}(x; \tilde{A})$ terminates and outputs some value.

Now one can define the reduction Θ based on φ_e as follows: $\Theta^{E,F}(x)$ takes $\varphi_e^E(x; F)$ in the case that this reduction terminates and 0 in the case that E does not extend θ_e or F does not extend σ_e or F is not an infinite branch of T; if both cases apply then the output follows that case which is determined first to apply. Since Θ is defined for all oracles E, F, Θ is actually a truth-table reduction. There is a recursive function $\tilde{p}$ mapping any input k to a program of size up to $\log(k) + c$ for some constant c independent of k such that $U(\tilde{p}(k))$ computes the use of the tt-reduction when requested to compute $\Theta^{E,F}(0)\ldots\Theta^{E,F}(k)$ and needs also this use as computation time. Thus, by the construction of the intervals I_n, whenever $|I_n| > 2 \cdot |\tilde{p}(k)|$ then the use of $\Theta^{E,F}(l)$ is below $\min(I_n)$ for $l = 0, 1, \ldots, k$.

Since $|\tilde{p}(|I_n|)| \leq \log(|I_n|) + c$ for all n, there is an interval I_n outside the domain of θ_e where $\tilde{p}(|I_n|) < |I_n|/2$. Now let $\tilde{A}$ be any extension of σ_e inside T. There is an extension $\tilde{B}$ of θ_e such that $\tilde{B}(I_n) = \Theta^{\tilde{A},\tilde{B}}(\{1, \ldots, |I_n|\})$; this can be obtained by first choosing $\tilde{B}$ arbitrarily and then changing $\tilde{B}_n$ on I_n which does not affect the value of $\Theta^{\tilde{A},\tilde{B}}(\{1, \ldots, |I_n|\})$ since $\tilde{B}$ is not queried on I_n for these computations.

Let θ_{e+1} be the extension of θ_e which takes the value $\tilde{B}(z)$ whenever z is in the domain of θ_e or $z \leq \max(I_n)$ for the I_n determined above. The string σ_{e+1} is the prefix of $\tilde{A}$ of length $|\sigma_e| + \max(I_n)$.

This completes the distinction of Cases 1, 2, 3 to build σ_{e+1} and θ_{e+1} in stage e in the simultaneous construction of A and B. The following properties of the construction can be verified.

- Every σ_e can be extended to an infinite branch of T.
- Each θ_e is a finite variant of ψ and thus a partial-recursive function.
- For every x there is an e such that x is in the domain of σ_e and θ_e: This follows from the fact that there are infinitely many indices e for the wtt-reduction mapping every set A to A and for each x there is such an index e with the additional property that the I_n picked satisfies $\max(I_n) \geq x$. The construction which has to go through the third case, causes that $\theta_{e+1}(x)$ and $\sigma_{e+1}(x)$ is defined.
- The limits A of the σ_e and B of the θ_e exist and extend all σ_e and θ_e. Furthermore A is an infinite branch of T and thus $\dim_H(A) = 1$. It follows from (c) that $\dim_H^B(A) \geq \frac{1}{4}$.
- If $D \leq_{wtt}^B A$ then this reduction has infinitely many indices e and the construction goes through Case 3 for these e. For each such e there is an interval I_n outside the domain θ_e on which B takes the value-sequence $D(\{1, \dots, |I_n|\})$. It follows from condition (d) that $\dim_H^B(D) \leq \frac{1}{2}$.

These properties complete the proof of the theorem. ⊣

Acknowledgments. The author wants to thank Sanjay Jain, Jan Reimann and Ted Slaman for helpful discussions on the subject. Furthermore, the anonymous referee of the Proceedings of the Logic Colloquium 2004 made several suggestions to improve the presentation of the paper.

REFERENCES

[1] Jin-Yi Cai and Juris Hartmanis, *On Hausdorff and topological dimensions of the Kolmogorov complexity of the real line*, ***Journal of Computer and System Sciences***, vol. 49 (1994), no. 3, pp. 605–619.

[2] Felix Hausdorff, *Dimension und äußeres Maß*, ***Mathematische Annalen***, vol. 79 (1918), no. 1-2, pp. 157–179.

[3] Bjørn Kjos-Hanssen, Wolfgang Merkle, and Frank Stephan, *Kolmogorov complexity and the recursion theorem*, Manuscript, 2005.

[4] Ming Li and Paul Vitányi, ***An Introduction to Kolmogorov Complexity and its Applications***, second ed., Springer-Verlag, New York, 1997.

[5] Jack H. Lutz, *The dimensions of individual strings and sequences*, ***Information and Computation***, vol. 187 (2003), no. 1, pp. 49–79.

[6] Per Martin-Löf, *The definition of random sequences*, ***Information and Control***, vol. 9 (1966), pp. 602–619.

[7] André Nies and Jan Reimann, ***A Lower Cone in the wtt Degrees of Non-Integral Effective Dimension***, Technical Report 63, Workgroup Mathematical Logic and Theoretical Computer Science, University of Heidelberg, 2005.

[8] Piergiorgio Odifreddi, ***Classical Recursion Theory***, North-Holland, Amsterdam, 1989.

[9] Jan Reimann, ***Computability and Fractal Dimension***, Doctoral thesis, Heidelberg, 2005.

[10] JAN REIMANN and THEODORE SLAMAN, *Randomness, entropy and reducibility*, Manuscript, 2005.

[11] ———, *Private communication*, 2005.

[12] BORIS RYABKO, *Noise-free coding of combinatorial sources, Hausdorff dimension and Kolmogorov complexity*, ***Problemy Peredachi Informatsii***, vol. 22 (1986), no. 3, pp. 16–26.

[13] ROBERT I. SOARE, ***Recursively Enumerable Sets and Degrees***, Springer-Verlag, Berlin, 1987.

[14] LUDWIG STAIGER, *Constructive dimension equals Kolmogorov complexity*, ***Information Processing Letters***, vol. 93 (2005), no. 3, pp. 149–153.

SCHOOL OF COMPUTING AND DEPARTMENT OF MATHEMATICS
NATIONAL UNIVERSITY OF SINGAPORE
3 SCIENCE DRIVE 2, SINGAPORE 117543
REPUBLIC OF SINGAPORE
E-mail: fstephan@comp.nus.edu.sg.

COMPUTATION AND THE EXPLANATION OF INTELLIGENT BEHAVIOURS: ETHOLOGICALLY MOTIVATED RESTART

GUGLIELMO TAMBURRINI AND EDOARDO DATTERI

Abstract. Computational theorizing is fruitfully pursued in the investigation of sensorimotor coordination mechanisms of simple biological systems, such as unicellular organism chemotaxis. These investigations undermine the sweeping claim according to which intelligent and adaptive behaviours in biological systems are to be accounted for in terms of continuous systems. Moreover, these investigations suggest the opportunity of developing a more fine-grained framework for analyzing the hierarchical interplay between computational, dynamical, and hybrid models of adaptive behaviours in both biological systems and machines. Key epistemological issues arising in this context of inquiry are clearly identified in Turing's and von Neumann's early reflections on the computational modelling of intelligent behaviours and brain functions.

§1. **Introduction.** A variety of sensorimotor coordination mechanisms are being successfully modelled on the basis of continuous dynamical system approaches (Beer [1997]; Steinhage and Bergener [2000]; Turvey and Carello [1995]). This work is invoked as empirical support for a sweeping "dynamicist" thesis: intelligent and adaptive biological behaviours are to be ultimately accounted for in terms of continuous (dynamical) systems; properly computational investigations make approximate simulation tools available for dynamical theories but play no essential theoretical role (Port and Gelder [1995]). Similar claims about mathematical theorizing in cognitive ethology and biology at large can be found in (Beer [1995], Steels [1995]) and (Longo [2003]), respectively. These claims are critically examined here, in the light of theoretical models of simple sensorimotor adaptive behaviours. These case studies are particularly suited to our purposes, for continuous dynamical approaches are supposedly at their best in the modelling of sensorimotor coordination mechanisms. Computational approaches, we submit, are being fruitfully pursued there too.

This critical examination suggests the opportunity of developing a more detailed analysis of interactions between (and respective roles of) computational

We gratefully thank an anonymous referee, Ernesto Burattini, and Franco Giorgi for valuable criticisms and suggestions.

Logic Colloquium '04
Edited by A. Andretta, K. Kearnes, and D. Zambella
Lecture Notes in Logic, 29

and continuous modelling of intelligent and adaptive behaviours. A key epistemological problem arising in this context of inquiry was already identified by Alan Turing, in his early reflections on computational intelligence. In particular, Turing distinguishes between "continuous" and "discrete" machinery:

> We may call a machine 'discrete' when it is natural to describe its possible states as a discrete set, the motion of the machine occurring by jumping from one state to another. The states of 'continuous' machinery on the other hand form a continuous manifold and the behaviour of the machine is described by a curve on this manifold (Turing [1992]).

Strictly speaking, Turing claims, there is no genuinely discrete machine: everything moves continuously, and thus every machine should be ultimately regarded as a "continuous" system. He is careful to add, however, that "there are many kinds of machine which can profitably be *thought of* as being discrete state machines"[1]. John von Neumann's open-minded speculations on computational modelling of brain functions reflect a strikingly similar viewpoint. Although analog processing is a pervasive feature of brain processing, for some functional modelling purposes this feature may turn out to be an *inessential complexity* (Von Neumann [1958]), p. 58.

Turing's and von Neumann's remarks point to the right kind of questions about the interplay discrete and continuous modelling approaches: When is the discrete-state machine viewpoint profitably adopted to theorize about mechanisms underlying intelligent and adaptive behaviours? When is analog processing an inessential complexity of such mechanisms, which may be disregarded without loss of empirical and explanatory adequacy?

Only a few years before Turing and von Neumann presented these reflections, pioneers of Cybernetics Arturo Rosenblueth and Norbert Wiener developed a sensible epistemological framework for an examination of the above questions (Rosenblueth and Wiener [1945]). One may attempt to locate within this broad framework a proper role for "discrete-state machine" theorizing, for in this framework relatively few requirements are set on kinds of machines to be used in the investigation of adaptive behaviours (apart from the familiar cybernetic emphasis on feedback). Moreover, the pioneers of Cybernetics emphasized the hierarchical organization and the explanatory role of "box-and-arrows" theoretical models of adaptive biological behaviours. Finally,

[1] See (Turing [1992]). See also (Turing [1950]) p. 439: "The digital computers considered in the last section may be classified amongst the 'discrete state machines'. These are the machines which move by sudden jumps or clicks from one quite definite state to another. These states are sufficiently different for the possibility of confusion between them to be ignored. Strictly speaking there are no such machines. Everything really moves continuously. But there are many kinds of machine which can profitably be *thought of* as being discrete state machines." For discussion of this distinction in connection with the closely related digital-analog distinction, see (Trautteur [2005]).

significant experimental uses of machines are suggested for the purpose of testing and revising these theoretical models.[2] Let's see.

§2. The Cybernetic legacy: mechanistic models and machine experiments. Significant machine-based modelling efforts are being carried out in cognitive ethology. A graphic example is the investigation, carried out by means of robotic models and experiments (Lambrinos *et al.* [2000]), of the homing behaviour of desert ant *Cataglyphis*, which is capable of reaching its nest following a straight path after wandering for dozens of meters in search of food.

Cybernetics addressed the problem of explaining this sort of adaptive system capacities (why is desert ant *Cataglyphis* capable of straight-line homing, even when the nest is not in the range of its sensory apparatus?) and behavioural events (why did ant #3 follow a certain trajectory on trial #6?) by means of special kinds of interconnected regularities, that we call here *mechanism schemata*[3]. Cybernetic mechanism schemata are theoretical models grounded into functional analyses of behaviour, to be pursued from an initial "closed box" view of the system S under investigation (Rosenblueth and Wiener [1945]). A closed box view of S is adopted when the relationships between a fixed finite number of its input and output variables are investigated irrespective of what processes S carries out in order to bring about these relationships (whence the closed box metaphor). Rosenblueth and Wiener are careful to point out that a closed-box approach presupposes that there is only "loose coupling" between the investigated variables on the one hand, and the other system variables on the other hand. Loose coupling ensures that the interactions of the selected variables with other system variables are negligible for approximate modelling purposes[4].

A closed box can be "opened" by iteration of closed box analysis. This iteration may enable one to identify a mechanism schema formed by functionally characterized and interconnected parts: one divides the original closed box into "smaller" compartments, thereby fixing their contribution to overall system behaviour by means of their mutual I/O interconnections. Deciding how far to pursue the "opening" of some closed box depends on various factors, including current theoretical interests and system variables that happen to be measurable by available techniques.

The explanatory adequacy of functional mechanism schemata depends on the following features. First, mechanism schemata enable one to provide a *causal* account for how some behavioural events are produced or why the

[2]See (Tamburrini and Datteri [2005]) for more extensive discussion of Cybernetic methodology and its import on contemporary biorobotics.

[3]This terminology dates back at least to (Nagel [1961]) Chapter 8.

[4]The notion of nearly decomposable system, in the sense of (Simon [1996], p. 197 ff.), provides a somewhat more precise understanding of Rosenblueth and Wiener's idea that "a number of variables are only loosely coupled with the rest of those belonging to the system".

system is endowed with some behavioural capacities. Second, overall system behaviour is accounted in terms of the behaviour of *simpler* functional parts. Third, *unified* understanding is achieved, insofar as the same causal factors account for behaviour observed in structurally different systems. Indeed, the characteristic unification promise of Cybernetics is based on the identification of mechanism schemata harnessing the behaviour of both living organisms and machines, that is, the behaviour of systems differing from each other in their "structural" properties (Rosenblueth *et al.* [1943]). In passing, let us notice that abstraction from material support of behaviour is also invoked by Turing in order to go beyond shallow analogies between computers and nervous systems[5].

Cybernetic mechanism schemata are naturally accommodated by the mechanistic model of explanation (Machamer *et al.* [2000]; Glennan [2002]), but are significantly related the covering-law model (Hempel and Oppenheim [1948]) of scientific explanation too, insofar as cybernetic schemata are theoretical descriptions taking the form of coordinated *regularities*. Ernest Nagel, a major advocate of the covering law model, went as far as to assert that Cybernetic accounts of homeostatic system behaviours are special cases of covering-law explanations: "all such systems can in principle be analyzed into parts which are causally relevant to the maintenance of some feature in those systems, and which stand to each other and to environmental factors in determinate relations capable of being formulated as general laws. The discovery and analysis of such detailed mechanisms is the task of specialized scientific inquiry" (Nagel [1961], p. 418).

The uniform functional analysis of both machines and living organisms sought for by Cybernetics paves the way to a principled use of machines in experiments about mechanism schemata. A *material model* is the representation of a complex system by another system "which is assumed simpler and which is also assumed to have some properties similar to those selected for study in the original complex system". (Rosenblueth and Wiener [1945], p. 317). The crucial relationship between the original complex system and its material model is one of similarity. In this context, distinct material systems are similar to the extent their functional organization is harnessed by the same abstract theoretical model. Accordingly, Rosenblueth and Wiener are careful to point out that the basis for passing similarity judgments between the original system and its material model is provided by some theoretical model, e.g., a mechanism schema, of the original system. Similarity is a theoretical

[5]"Importance is often attached to the fact that modern digital computers are electrical, and that the nervous system also is electrical.... In the nervous system chemical phenomena are at least as important as electrical. In certain computers the storage system is mainly acoustic. The feature of using electricity is thus seen to be only a very superficial similarity. If we wish to find such similarities we should look rather for mathematical analogies of function." (Turing [1950], p. 439).

notion, and similarity judgments are always relative to some sets of properties: the theoretical model is the yardstick in terms of which similarity judgments are issued.

A material model singled out on the basis of some mechanism schema can be used to test the empirical adequacy of the mechanism schema, insofar as it may enable one to carry out "experiments under more favorable conditions than would be available in the original system". The similarity between material model and original system is a crucial *empirical assumption* underlying these experimental enterprises. In addition to corroboration and falsification, material models may enable one to obtain results which "could not have been easily anticipated on the basis of the formal model alone" (Rosenblueth and Wiener [1945], p. 317).

In our view, this broad cybernetic picture of the relationship between theoretical modelling, explanation, and machine experiments provides a sensible methodological framework for much work in contemporary biorobotics, which is characteristically concerned with the problem of conducting experiments on adaptive biological behaviours by means of robotic material models (Tamburrini and Datteri [2005], Datteri and Tamburrini [2007]). Let us now turn to compare, within this overall methodological framework, "continuous state machines" with "discrete state machines" viewpoints in the theoretical modelling of simple sensorimotor coordination mechanisms.

§3. Production-rule mechanisms and simple organism chemotaxis.

3.1. Bacterial chemotaxis. Bacterial chemotaxis has been extensively studied since the end of the 19th century. It can be observed by pushing a capillary tube containing a solution of glucose into a suspension of *E. coli* on a slide; bacteria will massively go inside the tube (Adler [1969]). By considering only bacteria positions and chemical concentrations, the behaviour of E. coli is readily modelled as a gradient ascent: bacteria go towards the region of maximum chemical concentration (Staddon [2003], Chapter 2), where the quantities involved (positions and chemical gradients) are naturally represented as continuous state variables.

Further observation of *E. coli* behaviour reveals additional features of their chemotaxis capacities (Bourret and Stock [2002]; Sourjik [2004]; Webre *et al.* [2004]). *E. coli* exhibit *only two* actions: they swim in a straight line or tumble down. The relative frequencies of swimming and tumbling are the key to understanding how *E. coli* finds a chemical source. In solutions characterized by uniform, non-changing attractant concentrations, bacterial motility exhibits typical patterns: bacteria swim in fairly straight lines; occasionally they tumble and start swimming in completely new directions. When an attractant is inserted in the solution, bacterial motility appears much smoother and better co-ordinated than normal: the frequency of straight-line swimming increases with respect to tumbling and randomly changing direction.

This produces an overall motion upward along the concentration gradient. A few minutes after stopping the flow of attractant, the frequency of tumbling is restored to its steady-state value and bacterial motility shows the typical patterns described above. Negative attractant gradients produce opposite behaviours: tumbling increases over straight-line swimming, and bacterial motion appears erratic and non-coordinated (Macnab and Koshland [1972]). Chemotaxis is thus achieved by switching between two distinct actions, tumbling and swimming.

Let us consider current explanations of how these two actions are produced. Each *E. coli* is endowed with some flagella (five to ten per cell). Flagella can rotate clockwise or counterclockwise, producing tumbling and smooth swimming respectively. Transmembrane receptors bind to attractant molecules producing a chain of protein phosphorylation processes, that ultimately leads to variation in the concentration of phosporylated protein CheY (CheY-P). The latter binds to protein FliM, which is thought to be a "switch" protein, responsible for changes of flagellar motion direction (Bren and Eisenbach [2001]). The resulting behaviour appears to be governed by simple rules:

> The question of how the chemotaxis system is regulated [...] reduces to that of how the direction of flagellar rotation is controlled. The default direction of rotation of the flagellar motor is counterclockwise. A switch to rotation in the clockwise direction occurs in response to a sensory signal, transduced by the chemotaxis protein CheY (Eisenbach and Caplan [1998]).

Two rules are included for two distinct and mutually exclusive behaviours. This hypothesis is also at the core of a computer simulation of bacterial chemotaxis, which takes into account a wide variety of processes that are known to occur within the cell, from membrane receptors to motor activation via phosphorylation signalling pathways (Bray *et al.* [1993]). In their model, the state assumed by the flagellar motor complex M (run, pause, and tumble) depends on different binding possibilities between M and phosphorylated CheY: M can bind up to four different phosphorylated CheY molecules, thus resulting in five distinct molecules. Counterclockwise motion is elicited by two of these molecules (M alone, and M bound with just one CheY-P molecule); motion pause is elicited by the binding of M with two CheY-P molecule; the other two combinations elicit clockwise behaviour. Predictions obtained with this computer simulation match with observations in many significant cases[6].

In the simple and well-studied case of bacterial chemotaxis, the "same" system (*E. coli* exemplars) can be modelled either as a discrete or else as

[6]The two-actions behaviour of bacterial chemotaxis has served as an inspirational source for a recent robotic system, capable of locating gradient sources, and applied to the field of environmental monitoring (Dhariwal *et al.* [2004]).

a continuous-state mechanism. Which is chosen as the more suitable explanatory source depends on the character of the *explanandum*. If one wants to understand the relation holding between chemical gradients and *E. coli* positions, a "continuous state machine" approach may be preferred over a discrete one. Accordingly, quantization of gradients and positions in space — introduced, say, for practical reasons concerning the handling of simpler mathematical or simulation tools — is regarded only as an *approximation* of quantities that, in the theoretical account of the system, take values in the real field. Discrete-state theoretical models provide empirically supported explanations of the relations holding between the reception of chemical molecules by *E. coli* transmembrane proteins, and the direction of flagellar motion. In the model reviewed above, the flagellar behavioural repertoire of *E. coli* is represented as a discrete set of mutually exclusive actions; quantization is empirically (rather than practically) justified, insofar as the discrete theoretical model is supported by empirical testing. Further opening of the boxes into which this discrete model functionally decomposes (for example, by looking at the chemical dynamics involved in binding and reaction processes occurring within the cell), may induce one to switch to a continuous-state approach again. In specific empirical investigations, discrete state machine descriptions and continuous dynamic system descriptions may happen to be interleaved in hierarchical, multilevel mechanism schemata. Which level of the hierarchy is chosen as an appropriate source of explanations depends on the nature of explananda and pragmatic considerations about explanation needs.

Similarly, production-rule mechanisms operating on quantized variables may have to be combined with other mechanisms in a horizontal (i.e. non-hierarchical) fashion, in order to capture some biological action selection mechanism.[7] The opportunity of introducing such hybrid mechanisms may emerge even under the assumption - embedded into models of *E. coli* chemotaxis - that actions are mutually incompatible. For example, McFarland's hybrid models of animal behaviour (McFarland and Sibly [1975]) rest on a similar strong assumption:

> it is always possible to classify the behavioural repertoire of a species in such a way that the classes are mutually exclusive in the sense that members of different classes cannot occur simultaneously

Animals do *only* action x at time t, where x can be classified as an *unit of*

[7]"Imagine a zebra in the African savannah. At each moment in time this zebra has to weigh up alternative courses of action before deciding which will be most beneficial to it. For instance, it may want to graze because it is short of food, or it may want to head towards a water hole because it is short of water, or it may want to remain motionless in order to avoid detection by the predator it can see lurking nearby. This is an example of the problem of *action selection*: how to choose, at each moment in time, the most appropriate out of a repertoire of possible actions" (Tyrrell [1993]).

behaviour, in McFarland's words. Units of behaviours are identified on the basis of a number of their distinctive features. McFarland is not alone in this suggestion. Tyrrell, Hull, Tinbergen, Lorenz and Baerends propose models of action selection conforming to this oft-maintained assumption, insofar as the behavioural repertoire is given by a finite set of mutually incompatible actions. In these theoretical models, however, variables are included which range over some real interval and participate in various ways to the selection of a winning action. McFarland's causal factors are a significant case in point, as well as Hull's *drives*, construed as internal motivation variables associated to behaviours (Hull [1943]). Each behaviour has a drive, whose magnitude is determined by the strength of internal and external stimuli and by other factors that change with learning. One action is chosen from a set of competing actions on the basis of its drive strength. Baerends is more explicit in postulating a hierarchical organization of behaviours in his study of digger wasp and herring gull: the animal's 'higher-level instincts' activate some 'lower-level instinct', which in turn activates some 'fixed action pattern' (Baerends [1976]).

These various modelling enterprises suggest that an adequate epistemological analysis of the interplay between continuous and discrete state machine approaches requires a detailed study of the hierarchical interleaving of continuous and discrete mechanism descriptions as well as their "horizontal" combination into hybrid mechanism descriptions. This detailed study goes clearly beyond the limits of this paper, but we can here strengthen this motivation by reference to a historically prominent case study.

3.2. Stentor. In *Behaviour of the low organisms*, H. S. Jennings proposes a mechanism schema to explain the behaviour of Stentor, a unicellular Protozoan organism. Stentor resembles a "trumpet" with cilia surrounding its larger opening (Jennings [1906]). By moving the cilia, Stentor gives rise to a water current that carries food into its body. Jennings accounts for the behaviour of Stentor when a noxious chemical tray reaches the opening in terms of a simple adaptive mechanism. This mechanism is adequately described by means of a few production rules[8]. Each production rule encodes, in its action part, a different reaction to the noxious stimulus. And which action is chosen (which rule wins) depends on the organism's past history.

Let us consider Jennings' model in more detail. After sensing the noxious tray,

> [t]he animal bends to one side ... It thus as a rule avoids the cloud of particles, unless the latter is very large. [...] If the first reaction is not successful, it is usually repeated one or more times. If the

[8]Jenning's discrete-state model of Stentor has been discussed in (Cordeschi [2002]), in the context of machine-based investigations on adaptive biological behaviours.

> repeated turning to one side does not relieve the animal, so that the particles of carmine continue to come in a dense cloud, another reaction is tried. The ciliary movement is suddenly reversed in direction, so that the particles against the disk and in the pouch are thrown off. [...] If this fails to relieve the organism, the next reaction — contraction — usually supervenes.

Finally, if the stream reaches the cilia again, Stentor tries another very expensive strategy:

> it ceases to extend, but contracts repeatedly and violently while still enclosed in its tube. In this way the attachment of its foot to the object on which it is situated is broken, and the animal is free. Now it leaves its tube and swims away.

Jennings isolates a number of stereotyped actions (turning away; ciliary reversal; contract into tube; break and go away), all of them being possible responses to the same stimulus. The actions are tried sequentially (see also Staddon [2003], Chapter 3).

How does Stentor chooses the action to perform from its fixed repertoire $a_1, \ldots, a_n$ of actions when the noxious stimulus s is present? Jennings postulates the existence of some set of distinct physiological states, or "varying internal conditions of the organism". This is the reason why Stentor reacts to the same stimulus in different ways, depending on its current physiological status:

> After this [*the chemical stimulation*] has taken place once or twice we find that the Stentor no longer reacts as before. All the external conditions remain the same, the stimulus applied is the same. Nevertheless, the Stentor does not react. Therefore, we must conclude that the Stentor itself has changed. Its physiological condition is now different from what it was original.

A convenient production-rules representation of Jenning's account takes the form:

$$\mathrm{s} \wedge \mathrm{p}_1 \rightarrow \mathrm{a}_1$$

$$\ldots$$

$$\mathrm{s} \wedge \mathrm{p}_\mathrm{n} \rightarrow \mathrm{a}_\mathrm{n}$$

where each p_i, $1 \leq i \leq n$, is a physiological state taking into account Stentor's recording of recent interactions with the environment.

By means of this mechanism, Jennings can answer explanation requests about Stentor's behaviour, concerning, e.g., the observation that Stentor cannot perform more than one action at a time, or the causal factors determining its action selection. Whether this mechanism is a source of contextually appropriate explanations depends on the character of explanation requests. If one wants to understand, down to biochemical details, how sensory stimuli are processed by Stentor, and eventually affect its motor system, the functional accounts one may provide on the basis of this production-rule mechanism turn out to be unsatisfactory. A more satisfactory explanation will most likely involve a hierarchy of mechanisms, with some lower-level mechanism specified in terms of biochemical entities and processes which give rise to the behaviour functionally characterized by the higher-level production-rule mechanism. Once again, discrete and continuous state machine descriptions may turn out to be interleaved in a hierarchy of mechanism schemata.

The opportunity of developing a fine-grained framework for analyzing hierarchical and horizontal combinations of discrete and continuous models of adaptive behaviours emerges also from an examination of biologically inspired robotics. Sweeping dynamicist theses were advanced in this area of investigation too, especially building on the development of behaviour-based systems, which are both inspired to and provide heuristic insight for functional models of adaptive biological behaviours. Dynamicist theses stand in need of serious qualifications and amendments in this case too. Let's see.

§4. Biologically inspired robotics: discrete and continuous models.

4.1. Behaviour-based systems. The broad heuristic idea that many adaptive behaviours are adequately modelled and explained in terms of continuous state machines is a source of inspiration for much recent work in bio-robotics (Beer [1995], [1997]) and biologically inspired robotics (Steels [1995]). To exemplify, let us consider continuous state machine approaches in the framework of behaviour-based architectures. These are also used for modelling adaptive capabilities of many biological systems (from insect navigation to human limb control).

The functional structure of behaviour-based systems (Arkin [1998], Brooks [1986]) is usually described in terms of a set of m interconnected layers, called *behaviours*, each layer being associated to a function from sensor data to motor actions. At each time, t motor response r_i of behaviour b_i is

$$r_i = \gamma_i \cdot b_i(s_i),$$

where s_i is the state perceived by the sensor associated with behaviour b_i at time t, γ is a real-valued gain, and $0 \leq i \leq m$. The system is parallel and asynchronous. If n behaviours control the same actuator, a coordination function C, taking as input the outcomes of those n behaviours, outputs a command for the actuator (see Figure 1). A typical coordination function (termed cooperative) blends together (typically sums) multiple responses, previously weighted by means of some set of n real-valued gains. Function C, together with the various gains and mappings characterizing each behaviour, determines how the system acts in response to perceived environmental conditions. Let f be

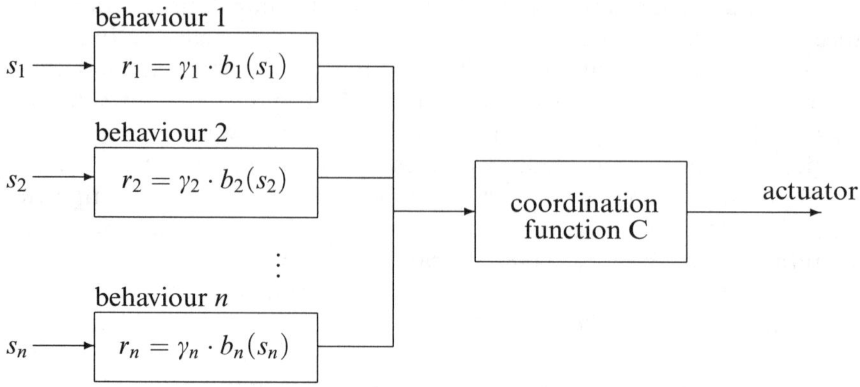

FIGURE 1. Functional structure of behaviour-based architectures.

the overall agent control function. Distinct sets of gains determine distinct reaction policies to environmental stimuli, i.e. distinct control functions.

A continuous-time, dynamical-system approach has been proposed as unifying treatment for behaviour-based systems. Notably, Steels (Steels [1995]) focuses on control programs in behaviour-based systems, and claims that these are best viewed as "dynamical systems which establish a continuous relation between the time varying data coming out of a sensor and a stream of values going to the actuators". Furthermore, he proposes to apply equilibrium analysis to the study of behaviour-based systems: a behaviour "reaches an equilibrium state if it exerts no influence on the dynamics" of the whole dynamical system, i.e. of the whole assemblage of behaviours. By means of equilibrium analyses one may identify, for example, whether a behaviour will reach equilibrium or the conditions under which a behavioural assemblage will reach some configurations in the state space.

Let us proceed to a critical examination of this approach to behaviour-based architectures in the light of adaptive robot navigation problems.

4.2. Symbol processing and potential field navigation. Adaptive navigation requires the capacity of orienting oneself to goals (whose position is directly perceived or known), while avoiding static or dynamic obstacles. Basic manifestations of these capacities, and their combination, are readily modelled in the framework of some behaviour-based system A, which is provided with the following behaviours:

- One behaviour, called *attractive*, is mathematically represented as a vector, whose direction points towards the goal, and whose magnitude is directly proportional to the distance between current point and goal or some sensory cue. For example, the attractive behaviour of a phototropic organism might be represented by a mapping from the perceived orientation of a light source to the attractive vector orientation, and by a mapping from light intensity to the attractive vector magnitude;

- another behaviour, called *repulsive*, is represented as a vector whose direction points away from perceived obstacles, and whose magnitude is inversely proportional to the distance between robot and obstacles. For example, if some distance sensors are available, their orientation and distance reading might be mapped to the two components of the repulsive vector.

A cooperative function sums the two vectors; at each given time, A's orientation is the orientation of the resulting vector, and its velocity is the vector magnitude. The contribution of the *repulsive* behaviour prevents A from colliding with obstacles, while the *attractive* behaviour directs it to the goal.

This modelling strategy has been more comprehensively formalized as the *potential field* method of navigation, according to which the system is represented as a point in space, acting under the influence of an *artificial potential field* U (Latombe [1991]). This potential field is typically defined as the vector sum of an attractive and a repulsive potential. More precisely, given an agent A (a single rigid object) and an environment E (a Euclidean space represented as R^2 or R^3), at each point q in space $U(q) = U_{att}(q) + U_{rep}(q)$. Typical equations for calculating the magnitude of attractive and repulsive potentials are

$$U_{att}(q) = \frac{1}{2}k\rho^2\, goal$$

$$U_{rep}(q) = \left\{ \begin{array}{cc} \frac{1}{2}j\left(\frac{1}{\rho}(q) - \frac{1}{\rho_0}\right) & \text{if } \rho(q) < \rho_0 \\ 0 & \text{if } \rho(q) \geq \rho_0 \end{array} \right\}$$

where ρ is the Euclidean distance between q and the goal, k and j are positive scaling factors, and ρ_0 is a positive constant called the *distance of influence* of the obstacles. The potential field approach is conceptually akin to Hull's goal-gradient hypothesis (Hull [1932]). Several computer simulation or robotic experiments have shown its significance in the modelling of adaptive navigation and adaptive interaction with the environment, over and above the robotic navigation domain (Khatib [1986]).

This navigation method shows that some navigation behaviours are trajectories predictable on the basis of some simple law; and in turn, such laws can be involved in the specification of a black-box mechanism schema, insofar as these laws identify the I/O behaviour of the attractive and repulsive modules

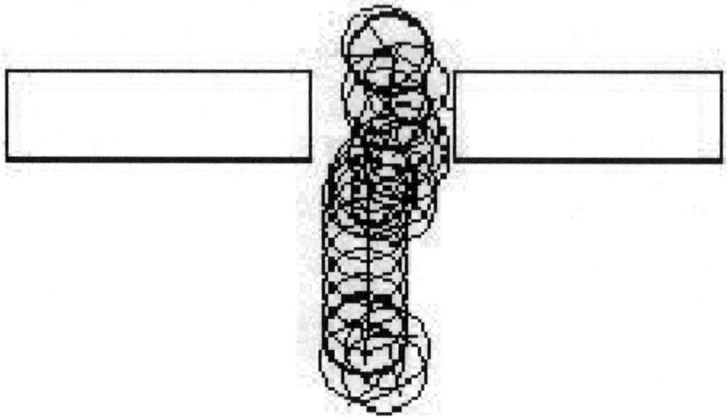

FIGURE 2. The robot gets stuck within the door aperture.

of the potential-field system. Let us now explore limitations of potential-field navigation methods and the opportunity of combining this method with symbol processing mechanisms, by reference to experiments involving robotic navigation behaviours.

Let us consider a potential-field based system, provided with some kind of distance sensor (like sonars or infrared). Koren and Borenstein (Koren and Borenstein [1991]) identify some of the main drawbacks of potential-fields models in cyclic behaviour due to local minima, no passage between closely spaced obstacles, oscillations in the presence of obstacles, and oscillations in narrow passages. These behaviours are claimed to be independent of implementations. And Latombe (Latombe [1991]) claims that

> [...] local minima remain an important cause of inefficiency for potential field methods. Hence, dealing with local minima is *the* major issue that one has to face in designing a planner based on this approach.

Even doors constitute a difficulty for behaviour-based agents based on potential fields. As Koren and Borenstein explain, depending on the size of the aperture (assuming it is larger than the robot body), on the depth of the walls, and on the position of distance sensors on the system, doors are perceived as walls, and the robot cannot go through them. Figure 2 shows superimposed snapshots from a simulation of a sonar-based robot which is commonly used in robotic investigations. This robot, represented by a circle, gets stuck within a door aperture.

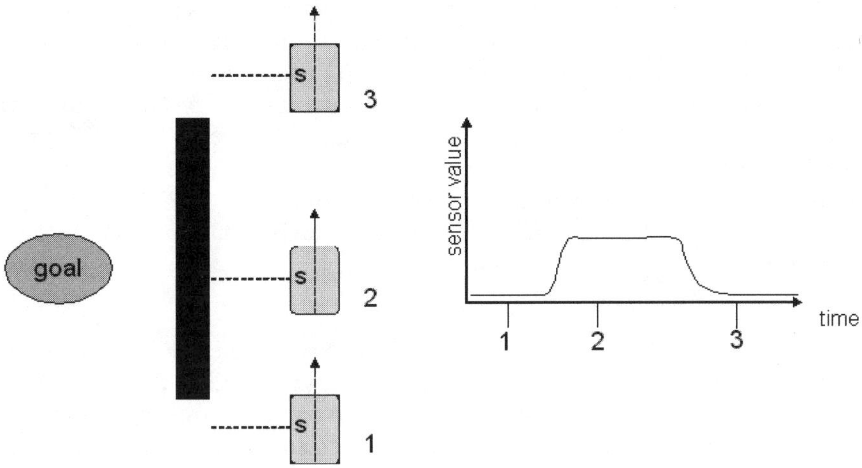

FIGURE 3. Values of sonar readings when the robot walks close to a wall.

To overcome problems of this sort, one may try and determine action on the basis of data that go beyond instantaneous sensory readings. A promising approach requires fitting the robotic system with an event detection mechanism based on sensor histories. Such event-detection mechanisms can give rise to genuine symbol-processing and more sophisticated planning behaviour. Let's see.

The graph on the right-hand side of Figure 3 shows the values signalled over time by a sonar of a behaviour-based robot. By considering sensor history, the presence of a wall can be hypothesized and distinguished from the presence of some, say, round obstacle: in the case of walls, the sensor curve typically shows three distinct periods, with a time lag of "high" value interleaved between two periods of "low" value.

This observation is the basis for devising a history-based mechanism, allowing the robot to perceive, in addition to local (current) potential fields, the presence of particular obstacles configurations, like walls or doors. Behaviour-based systems provided with this mechanism can modify their behavioural arrangement by adjusting their behavioural gains or even by completely suppressing behaviours[9] in order to cope with doors or walls. For example, as soon as the perceptual apparatus detects the presence of a wall on the left, by

[9]Inhibition of behaviours is characteristic of a particular coordination function (see Section 4.1), termed *competitive* control function.

detecting the characteristic sensor curve shown in the figure above, the system may adopt the following strategy in order to go around the wall: suppress for a while the attractive behaviour (by setting to 0 the corresponding gain), and activate at the same time a *wander* behaviour, directing the system in random directions; reactivate, after some time lag, the attractive behaviour. The *wander* behaviour is a source of noise, allowing the robot to escape from the local minimum induced by the wall-target configuration.

One can encode the first segment of the new navigation strategy as a production rule:

$$\text{wall\_left} \rightarrow (\text{on}(\text{wander}) \wedge \text{off}(\text{attractive})).$$

Similarly, if the perceptual system has the capacity of detecting doors, it can decrease the strength of the repulsive behaviour and increase the strength of the attractive behaviour, thus allowing the robot to escape from stalemate.

This strategy has been expanded and implemented in (Schönherr *et al.* [2001]), where histories of behavioural activation values are tracked and grouped in stereotyped patterns, corresponding to particular environmental situations. Behaviours are defined as in Section 4.1, and activation values correspond to the γs in the definition: behaviour activation values indicate the strength of their motor actions in relation to the magnitude of sensory inputs, and can be changed over time by other system modules (or behaviours), in order to inhibit, increase or completely suppress activity. The approach of (Schönherr *et al.* [2001]) is only apparently different from tracking *sensor values*, insofar as behaviour activation values strictly depend on sensor values, and are changed according to the following differential equation[10]:

$$\dot{\gamma}_b = g_b(\gamma_b, On_b, Off_b),$$

where γ_b is the activation value for behaviour b, g_b is a nonlinear function, On_b and Off_b represent the contribution of conditions positively and negatively related with the activation of b. The latter two parameters depend on the behaviour sensor values.

The usefulness of monitoring activation values for achieving adaptive behaviour is illustrated by an example, involving a behaviour-based system. Six behaviours are used (*turn-to-door*, *go-through-door*, *follow-wall*, *avoid-collision*, *wander*); three additional, "higher-level" behaviours are taken into account, whose outputs do not affect motor commands, but rather change the value of some other global variables. The experimental task is to pass through a door. The vector of the six activation values shows typical patterns in time, in correspondence with specific robot/environment configurations.

[10]This is a slightly simplified version, omitting variables connected to higher-level modules of the system.

For example, the sequence of behaviours activation values when the robot is in the vicinity of the door, goes through it and then exits, is characterized by

- a initial increase of one of the *follow-wall* behaviour;
- a subsequent increase of the *close-to-door* behaviour;
- during high activation of the *close-to-door* behaviour, a brief activation of *turn-to-door* followed by the simultaneous activation of *go-through-door* and *avoid-collision*.

Behavioural sequences of this kind are adequately represented by means of a suitable first-order formalism, and even simpler production rules such as

$$(\text{HIGH(wall-left) or HIGH(wall-right)}) \rightarrow (\text{ON(wander)} \wedge \text{OFF(attractive)})$$

suffice to overcome local minima difficulties arising with doors and walls. During robot motion, behavioural activity is logged continuously on in order to detect the occurrence of certain patterns. This pattern matching mechanism allows the system to extract "facts" from activation values, and to select actions on this basis. Note, however, that more than mere pattern recognition is required to qualify a perceived chronicle as a fact:

> The fact extraction technique does not presume or guarantee anything about the consistency of the facts that get derived over time. Achieving and maintaining consistency, and determining the ramifications of newly emerged facts remain issues that go beyond fact extraction. (Schönherr *et al.* [2001]).

Perception in these behaviour-based systems naturally takes the form of a hypothesize-and-test process: incoming chronicles are tested against the database of accepted facts and, possibly, against other facts that are expected in virtue of accepted facts. As long as temporal constraints characterizing adaptive action in the world can be properly met by the required reasoning systems, one has here a clear indication of the opportunity of combining symbol processing with continuous state machines in the modelling of behaviour-based architectures.

Various modes of combination are suggested by a closer examination of the examples provided so far. The event-detection module of the behaviour-based system described in this section is appended, in a cascade, to sensory modules whose state variables are assumed to range over some real interval. The cases of *E. coli* and *Stentor* suggest that continuous and discrete state mechanisms can be used as theoretical models of the same system, at different levels of the "box-and-arrows" modelling hierarchy[11]. In general, biologically inspired robotics provides additional motives to relinquish sweeping dynamicist theses,

[11] A neuro-symbolic language for combining continuous-state neural networks with production rules in a hierarchical fashion is described in (Burattini *et al.* [2005]).

and compelling reasons for developing a more fine-grained analysis of both hierarchical and horizontal combinations of discrete and continuous state machine approaches in the modelling of adaptive and intelligent behaviours. The selection of an appropriate explanatory source from these hierarchies of models (differing from each other in terms of functional primitives and organizations) involves pragmatic decisions that are motivated by particular explanation needs driving scientific inquiry.

REFERENCES

J. Adler [1969], *Chemoreceptors in bacteria*, ***Science***, vol. 166, no. 3913, pp. 1588–1597.

R. C. Arkin [1998], ***Behavior-Based Robotics***, The MIT Press.

G. Baerends [1976], *The functional organisation of behaviour*, ***Animal Behaviour***, vol. 24, pp. 726–735.

R. D. Beer [1995], *A dynamical systems perspective on agent-environment interaction*, ***Artificial Intelligence***, vol. 72, pp. 173–215.

R. D. Beer [1997], *The dynamics of adaptive behavior: A research program*, ***Robotics and Autonomous Systems***, vol. 20, pp. 257–289.

R. B. Bourret and A. M. Stock [2002], *Molecular information processing: Lessons from bacterial chemotaxis*, ***The Journal of Biological Chemistry***, vol. 277, no. 12, pp. 9625–9628.

D. Bray, R. B. Bourret, and M. I. Simon [1993], *Computer simulation of the phosphorylation cascade controlling bacterial chemotaxis*, ***Molecular Biology of the Cell***, vol. 4, pp. 469–482.

A. Bren and M. Eisenbach [2001], *Changing the direction of flagellar rotation in bacteria by modulating the ratio between the rotational states of the switch protein flim*, ***Journal of Molecular Biology***, vol. 312, pp. 699–709.

R. A. Brooks [1986], *A robust layered control system for a mobile robot*, ***IEEE Journal of Robotics and Automation***, pp. 14–23.

E. Burattini, E. Datteri, and G. Tamburrini [2005], *Neuro-symbolic programs for robots*, forthcoming in the *Proceedings of NeSy'05 Neural-Symbolic Learning and Reasoning*, Workshop at *IJCAI-05*, Edinburgh, Scotland, August 1st, 2005.

R. Cordeschi [2002], ***The Discovery of the Artificial. Behavior, Mind and Machines Before and Beyond Cybernetics***, Kluwer Academic Publishers, Dordrecht, The Netherlands.

E. Datteri and G. Tamburrini [2007], *Biorobotic experiments for the discovery of biological mechanisms*, forthcoming in Philosophy of Science.

A. Dhariwal, G. S. Sukhatme, and A. A. G. Requicha [2004], *Bacterium-inspired robots for environmental monitoring*, ***Proceedings of the IEEE International Conference on Robotics and Automation (ICRA 2004)***, pp. 1436–1443.

M. Eisenbach and S. R. Caplan [1998], *Bacterial chemotaxis: Unsolved mystery of the flagellar switch*, ***Current Biology***, vol. 8, pp. R444–R446.

S. Glennan [2002], *Rethinking mechanistic explanation*, ***Philosophy of Science***, no. 69, pp. S342–S353.

C. G. Hempel and P. Oppenheim [1948], *Studies in the logic of explanation*, ***Philosophy of Science***, vol. 15, no. 2, pp. 135–175.

C. L. Hull [1932], *The goal-gradient hypothesis and maze learning*, ***Psychological Review***, vol. 39, pp. 25–43.

C. L. Hull [1943], ***Principles of Behaviour: an Introduction to Behaviour Theory***, D. Appleton-Century Company.

H. S. Jennings [1906], ***Behavior of the Lower Organisms***, Columbia University Press, New York.

O. Khatib [1986], *Real-time obstacle avoidance for manipulators and mobile robots*, ***The International Journal of Robotics Research***, vol. 5, no. 1, pp. 90–98.

Y. Koren and J. Borenstein [1991], *Potential field methods and their inherent limitations for mobile robot navigation*, ***Proceedings of the IEEE Conference on Robotics and Automation***, Sacramento, California, pp. 1398–1404.

D. Lambrinos, R. Möller, T. Labhart, R. Pfeifer, and R. Wehner [2000], *A mobile robot employing insect strategies for navigation*, ***Robotics and Autonomous Systems***, vol. 30, pp. 39–64.

J. C. Latombe [1991], ***Robot Motion Planning***, Kluwer Academic Publishers, Norwell, MA.

G. Longo [2003], *Computer modelling and natural phenomena*, ***Proceedings of the European Software Engineering Conference and ACM SIGSOFT Symposium***, Helsinki.

P. Machamer, L. Darden, and C. F. Craver [2000], *Thinking about mechanisms*, ***Philosophy of Science***, vol. 67, pp. 1–25.

R. M. Macnab and D. E. Koshland [1972], *The gradient-sensing mechanism in bacterial chemotaxis*, ***Proceedings of the National Academy of Science USA***, vol. 69, no. 9, pp. 2509–2512.

D. J. McFarland and R. M. Sibly [1975], *The behavioural final common path*, ***Philosophical Transactions of the Royal Society of London Series B, Biological Sciences***, vol. 270, no. 907, pp. 265–293.

E. Nagel [1961], ***The Structure of Science: Problems in the Logic of Scientific Explanation***, Routledge and Kegan Paul, London.

R. F. Port and T. van Gelder (editors) [1995], ***Mind as Motion: Explorations in the Dynamics of Cognition***, Bradford Books, The MIT Press.

A. Rosenblueth and N. Wiener [1945], *The role of models in science*, ***Philosophy of Science***, vol. 12, pp. 316–321.

A. Rosenblueth, N. Wiener, and J. Bigelow [1943], *Behavior, purpose and teleology*, ***Philosophy of Science***, vol. 10, pp. 18–24.

F. Schönherr, M. Cistelecan, J. Hertzberg, and T. Christaller [2001], *Extracting situation facts from activation value histories in behavior-based robots*, ***KI-2001: Advances in Artificial Intelligence, LNAI 2174***, pp. 305–319.

H. Simon [1996], ***The Sciences of the Artificial***, The MIT Press, Cambridge, MA.

V. Sourjik [2004], *Receptor clustering and signal processing in E. coli chemotaxis*, ***Trends in Microbiology***, vol. 12, no. 12, pp. 569–576.

J. E. R. Staddon [2003], *Adaptive behavior and learning*, Internet Edition (`http://psychweb.psych.duke.edu/department/jers/abl/TableC.htm`).

L. Steels [1995], *Mathematical analysis of behavior systems*, ***Proceedings of the Conference "From Perception to Action" (Perarc)***, pp. 88–95.

A. Steinhage and T. Bergener [2000], *Learning by doing: A dynamic architecture for generating adaptive behavioral sequences*, ***Proceedings of the Second ICSC Symposium on Neural Computation NC2000, Berlin***, pp. 813–820.

G. Tamburrini and E. Datteri [2005], *Machine experiments and theoretical modelling: from cybernetic methodology to neuro-robotics*, forthcoming in ***Minds and Machines***.

G. Trautteur [2005], *Beyond the super-turing snare: analog computation and digital virtuality*, ***New Computational Paradigms*** (S. B. Cooper, B. Löwe, and L. Torenvliet, editors), LNCS, vol. 3526, Springer, Berlin, pp. 507–514.

A. M. Turing [1950], *Computing machinery and intelligence*, ***Mind***, vol. 59, pp. 433–460.

A. M. Turing [1992], *Intelligent machinery*, ***Collected Works of A. M. Turing: Mechanical Intelligence*** (D. C. Ince, editor), Elsevier Science Publishers.

M. T. Turvey and C. Carello [1995], *Some dynamical themes in perception and action*, (R. F. Port and T. van Gelder, editors), pp. 374–401.

T. Tyrrell [1993], ***Computational Mechanisms for Action Selection***, Ph.D. thesis, University of Edinburgh, Centre for Cognitive Science.

J. Von Neumann [1958], ***The Computer and the Brain***, Yale University Press, New Haven, CT.

D. J. Webre, P. M. Wolanin, and J. B. Stock [2004], *Modulated receptor interactions in bacterial transmembrane signalling*, ***Trends in Cell Biology***, vol. 14, no. 9, pp. 478–482.

DIPARTIMENTO DI SCIENZE FISICHE
UNIVERSITÀ DI NAPOLI FEDERICO II

DIPARTIMENTO DI FILOSOFIA
UNIVERSITÀ DI PISA

CONSTRUCTIBLE SHEAVES AND DEFINABILITY

IVAN TOMAŠIĆ

§1. Introduction. Several applications of model-theoretic methods in the theory of cohomology have appeared recently, most probably influenced by ideas of Macintyre. In his programmatic paper [10], he shows that, after expanding the language of rings by certain sorts and predicates for cohomology, the axioms of Weil cohomology theories are first order and one can form new cohomology theories as ultraproducts of already existing ones.

The present author has remarked that, although the cohomologies with torsion coefficients do not satisfy the axioms for a Weil cohomology theory individually, they do so "on average", and one can obtain cohomologies with coefficients in pseudofinite fields of characteristic zero by taking ultraproducts [13]. He shows that this "pseudofinite cohomology" is at least as good as the l-adic theory when dealing with issues around the Weil conjectures. In parallel, Brunjes and Serpe developed the theory of nonstandard sheaves systematically, and they even show that the pseudofinite cohomology is better behaved than the l-adic one, the former being a derived functor cohomology [1].

The purpose of this short note is to clarify which aspects and invariants of the theory of (étale) constructible sheaves and cohomologies are definable *in the language of rings*. It should serve as a bridge between the algebraic-geometric and model-theoretic language and should encourage model-theorists to use the sophisticated techniques already developed by geometers. We show that, in case one needs to consider an invariant defined in terms of constructible sheaves over a finite or a pseudofinite ground field, there is a good chance that it is definable.

We use the standard notation of algebraic geometry. For a field k, $\bar{k}$ denotes the algebraic closure of k, and for a scheme X over k, $\bar{X}$ denotes $X \times_{\mathrm{Spec}(k)} \mathrm{Spec}(\bar{k})$. A *variety* over S is a reduced and separated scheme of finite type over S.

§2. Fundamental group and constructible sheaves. We recall the basic definitions and facts from algebraic geometry in an effort to keep the presentation

This research was supported by a Marie Curie Fellowship.

Logic Colloquium '04
Edited by A. Andretta, K. Kearnes, and D. Zambella
Lecture Notes in Logic, 29

relatively self-contained. For more details we refer the reader to [5], [11] or the widely available manuscript [12].

Given a connected scheme X and a geometric point ξ of X (i.e., a point with values in some algebraically closed field), we have the profinite étale fundamental group $\pi_1(X, \xi)$, which classifies finite étale coverings of X [12, Section 3]. This gives a covariant functor on the category of pointed schemes (X, ξ). As in topology, varying ξ just changes $\pi_1(X, \xi)$ up to an inner automorphism. Thus we shall usually omit the base point, writing $\pi_1(X)$, when we only require calculations up to conjugacy.

In the special case $X = \mathrm{Spec}(k)$, where k is a field, a geometric point ξ is just a choice of an algebraically closed overfield L of k, and $\pi_1(X, \xi)$ is just the Galois group $\mathrm{Gal}(k^{\mathrm{sep}}/k)$, where k^{sep} is the separable closure of k in L. Another interesting special case is that of normal X. If a connected variety X is normal, it is irreducible, say with generic point η. Its function field K is the residue field $\boldsymbol{k}(\eta)$. If we view an algebraic closure $\bar{K}$ of K as a geometric generic point $\bar{\eta}$ of X, the group $\pi_1(X, \bar{\eta})$ is the quotient of $\mathrm{Gal}(K^{\mathrm{sep}}/K)$ which classifies those finite separable extensions L/K with the property that the normalisation of X in L is finite étale over X (i.e., unramified).

Given a connected scheme X, a field k and a k-valued point $x \in X(k)$, the associated morphism $\mathrm{Spec}\, k \to X$ induces a group homomorphism

$$\pi_1(\mathrm{Spec}(k)) = \mathrm{Gal}(k^{\mathrm{sep}}/k) \to \pi_1(X),$$

well-defined up to conjugacy. The image of this map is called the *Artin symbol* at x and is denoted by $\mathrm{Ar}(x)$. We shall discuss several incarnations of it in Section 3.

If X is geometrically irreducible, we have the short exact sequence for the fundamental group:

$$1 \longrightarrow \pi_1^{\mathrm{geom}}(X, \bar{\eta}) \longrightarrow \pi_1(X, \bar{\eta}) \longrightarrow \mathrm{Gal}(k^{\mathrm{sep}}/k) \longrightarrow 1,$$

where $\pi_1^{\mathrm{geom}}(X, \bar{\eta}) = \pi_1(X \times_k \bar{k}, \bar{\eta})$ is the *geometric* fundamental group. As a contrast, π_1 is sometimes called the *arithmetic* fundamental group.

Let us denote by $\mathrm{\acute{E}}(X)$ the category of étale extensions of a scheme X. An (étale) *presheaf* $\mathcal{F}$ on X in a category $\mathcal{C}$ (of sets, abelian groups, rings ...) is a contravariant functor

$$\mathcal{F} : \mathrm{\acute{E}}(X) \to \mathcal{C}.$$

A presheaf $\mathcal{F}$ is an (étale) *sheaf,* if it satisfies these additional properties:

(1) $\mathcal{F}(U \dot{\cup} V) = \mathcal{F}(U) \times \mathcal{F}(V)$;

(2) when $U \to V$ is surjective, the following sequence is exact:

$$\mathcal{F}(V) \longrightarrow \mathcal{F}(U) \rightrightarrows \mathcal{F}(U \times_V U)$$

For an object F in $\mathcal{C}$, we can define the "constant sheaf" F_X on X by putting

$$F_X(U) = F \times \cdots \times F = F^{n(U)}, \quad U \in \mathrm{\acute{E}}(X),$$

where $n(U)$ denotes the number of connected components of U. Any sheaf isomorphic to a sheaf of this form is called *constant*.

DEFINITION 2.1 ([12, Section 6], [5, Chapter I.4]). (1) A sheaf $\mathcal{F}$ on a scheme X is called *locally constant*, if there is an étale covering $U_i \to X$ such that each $\mathcal{F} \restriction U_i$ is constant. If the $\mathcal{F}(U_i)$ are all finite, we shall say that $\mathcal{F}$ is *finite locally constant*.

(2) A sheaf $\mathcal{F}$ on X is *constructible*, if X can be written as a union of finitely many locally closed subschemes $Y \subseteq X$ such that $\mathcal{F} \restriction Y$ is finite locally constant.

The following characterisation of locally constant sheaves is of great interest for our purpose.

PROPOSITION 2.2 ([12, Proposition 6.16], [5, Proposition A I.7]). *Let $\mathcal{F}$ be a finite locally constant sheaf of abelian groups on a connected and normal scheme X. For any geometric point ξ of X, the stalk $\mathcal{F}_\xi$ is in a natural way a continuous $\pi_1(X,\xi)$-module. The assignment $\mathcal{F} \mapsto \mathcal{F}_\xi$ establishes an equivalence between the category of finite locally constant sheaves of abelian groups and the category of finite continuous $\pi_1(X,\xi)$-modules.*

In this context, *continuity* clearly implies that the action comes from a finite quotient of $\pi_1(X,\xi)$.

Let X be a connected and normal scheme and l a prime invertible in X. By analogy, we define a *lisse $\bar{\mathbb{Q}}_l$-sheaf* $\mathcal{F}$ of rank r on X as an r-dimensional continuous $\bar{\mathbb{Q}}_l$-representation of $\pi_1(X,\bar{\eta})$ (cf. [12, Section 19], [5, Proposition A I.8], [9, Appendix A]). A *constructible $\bar{\mathbb{Q}}_l$-sheaf* $\mathcal{F}$ on X is given when X can be written as a union of finitely many locally closed subschemes U_i such that each $\mathcal{F} \restriction U_i$ is lisse. Note that in this case the action need not factor through a finite quotient. In our study of definability, however, we mostly consider the $\bar{\mathbb{Q}}_l$-sheaves with the *(FQ)-property*; a lisse $\bar{\mathbb{Q}}_l$-sheaf $\mathcal{F}$ on X is labelled (FQ) if it factors through a finite quotient of $\pi_1(X)$, while a constructible sheaf is called (FQ) if it is such on each piece of X where it is lisse.

Let $\mathcal{F}$ be a constructible or a constructible $\bar{\mathbb{Q}}_l$-sheaf on X, let $x : \mathrm{Spec}(k) \to X$ be a k-valued point of X and let $\bar{x} : \mathrm{Spec}(k^{\mathrm{sep}}) \to X$ be a geometric point lying over x. Since the underlying module of the *stalk* $\mathcal{F}_{\bar{x}}$ coincides with that of the pullback $x^*\mathcal{F}$, which can (tautologically) be considered as a sheaf on $\mathrm{Spec}(k)$, the stalk $\mathcal{F}_{\bar{x}}$ is naturally endowed with the action of $\mathrm{Gal}(k^{\mathrm{sep}}/k) = \pi_1(x,\bar{x})$.

§3. Galois formulae vs. formulae. We start with a general consideration of Galois stratification, developed in [7], [6], but adopting the more geometric language as in [4].

Let A be an integral and normal variety. A morphism of varieties $C \to A$ is a *Galois cover*, if C is integral, h is étale, and there is a finite group $G = G(C/A)$ acting on C such that h induces the isomorphism $C/G \simeq A$.

A Galois cover $C \to A$ is *coloured*, if $G(C/A)$ is equipped with a family Con of subgroups stable by conjugation. Let S be an integral normal scheme and let $X \to S$ be a variety over S. A *normal stratification* of X,

$$\langle X, C_i/A_i : i \in I \rangle,$$

is a partition of X into a finite set of integral and normal locally closed S-subschemes A_i, each equipped with a Galois cover $C_i \to A_i$. A *Galois stratification*

$$\mathcal{A} = \langle X, C_i/A_i, \mathrm{Con}(A_i) | i \in I \rangle$$

consists of a normal stratification in which each Galois cover C_i/A_i is coloured (by $\mathrm{Con}(A_i)$).

Let S be an integral and normal scheme and let $X \to S$ be a variety over S. For a field k, a point $s \in S$, associated with a morphism $\operatorname{Spec} k \to S$, we denote by X_s the fibre of X over s ($X_s = X \times_S \boldsymbol{k}(s)$). Let $\mathcal{A} = \langle X, C_i/A_i, \mathrm{Con}(A_i) | i \in I \rangle$ be a Galois stratification of X, let $s \in S(k)$, and let $x \in A_{i,s}$. The *Artin symbol*, $\mathrm{Ar}(C_i/A_i, s, x)$ is the conjugacy class of subgroups of $G(C_i/A_i)$ consisting of the decomposition subgroups at x. More precisely, considering the map corresponding to x, $\mathrm{Spec}(k) \to A_{i,s} \to A_i$, we have the induced map

$$\mathrm{Gal}(k^{\mathrm{sep}}/k) \to \pi_1(A_{i,s}) \to \pi_1(A_i) \to G(C_i/A_i),$$

and $\mathrm{Ar}(C_i/A_i, s, x)$ is its image, defined up to conjugacy.

Let $\mathcal{A}$ be a Galois stratification over S, as above. We will call an expression of the form

$$A := \{x \in X : \mathrm{Ar}(x) \subseteq \mathrm{Con}(\mathcal{A})\}$$

a *Galois formula* over S. For a field k, $s \in S(k)$ and $x \in A_{i,s}(k)$, we write $\mathrm{Ar}(x) \subseteq \mathrm{Con}(\mathcal{A})$ for $\mathrm{Ar}(C_i/A_i, s, x) \subseteq \mathrm{Con}(C_i/A_i)$, and we consider the *set of k-valued points of $\mathcal{A}_s$*,

$$\mathcal{A}_s(k) := \{x \in X_s(k) : \mathrm{Ar}(x) \subseteq \mathrm{Con}(\mathcal{A})\}.$$

Formulae as above will be referred to as *Galois formulae*. The following result shows that every Galois formula corresponds to a formula. The converse is true only under additional assumptions on the base field, cf. [7].

PROPOSITION 3.1. *Let $Y \to X$ be a Galois cover over S with group G, and let $\mathcal{D} \subseteq G$ be a conjugacy domain of subgroups. There exists a formula $\theta^{\mathcal{D}}$ over S such that for every field k, and every $s \in S(k)$,*

$$\{x \in X_s(k) : \mathrm{Ar}(x) \subseteq \mathcal{D}\} = \{x \in X_s(k) : k \models \theta_s^{\mathcal{D}}(x)\}.$$

This result is stated under more restrictive hypotheses in [6], but they are not needed in the proof. For the convenience of the reader, we repeat the proof in our notation.

PROOF. We can reduce to the case of affine varieties and we may assume that $\mathcal{D}$ is *full*, i.e., that it contains all subgroups of each group in $\mathcal{D}$. Indeed, if $\mathcal{D}$ is a single conjugacy class of groups, then $\mathcal{D} = \mathcal{D}' \setminus \mathcal{D}''$, where $\mathcal{D}'$ is the conjugacy domain of all subgroups of the groups in $\mathcal{D}$ and $\mathcal{D}''$ is the conjugacy domain of all proper subgroups of the groups in $\mathcal{D}$. Then $\theta^{\mathcal{D}} \equiv \theta^{\mathcal{D}'} \wedge \neg\theta^{\mathcal{D}''}$. In general, a conjugacy domain is a union of conjugacy classes and the required formula will be a disjunction of formulae as above.

Consider the Galois extension F/E of function fields of Y and X. For each subgroup H of G, let E_H be the fixed field of H in F and let X_H be the normalisation of X in E_H. Clearly, Y/X_H is a Galois cover with group H.

Let $x \in X_s(k)$, and let $\bar{x}$ be a geometric point over x. We claim that for every H, the image of $\pi_1(x,\bar{x}) \to \pi_1(X,\bar{x}) \to G$ is a subgroup of H if and only if there exists a $z \in X_H(k)$ that maps onto x. This will then imply (when we forget to specify the base point $\bar{x}$) that $\mathrm{Ar}(x) \subseteq \mathcal{D}$ is equivalent to

$$\bigvee_{H \text{ max. in } \mathcal{D}} \exists z \in X_H(k)\ z \mapsto x,$$

the latter condition being definable by a formula over S.

To prove the claim, let us choose geometric points $\bar{y}$ in Y and $\bar{z}$ in X_H so that $\bar{y} \mapsto \bar{z} \mapsto \bar{x}$. If the image of $\pi_1(x,\bar{x})$ is contained in H and H fixes Z, then $\mathrm{Gal}(k)$ fixes $\bar{z}$ so we have found a k-rational point of X_H mapping onto x.

Conversely, if there is such a z, we have the diagram (the last column is obtained by quotienting the middle column by $\pi_1(Y,\bar{y})$):

$$\begin{array}{ccccc}
\pi_1(z,\bar{z}) & \longrightarrow & \pi_1(X_H,\bar{z}) & \longrightarrow & H \\
\downarrow & & \downarrow & & \downarrow \\
\pi_1(x,\bar{x}) & \longrightarrow & \pi_1(X,\bar{x}) & \longrightarrow & G
\end{array}$$

Since the left vertical arrow is an isomorphism, the image of $\pi_1(x,\bar{x})$ sits in H. ⊣

COROLLARY 3.2. *Let $\mathcal{F}$ be a constructible sheaf (resp. a constructible $\bar{\mathbb{Q}}_l$-sheaf with (FQ)) on a variety X over S. There is a finite number of finite modules (resp. finite-dimensional $\bar{\mathbb{Q}}_l$-vector spaces) M_i endowed with group actions such that for every field k, $s \in S(k)$, and $x \in X_s(k)$, the action of $\mathrm{Ar}(x)$ on $\mathcal{F}_x$ is isomorphic to some M_i. Moreover, for each M_i there is a formula θ^{M_i} over S*

such that

$$\{x \in X_s(k) : \textit{the action of } \mathrm{Ar}(x) \textit{ on } \mathcal{F}_x \textit{ is isomorphic to } M_i\} = \{x \in X_s(k) : k \models \theta_s^{M_i}(x)\}.$$

PROOF. Since $\mathrm{Ar}(x)$ is a conjugacy class of subgroups, it does make sense to talk about the isomorphism class of the action as above and it is clear that the isomorphism type is fixed on a union of conjugacy classes (a conjugacy domain) of subgroups. The sheaf has (FQ) using either assumption and on each stratum where it is locally constant (resp. lisse) we only need to worry about the actions obtained by restriction of the action of a finite quotient of the fundamental group to its subgroups. Definability follows directly from the proposition. ⊣

In model theory of fields, one often studies a class of fields with a specific structure of their absolute Galois groups. Although 3.2 looks quite interesting, it cannot be used to, say, compute traces or characteristic polynomials of some distinguished element of the Galois group of the ground field on the stalk $\mathcal{F}_{\bar{x}}$, unless one expands the language with symbols which allow it to be mentioned, as in [8]. To overcome this weakness of the language of rings, one is forced to restrict the class of acceptable sheaves, as expounded in the following definition.

DEFINITION 3.3. Let $\mathcal{K}$ be a class of fields containing a given field k_0. Let $\mathcal{F}$ be a constructible or a constructible $\bar{\mathbb{Q}}_l$-sheaf on a variety X over k_0. We shall say that $\mathcal{F}$ is *$\mathcal{K}$-invariant*, if for every $k \in \mathcal{K}$, for every k-rational point $x \in X(k)$, any two geometric points $\bar{x}_1, \bar{x}_2 \in X(k^{\mathrm{sep}})$ lying over x, and every group isomorphism $\varphi : \pi_1(x, \bar{x}_1) \to \pi_1(x, \bar{x}_2)$, we have the isomorphism of $\mathrm{Gal}(k) = \pi_1(x, \bar{x}_1)$-modules

$$\varphi^* \mathcal{F}_{\bar{x}_2} \simeq \mathcal{F}_{\bar{x}_1}.$$

Motivated by a part of the above definition, when two isomorphic groups G and G' act on modules M and M', we shall say that these actions are *strongly isomorphic* if, for *every* isomorphism $\varphi : G \to G'$, the G-modules $\varphi^* M'$ and M are isomorphic.

THEOREM 3.4. *Let $\mathcal{K}$ be a class of fields. Let $\mathcal{F}$ be a $\mathcal{K}$-invariant constructible sheaf (resp. constructible $\bar{\mathbb{Q}}_l$-sheaf with (FQ)) on a variety X over S. There is a finite number of finite modules (resp. finite-dimensional vector spaces over $\bar{\mathbb{Q}}_l$) M_i endowed with group actions such that for every $k \in \mathcal{K}$, $s \in S(k)$, and $x \in X_s(k)$, the action of* $\mathrm{Gal}(k)$ *on $\mathcal{F}_x$ is strongly isomorphic to some M_i. Moreover, for each M_i there is a formula θ^{M_i} over S such that*

$$\{x \in X_s(k) : \textit{the action of } \mathrm{Gal}(k) \textit{ on } \mathcal{F}_x \textit{ is strongly isomorphic to } M_i\} = \{x \in X_s(k) : k \models \theta_s^{M_i}(x)\}.$$

PROOF. Having in mind that the action of Gal(k) on $\mathcal{F}_{\bar{x}}$ is the pullback of the action of Ar(x), we argue as in 3.2. By k-invariance, on each stratum Y where $\mathcal{F}$ is locally constant, the left hand side of the above equality is of the form

$$\{x \in Y_s(k) : Ar(x) \subseteq \mathcal{D}\},$$

for some conjugacy domain $\mathcal{D}$. In other words, this set is given by a Galois formula over S. By 3.1, there is a formula over S equivalent to it. ⊣

REMARK 3.5. At the end of Section 2, we mentioned that there exist lisse $\mathbb{Q}_l$-sheaves which do not factor through a finite quotient of the fundamental group. It is intuitively clear that such sheaves encode information contained in an infinite sequence of constructible sheaves and that we cannot expect their invariants to be definable, but only possibly ∞-definable.

For example, let k be a field of characteristic different from a prime number l. For each n, consider the group T_n of l^n-th roots of unity in $\bar{k}$, together with the action of the absolute Galois group G of k. Then, $T := \varprojlim_n T_n \simeq \mathbb{Z}_l$ is a continuous G-module in its l-adic topology, and we obtain a lisse $\bar{\mathbb{Q}}_l$-sheaf as the representation of G on $T \otimes_{\mathbb{Z}_l} \bar{\mathbb{Q}}_l$. In particular, this sheaf "knows" which l^n-th roots of unity are contained in k for all n, and this is clearly not something expressible by a single formula.

§4. Applications.

4.1. Cohomology. Let us recall some facts about étale cohomology groups, as one of the most important classes of locally constant sheaves. Let X be connected over a finite field k and let $\mathcal{F}$ be a constructible sheaf. We have *étale cohomology groups with compact support* (cf. [12, Section 18], [5, Chapter I.8]),

$$H_c^i(\bar{X}, \mathcal{F}),$$

which are finite continuous Gal(k^{sep}/k)-modules, and which vanish unless $i \in \{0, \dots, 2d\}$. We consider them as locally constant sheaves on k.

Suppose now that S is a connected normal variety over $\mathbb{Z}$, and let $\pi : X \to S$ be a normal and connected variety over S. Let $\mathcal{F}$ be a constructible sheaf on X. It is shown in algebraic geometry [12, Proposition 18.4], [5, Theorem I.8.10] that the *higher direct images of $\mathcal{F}$ with compact support,* $R^i\pi_!\mathcal{F}$, are constructible sheaves on S. Given an algebraically closed field field Ω and any geometric point $\bar{s} \in S(\Omega)$, by the proper base change theorem [12, Theorem 17.10], [5, Theorem I.6.1, Theorem I.8.7] we have the *specialisation* property:

$$(R^i\pi_!\mathcal{F})_{\bar{s}} = H_c^i(X_{\bar{s}}, \mathcal{F}_{\bar{s}}),$$

where $X_{\bar{s}}$ is the geometric fibre over $\bar{s}$, and $\mathcal{F}_{\bar{s}}$ is the restriction (pullback) of $\mathcal{F}$ to $X_{\bar{s}}$.

THEOREM 4.1. *Let $\mathcal{F}$ be a (torsion!) constructible sheaf on a variety X over S. There are finitely many finite modules M_i^r endowed with group actions such that for every field k, $s \in S(k)$ and every integer r, there exists an i such that $H_c^r(\bar{X}_s, \mathcal{F})$ is isomorphic to M_i^r. Moreover, for all i and r, there is a formula θ_i^r such that*

$$\{s \in S(k) : H_c^r(\bar{X}_s, \mathcal{F}) \simeq M_i^r\} = \{s \in S(k) : k \models \theta_i^r(s)\}.$$

PROOF. The theorem follows from the étale cohomology facts mentioned above, namely the constructibility of higher direct images with proper supports and the specialisation property, together with 3.2. ⊣

4.2. Finite and pseudofinite fields. Let us consider a property related to $\mathcal{K}$-invariance, which turns out to be particularly well-behaved in the case of finite and pseudofinite fields. The absolute Galois group of a finite or a pseudofinite field being isomorphic to $\hat{\mathbb{Z}}$, our attention is drawn to the consideration of properties invariant of the choice of a topological generator of the Galois group.

DEFINITION 4.2. Let G be a profinite group. We shall call a function f on G $\mathbb{Q}$-central, if $f(x) = f(x')$ whenever x and x' topologically generate conjugate subgroups of G.

REMARK 4.3. Let $\mathcal{K}$ be the class of finite or pseudofinite fields. Suppose $\mathcal{F}$ is a lisse $\bar{\mathbb{Q}}_l$-sheaf on a scheme S. Then $\mathcal{F}$ is $\mathcal{K}$-invariant if and only if its character is $\mathbb{Q}$-central.

For a (torsion) constructible sheaf, it can be rather intricate to talk about its character. For a commutative ring Λ (usually $\Lambda = \mathbb{Z}/n\mathbb{Z}$), it is clearly possible when it is a sheaf of free Λ-modules, but it can be shown that traces can be taken even in the case of projective Λ-modules or even perfect complexes of Λ-modules [5, Chapter II.3]. By a slight abuse of notation, we will call a constructible (torsion) sheaf $\mathbb{Q}$-central if it is $\mathcal{K}$-invariant.

We have isolated the property of $\mathbb{Q}$-centrality precisely for the reason that, given a $\mathbb{Q}$-central sheaf $\mathcal{F}$, the trace or the characteristic polynomial of σ on $\mathcal{F}_{\bar{x}}$ does not depend on the choice of the topological generator σ of the absolute Galois group of the ground field. Moreover, it is *hereditary* in the sense that it is preserved under taking direct images with proper supports.

PROPOSITION 4.4. *Let $\mathcal{F}$ be a constructible (resp. constructible $\bar{\mathbb{Q}}_l$ with (FQ)) $\mathbb{Q}$-central sheaf on a variety $X \xrightarrow{\varphi} S$. Then the direct image with compact support, $\varphi_! \mathcal{F}$, is also $\mathbb{Q}$-central (resp. $\mathbb{Q}$-central and (FQ)).*

PROOF. Firstly, we reduce to the case where φ is a proper map. Stein factorisation tells us that φ can be written as a composition of a finite map and a map with geometrically connected fibres. In both cases we have an explicit description of the stalks of the direct image and it is straightforward to verify that $\mathbb{Q}$-centrality (resp. the property (FQ)) is inherited. ⊣

The above concepts are best applied in conjunction with the Grothendieck-Lefschetz fixed point formula and Deligne's theory of weights [3]. Using these techniques, the author shows in [14] the following conceptual improvement of the results from [6] and [2].

THEOREM 4.5. *Let S be a connected normal variety over $\mathbb{Z}$ and let $\mathcal{F}$ be a constructible $\bar{\mathbb{Q}}_l$-sheaf on a variety X over S which is $\mathbb{Q}$-central and (FQ). Let χ be its character and fix a non-canonical embedding $\iota : \bar{\mathbb{Q}}_l \to \mathbb{C}$. Then there is a localisation $\mathbb{Z}[1/n]$ of $\mathbb{Z}$, a finite number of (continuous) $\mathbb{Q}$-central characters $\alpha_i : \hat{\mathbb{Z}} \to \mathbb{C}$ and a constant $C > 0$ such that, writing S' for the restriction of S to $\mathbb{Z}[1/n]$, for every finite field k, its extension k_n of degree n, and every parameter $s \in S'(k)$, there exists an i with*

$$\left| \sum_{x \in X_s(k_n)} \iota\chi_s(F_{k_n,x}) - \alpha_i(n)|k_n|^{\dim(X_s)} \right| \leq C|k_n|^{\dim(X_s)-1/2},$$

where F_k denotes the Frobenius $x \mapsto x^{|k|}$ and $F_{k,x}$ its image in $\mathrm{Ar}(x)$, *the local Frobenius element.*

Moreover, for every i there is a formula θ_{α_i} in the language of rings which defines, in each finite field k, the set of $s \in S'(k)$ for which the above estimate holds.

For further applications in pseudofinite fields, such as showing the near-rationality of pseudofinite L-functions, results regarding pseudofinite version of Dirichlet density, we refer the reader to [14].

QUESTION 4.6. Is $\mathbb{Q}$-centrality preserved under *higher* direct images with proper supports? What is the most general class of fields $\mathcal{K}$ (with respect to properties of their absolute Galois groups) such that $\mathcal{K}$-invariance is preserved under taking higher direct images?

REFERENCES

[1] L. BRÜNJES and C. SERPÉ, *Nonstandard étale cohomology*, Preprint, 2003.

[2] ZOÉ CHATZIDAKIS, LOU VAN DEN DRIES, and ANGUS MACINTYRE, *Definable sets over finite fields*, ***Journal für die Reine und Angewandte Mathematik***, vol. 427 (1992), pp. 107–135.

[3] PIERRE DELIGNE, *La conjecture de Weil. II*, ***Institut des Hautes Études Scientifiques. Publications Mathématiques***, (1980), no. 52, pp. 137–252.

[4] JAN DENEF and FRANÇOIS LOESER, *Definable sets, motives and p-adic integrals*, ***Journal of the American Mathematical Society***, vol. 14 (2001), no. 2, pp. 429–469 (electronic).

[5] EBERHARD FREITAG and REINHARDT KIEHL, ***Étale Cohomology and the Weil Conjecture***, Ergebnisse der Mathematik und ihrer Grenzgebiete (3), vol. 13, Springer-Verlag, Berlin, 1988, Translated from the German by Betty S. Waterhouse and William C. Waterhouse, With an historical introduction by J. A. Dieudonné.

[6] MICHAEL D. FRIED, DAN HARAN, and MOSHE JARDEN, *Effective counting of the points of definable sets over finite fields*, ***Israel Journal of Mathematics***, vol. 85 (1994), no. 1-3, pp. 103–133.

[7] MICHAEL D. FRIED and MOSHE JARDEN, ***Field Arithmetic***, Ergebnisse der Mathematik und ihrer Grenzgebiete (3), vol. 11, Springer-Verlag, Berlin, 1986.

[8] Dan Haran and Moshe Jarden, *Bounded statements in the theory of algebraically closed fields with distinguished automorphisms*, ***Journal für die Reine und Angewandte Mathematik***, vol. 337 (1982), pp. 1–17.

[9] Reinhardt Kiehl and Rainer Weissauer, ***Weil Conjectures, Perverse Sheaves and*** l***-Adic Fourier Transform***, Ergebnisse der Mathematik und ihrer Grenzgebiete (3), vol. 42, Springer-Verlag, Berlin, 2001.

[10] Angus Macintyre, *Weil cohomology and model theory*, ***Connections Between Model Theory and Algebraic and Analytic Geometry***, Quaderni di Matematica, vol. 6, Deptartment of Mathimatics, Seconda University Napoli, Caserta, 2000, pp. 179–199.

[11] James S. Milne, ***Étale Cohomology***, Princeton Mathematical Series, vol. 33, Princeton University Press, Princeton, N.J., 1980.

[12] ———, *Lectures on Étale Cohomology*, 1998, Available online at `www.jmilne.org`.

[13] Ivan Tomašić, *A new Weil cohomology theory*, ***The Bulletin of the London Mathematical Society***, vol. 36 (2004), no. 5, pp. 663–670.

[14] ———, *Independence, measure and pseudofinite fields*, To appear in ***Selecta Mathematica***, 2004.

PMF-MATEMATIČKI ODJEL
BIJENIČKA 30
10000 ZAGREB, CROATIA
E-mail: tomasic@math.hr

AN OVERVIEW OF MODERN UNIVERSAL ALGEBRA

ROSS WILLARD

Abstract. This article, aimed specifically at mathematical logicians, gives a gentle introduction to some of the central achievements and problems of universal algebra during the last 25 years.

I intend in this article to introduce nonspecialists to the fact that there are deep results in contemporary universal algebra. The first four sections give the context in which universal algebra has something to say, and describe some of the basic results upon which much of the work in the field is built. Section 5 covers the highlights of tame congruence theory, a sophisticated point of view from which to analyze locally finite algebras. Section 6 describes some of the field's "big" results and open problems concerning finite algebras, notably the undecidability of certain finite axiomatizability problems and related problems, and the so-called "RS problem," currently the most important open problem in the field.

This article presents a personal view of current universal algebra, one which is limited both by my ignorance of large parts of the field as well as the likely interests of the intended audience and the need to keep the article focused. For example, I do not mention natural duality theory, one of the most vigorous subdisciplines of the field, nor algebraic logic or work that is motivated by and serves computer science. Therefore the views expressed in this article should not be taken to be a comprehensive statement of "what universal algebra is."

The text is frequently imprecise and proofs, when present, are merely sketched. Credit is not always given where due. My aim is to give the reader an impression of the field. Resources for further reading are provided in the final section.

I wish to thank Keith Kearnes and Emil Kiss for helpful criticisms of the penultimate version of this paper.

2000 *Mathematics Subject Classification.* Primary 08-02; Secondary 03C05.
Key words and phrases. universal algebra, survey.
Work supported by the NSERC of Canada.

Logic Colloquium '04
Edited by A. Andretta, K. Kearnes, and D. Zambella
Lecture Notes in Logic, 29

§1. Algebras and equational classes. The fundamental objects of study in universal algebra are *algebraic structures*, or *algebras* for short. These are structures in the traditional sense of model theory but with no relations. The algebras in this article can be assumed with no loss of generality to have a countable language, and often the language will be finite. Notationally I will use $\boldsymbol{A}$ to denote an algebra with universe A, and will write $\boldsymbol{A} = \langle A; f_0, f_1, \dots \rangle$. Unlike model theory, which is primarily interested in the structure $\langle A; \{\text{all definable sets of } \boldsymbol{A}\}\rangle$, universal algebra might be described as the study of the more mundane $\langle A; \{\text{all term functions of } \boldsymbol{A}\}\rangle$.

Universal algebra from the very beginning has focused on the *equations* $\forall \overline{x} : t_1(\overline{x}) = t_2(\overline{x})$ (also called *identities*) modeled by an algebra. This focus immediately gives rise in the usual fashion to the galois connection between *equational theories* (set of all equations modeled by an algebra, or by a class of algebras) and *equational classes* (class of all models of an equational theory). The latter are usually called *varieties* by universal algebraists but I will avoid this usage in this article.

While the basic constructions for model theory might be ultraproducts and elementary substructures, the basic constructions for universal algebra are simpler ones which preserve equations, including:

(1) *Homomorphic images* (i.e., images of surjective maps which preserve atomic formulas).
(2) *Subalgebras* (i.e., substructures).
(3) *Products* (i.e., full cartesian products over arbitrary index sets).

$\boldsymbol{H}$, $\boldsymbol{S}$ and $\boldsymbol{P}$ denote closure (of a class of algebras in the same language) under homomorphic images, subalgebras and products respectively. Two basic theorems concerning these closure operators are:

THEOREM 1.1 (G. Birkhoff). *A class $\mathcal{K}$ of algebras in the same language is an equational class if and only if it is closed under $\boldsymbol{H}$, $\boldsymbol{S}$ and $\boldsymbol{P}$.*

THEOREM 1.2 (A. Tarski). *Suppose $\mathcal{K}$ is a class of algebras in the same language. Then the smallest equational class containing $\mathcal{K}$ is $\boldsymbol{HSP}(\mathcal{K})$.*

There has been long-standing fascination with equational classes $\boldsymbol{HSP}(\boldsymbol{A})$ generated by a finite algebra $\boldsymbol{A}$. Perhaps the fascination stemmed from particular equational classes that were seen to be significant from the early days of universal algebra, such as:

(1) The class of boolean algebras. This class is known to be the equational class generated by the 2-element boolean algebra:

$$\{\text{boolean algebras}\} = \boldsymbol{HSP}(\langle 2; \wedge, \vee, {}', 0, 1\rangle).$$

(2) The class of distributive lattices. This class is known to be the equational class generated by the 2-element (distributive) lattice:

$$\{\text{distributive lattices}\} = \boldsymbol{HSP}(\langle 2; \wedge, \vee\rangle).$$

If an equational class has a finite language, one can ask if the class is finitely axiomatizable. Universal algebraists have been particularly interested in asking this question when the class is generated by a finite algebra. Two enduring problems are:

Finite Basis Problem (1950's). For which finite algebras $\boldsymbol{A}$ with a finite language is $\boldsymbol{HSP}(\boldsymbol{A})$ finitely axiomatizable?

Tarski's Finite Basis Problem (1960's). Is the answer to the Finite Basis Problem recursive?

The first problem has motivated a great deal of work over the last 50 years and is considered to be intractable in general but possibly solvable when restricted to well-behaved classes of algebras. I will return to these problems later in this article.

More recent work, meaning since 1980 or so, has largely focused on structural or "representational" questions. In general terms, the emphasis is on the study and classification of equational classes, particularly those generated by a finite algebra, according to the nature of the algebraic and combinatorial structure of their members.

§2. Congruences. Let h be a homomorphism from $\boldsymbol{A}$ to $\boldsymbol{B}$, i.e., a map $h : A \to B$ which preserves atomic formulas. h partitions the universe of A into blocks $h^{-1}(b), b \in h(A)$. Such a partition induced by a homomorphism is called a *congruence* of $\boldsymbol{A}$. Natural examples of congruences include the partition of a group given by the cosets of a normal subgroup, or the partition of a ring given by the cosets of an ideal. As in the case of groups and rings, there is an intrinsic definition of congruence which does not make reference to homomorphisms.

Given a congruence α of an algebra $\boldsymbol{A}$, there is a natural way to define a *quotient algebra* $\boldsymbol{A}/\alpha$; the universe of this algebra is the set of blocks of α, and $\boldsymbol{A}/\alpha$ itself is a homomorphic image of $\boldsymbol{A}$. The standard homomorphism and isomorphism theorems of group and ring theory have their counterparts in this general setting.

Fix an algebra $\boldsymbol{A}$. We shall consider the set $\operatorname{Con}\boldsymbol{A}$ of all congruences of $\boldsymbol{A}$. For congruences α, β of $\boldsymbol{A}$, we write $\alpha \leq \beta$ if α is a refinement of β, i.e., each α-block is contained in a single β-block. It turns out that $\langle \operatorname{Con}\boldsymbol{A}; \leq\rangle$ is a *bounded lattice*, i.e., with respect to the ordering $\leq$, $\operatorname{Con}\boldsymbol{A}$ has a greatest element 1 and a least element 0, and any two congruences α, β of $\boldsymbol{A}$ have a least upper bound $\alpha \vee \beta \in \operatorname{Con}\boldsymbol{A}$ and a greatest lower bound $\alpha \wedge \beta \in \operatorname{Con}\boldsymbol{A}$. The greatest element 1 is the partition of A having only the one block A, while the least element 0 is the partition of A in which every block has only one element.

The congruence lattice of an algebra is typically the first point of reference when a universal algebraist seeks to understand the structure of the algebra.

Just as a ring theorist might be interested in the maximal ideals of a ring, a universal algebraist might need to understand the maximal congruences of an algebra. Where a group theorist might consider a composition series of a finite group, a universal algebraist will consider a maximal increasing chain in the congruence lattice of a finite algebra.

Something that seems unique to universal algebra, however, is the observation that the possession by an algebra of term functions satisfying certain equations is connected to restrictions on the *shape* of the congruence lattice of the algebra. To illustrate, I offer two samples. Recall that the *distributive law* for lattices is

$$x \vee (y \wedge z) = (x \vee y) \wedge (x \vee z)$$

while the *modular law* is the restriction of this law to those instances with $x \leq y$.

THEOREM 2.1. *Suppose* $\boldsymbol{A}$ *has a Mal'cev term function, i.e., a ternary term function* $p(x, y, z)$ *such that* $\boldsymbol{A} \models \forall x, y : \ p(x, x, y) = p(y, x, x) = y$. *Then* Con $\boldsymbol{A}$ *satisfies the modular law.*

As an example, the term $(x \cdot y^{-1}) \cdot z$ defines a Mal'cev term function in any group; hence the congruence lattice (= normal subgroup lattice) of any group is modular.

THEOREM 2.2. *Suppose* $\boldsymbol{A}$ *has a majority term function, i.e., a ternary term function* $m(x, y, z)$ *such that* $\boldsymbol{A} \models \forall x, y : m(x, x, y) = m(x, y, x) = m(y, x, x) = x$. *Then* Con $\boldsymbol{A}$ *satisfies the distributive law.*

As an example, the term $(x \wedge y) \vee (y \wedge z) \vee (z \wedge x)$ defines a majority term function in any boolean algebra, and hence the congruence lattice (= lattice of filters) of any boolean algebra is distributive.

§3. Irreducible algebras. A *representation* of an algebra $\boldsymbol{A}$ is a collection $\{h_i : i \in I\}$ of homomorphisms with domain $\boldsymbol{A}$ which collectively separate the points of A. That is, if $a, b \in A$ with $a \neq b$ then there must exist $i \in I$ such that $h_i(a) \neq h_i(b)$.

The point of a representation is that it provides a way to embed the algebra into a product of potentially smaller algebras. For suppose $\{h_i : i \in I\}$ is a representation of $\boldsymbol{A}$ with $h_i : \boldsymbol{A} \to \boldsymbol{B}_i$. Then the map $a \mapsto (h_i(a))_{i \in I}$ embeds $\boldsymbol{A}$ into $\prod_{i \in I} \boldsymbol{B}_i$. If each h_i is surjective (which can be assumed), then each $\boldsymbol{B}_i$ is a homomorphic image of $\boldsymbol{A}$. Thus a representation of an algebra gives a way to realize it as a subalgebra of a product of (some of) its homomorphic images. One would like these images to be simpler than $\boldsymbol{A}$ itself. This will certainly not be the case if one of the images is $\boldsymbol{A}$ itself, i.e., if some h_i is injective. Thus we say that a representation $\{h_i : i \in I\}$ of $\boldsymbol{A}$ is *trivial* if some h_i is injective.

An algebra is *irreducible* (or *subdirectly irreducible*) if every representation of it is trivial. For example, the ring $\mathbb{Z}_n$ is irreducible if and only if n is a prime power. More generally, a ring is irreducible if and only if it has a least nonzero ideal, i.e., a nonzero ideal which is contained in every nonzero ideal. This generalizes beyond rings:

THEOREM 3.1. *An algebra having more than one element is irreducible if and only if it has a least nonzero congruence.*

Equational classes are generated by their irreducible members, as a consequence of Theorem 1.2 and the following theorem.

THEOREM 3.2 (G. Birkhoff). *For every algebra $\boldsymbol{A}$ having more than one element there exists a representation $\{h_i : i \in I\}$ of $\boldsymbol{A}$ such that each h_i is a surjective homomorphism from $\boldsymbol{A}$ onto an irreducible homomorphic image of $\boldsymbol{A}$.*

To close this section, we connect the concepts introduced so far by stating and proving the following theorem, whose proof (except for the first paragraph) was given by A. Foster and A. Pixley in the 1960's.

THEOREM 3.3. *Let $\boldsymbol{A}$ be a finite algebra. If* $\operatorname{Con} \boldsymbol{B}$ *is distributive for every $\boldsymbol{B} \in \boldsymbol{HSP}(\boldsymbol{A})$, then every irreducible member of $\boldsymbol{HSP}(\boldsymbol{A})$ has cardinality at most $|A|$.*

PROOF SKETCH. It can be shown that every irreducible algebra $\boldsymbol{D}$ can be embedded in an ultraproduct of finitely generated irreducible members of $\boldsymbol{HS}(\boldsymbol{D})$. Hence it suffices to restrict our attention to the finitely generated irreducible members of $\boldsymbol{HSP}(\boldsymbol{A})$.

Let $n = |A|$ and let $\boldsymbol{D}$ be a finitely generated irreducible member of $\boldsymbol{HSP}(\boldsymbol{A})$. Because $\boldsymbol{A}$ is finite, $\boldsymbol{HSP}(\boldsymbol{A})$ is *locally finite* (i.e., every finitely generated subalgebra of a member of $\boldsymbol{HSP}(\boldsymbol{A})$ is finite), so in particular $\boldsymbol{D}$ is finite.

We have $\boldsymbol{D} \in \boldsymbol{HSP}(\boldsymbol{A})$, which means there exist a nonempty set I, a subalgebra $\boldsymbol{B}$ of $\boldsymbol{A}^I$, and a congruence δ of $\boldsymbol{B}$ with $\boldsymbol{D} \cong \boldsymbol{B}/\delta$. We may assume $\boldsymbol{D} = \boldsymbol{B}/\delta$. Because $\boldsymbol{D}$ is finite, we may also assume that $\boldsymbol{B}$ is finitely generated and hence is finite.

Because $\boldsymbol{D}$ is irreducible, it has a least nonzero congruence μ. Because $\boldsymbol{D} = \boldsymbol{B}/\delta$, the congruences of $\boldsymbol{D}$ correspond exactly to the congruences of $\boldsymbol{B}$ above δ. Thus $\boldsymbol{B}$ has a least congruence ν strictly above δ. Since $\operatorname{Con} \boldsymbol{B}$ is a distributive lattice by assumption, it follows that δ is *meet-prime* in $\operatorname{Con} \boldsymbol{B}$; that is, if $\rho_0, \dots, \rho_{N-1}$ are congruences of $\boldsymbol{B}$ and $\rho_0 \wedge \cdots \rho_{N-1} \leq \delta$, then there exists $i < N$ such that $\rho_i \leq \delta$.

(Proof: assuming $\rho_0 \wedge \cdots \rho_{N-1} \leq \delta$ we have

$$
\begin{aligned}
\delta &= \delta \vee (\rho_0 \wedge \cdots \wedge \rho_{N-1}) \\
&= (\delta \vee \rho_0) \wedge \cdots \wedge (\delta \vee \rho_{N-1})
\end{aligned}
$$

where the last line is an application of the distributive law in $\operatorname{Con} \boldsymbol{B}$. Recall that ν is the least congruence of $\boldsymbol{B}$ strictly above δ. Thus for each $i < N$, since $\delta \leq \delta \vee \rho_i$, either $\delta = \delta \vee \rho_i$ or $\nu \leq \delta \vee \rho_i$. If $\delta \neq \delta \vee \rho_i$ for every $i < N$, then

ν would be a common lower bound to all the $\delta \vee \rho_i$ and hence

$$\delta < \nu \leq (\delta \vee \rho_0) \wedge \cdots \wedge (\delta \vee \rho_{N-1})$$

which would contradict the previous displayed equations. Therefore there exists $i < N$ such that $\delta = \delta \vee \rho_i$, which is equivalent to $\rho_i \leq \delta$.)

Because $\boldsymbol{B}$ is a subalgebra of $\boldsymbol{A}^I$, there is a representation $\{h_i\}$ of $\boldsymbol{B}$ consisting of homomorphisms from $\boldsymbol{B}$ into $\boldsymbol{A}$. As $\boldsymbol{B}$ is finite, the set $\{h_i\}$ can be taken to be finite, say $\{h_i\} = \{h_i : i < N\}$. For each $i < N$ let ρ_i be the congruence of $\boldsymbol{B}$ induced by h_i. As $\{h_i : i < N\}$ is a representation of $\boldsymbol{B}$, we have $\rho_0 \wedge \cdots \wedge \rho_{N-1} = 0$ in $\operatorname{Con} \boldsymbol{B}$. Hence $\rho_0 \wedge \cdots \rho_{N-1} \leq \delta$. As δ is meet-prime, there exists $i < N$ such that $\rho_i \leq \delta$. Hence ρ_i is a refinement of δ, so δ has no more blocks than ρ_i. As the number of blocks of δ is equal to $|D|$ (because $\boldsymbol{D} = \boldsymbol{B}/\delta$) while the number of blocks of ρ_i is equal to $|h_i(B)|$ and hence is at most n (because $h_i : \boldsymbol{B} \to \boldsymbol{A}$ and $|A| = n$), it follows that $|D| \leq n$. ⊣

§4. Abelian congruences and congruence modular equational classes. An algebra $\boldsymbol{A}$ is said to be *abelian* if it models the following infinite family of universal Horn sentences: for every term $f(\overline{x}, \overline{y})$ in the language of $\boldsymbol{A}$,

$$\forall \overline{a}, \overline{b}, \overline{c}, \overline{d} : f(\overline{a}, \overline{c}) = f(\overline{a}, \overline{d}) \leftrightarrow f(\overline{b}, \overline{c}) = f(\overline{b}, \overline{d}).$$

For example:

(1) Sets are abelian.
(2) R-modules (construed as abelian groups with endomorphisms named by the elements of R) are abelian.
(3) A group is abelian in the above sense if and only if it is abelian in the usual sense.
(4) No ring with identity having more than one element is abelian.
(5) In equational classes in which every member has a distributive congruence lattice, no algebra having more than one element is abelian.

This definition of "abelian" is not entirely well-behaved. For example, it is not generally true that homomorphic images of abelian algebras are abelian; in fact, *every* algebra is a homomorphic image of some abelian algebra. Nonetheless, it turns out to be a surprisingly useful notion in some contexts, as we shall see.

One can relativize the notion of abelianness to congruences. One does this by restricting the quantified parameters $\overline{a}, \overline{b}, \overline{c}, \overline{d}$ to those instances in which each pair a_i, b_i belong to a common block of the congruence, and likewise each pair c_j, d_j belong to a common block of the congruence. Then:

(1) Every congruence of an abelian algebra is abelian.
(2) A congruence of a group is abelian if and only if the normal subgroup corresponding to it is abelian in the usual sense.

(3) A congruence of a ring is abelian if and only if the ideal corresponding to it annihilates itself.
(4) In congruence distributive equational classes, only the zero congruence is abelian.

I have already defined the modular law for lattices and observed that it is a weaker than the distributive law. An equational class is *congruence modular* if Con $\boldsymbol{A}$ satisfies the modular law for every algebra $\boldsymbol{A}$ in the class. It turns out that essentially all equational classes of "classical" algebras (e.g., groups, rings, modules, boolean algebras, lattices) are congruence modular. Thus it was with some excitement in the 1980's that the community uncovered a rich and deep theory of the good behavior of the notion of abelianness in congruence modular equational classes [34, 10, 9, 8]. (For example, abelian algebras in this context are definitionally equivalent to R-modules, and homomorphic images of abelian algebras are abelian.) This theory is called *modular commutator theory*. At its most basic level it reveals how, in congruence modular equational classes, the abelian/nonabelian dichotomy mediates tight connections between (1) the term functions of $\boldsymbol{A}$, and (2) the local shape of Con $\boldsymbol{A}$. Somewhat more precisely,

(1) Abelian congruences restrict the complexity of term functions: the blocks support translations of linear functions over modules only.
(2) Nonabelian congruences enforce consequences of distributivity near the congruence.

Here is an example of what I mean by "enforcing consequences of distributivity" in the nonabelian case. Suppose $\boldsymbol{D}$ is an irreducible algebra with least nonzero congruence μ. Suppose moreover that $\boldsymbol{D} = \boldsymbol{B}/\delta$ for some algebra $\boldsymbol{B}$ and some congruence $\delta \in$ Con $\boldsymbol{B}$. Then in this situation, if $\boldsymbol{B}, \boldsymbol{D}$ belong to a congruence modular equational class and μ is nonabelian in Con $\boldsymbol{D}$, then δ is meet-prime in Con $\boldsymbol{B}$. Since this fact was the only consequence of distributivity used in the proof of Theorem 3.3, we have the following corollary.

COROLLARY 4.1. *Let* $\boldsymbol{A}$ *be a finite algebra. If* $\boldsymbol{HSP}(\boldsymbol{A})$ *is congruence modular, then every irreducible member of* $\boldsymbol{HSP}(\boldsymbol{A})$ *whose least nonzero congruence is nonabelian has cardinality at most* $|A|$.

The previous result is the first step in proving the following deep result about the irreducible members of finitely generated congruence modular varieties.

THEOREM 4.2 (R. Freese, R. McKenzie [7]). *If* $\boldsymbol{A}$ *is a finite algebra and* $\boldsymbol{HSP}(\boldsymbol{A})$ *is congruence modular, then*

(1) (*Gap*) *Either there is a finite bound to the cardinalities of the irreducibles in* $\boldsymbol{HSP}(\boldsymbol{A})$, *or there is no bound at all.*
(2) (*Recursive*) *If the language of* $\boldsymbol{A}$ *is finite, then one can effectively determine from* $\boldsymbol{A}$ *which alternative holds.*

Proof sketch. By the previous result, it suffices to consider only the irreducibles in $\boldsymbol{HSP}(\boldsymbol{A})$ with abelian least nonzero congruence. If such an irreducible has a nonabelian congruence "independent from" the least nonzero congruence (in a certain technical sense), then a uniform construction produces irreducibles in $\boldsymbol{HSP}(\boldsymbol{A})$ of every infinite cardinality. If no irreducible has such an independent nonabelian congruence, then the structure of each irreducible is relatively well understood. In particular, on the blocks of a large abelian congruence of the irreducible one can build a many-sorted irreducible module (each block an abelian group, each operation linear) in which (i) the number of sorts and (ii) the number of operations are finite and uniformly bounded. This implies a finite uniform bound to the size of these irreducibles, which moreover is computable from $\boldsymbol{A}$ if the language of $\boldsymbol{A}$ is finite. ⊣

This theorem and other data convinced D. Hobby and McKenzie to formulate in the 1980's what we call here the RS (gap) conjecture; its recursive companion was implicitly accepted as an inevitable consequence of the eventual proof of the conjecture.

RS Conjecture 4.3. For any finite algebra $\boldsymbol{A}$,

(1) (Gap) Either there is a finite bound to the cardinalities of the irreducibles in $\boldsymbol{HSP}(\boldsymbol{A})$, or there is no bound at all.
(2) (Recursive) If the language of $\boldsymbol{A}$ is finite, then one can effectively determine from $\boldsymbol{A}$ which alternative holds.

§5. Elements of tame congruence theory.

Also in the 1980's, Hobby and McKenzie developed a remarkable, exotic, deep tool for understanding fine structure in finite algebras. The analysis is essentially combinatorial and is built on the idea of "localization at a minimal nonzero congruence." This section introduces the reader to the basic definitions and ideas of the theory.

5.1. Polynomial subreducts. Let $\boldsymbol{A}$ be an algebra. A *polynomial function* of $\boldsymbol{A}$ is a term function allowing parameters from A. For any nonempty subset $S \subseteq A$ we define a new algebra $\boldsymbol{A}|_S$ (in a new language) as follows:

(1) The universe of $\boldsymbol{A}|_S$ is S.
(2) The functions of $\boldsymbol{A}|_S$ are all restrictions to S of n-ary polynomial functions f of $\boldsymbol{A}$ (all n) which satisfy $f(S^n) \subseteq S$.

This new algebra $\boldsymbol{A}|_S$ is called the *polynomial algebra induced on S* (by $\boldsymbol{A}$).

Example 5.1. Let $\boldsymbol{V}$ be a 1-dimensional vector space over $GF(4)$. The polynomial functions of $\boldsymbol{V}$ are all n-ary functions of the form

$$\sum_{i=1}^{n} \lambda_i x_i + a, \ \lambda_i \in GF(4), \ a \in V.$$

Let $S = V \setminus \{0\}$. Then every function of $\boldsymbol{V}|_S$ is either constant or depends on only one variable, and in the latter case is either the identity map or a cyclic permutation of S applied to that variable. Let $\boldsymbol{S}_1$ be the algebra $\langle S;\ g(x)\ (g \in C_3)\rangle$ where C_3 is the cyclic permutation group of order 3 acting on S. $\boldsymbol{S}$ is an example of a G-set (with $G = C_3$). Then $\boldsymbol{V}|_S$ is *polynomially equivalent* to $\boldsymbol{S}$, meaning that $\boldsymbol{V}|_S$ and $\boldsymbol{S}$ have the same universe and the same polynomial functions.

EXAMPLE 5.2. Start with the alternating group $\boldsymbol{A}_4$ and let V be its unique 4-element normal subgroup. We shall describe the polynomial algebra induced on V by $\boldsymbol{A}_4$. Let $\boldsymbol{V}$ be the abelian group V expanded to become a 1-dimensional vector space over $GF(4)$. The polynomial functions of $\boldsymbol{V}$ were described in the previous example. It can be shown that $\boldsymbol{A}_4|_V$ is polynomially equivalent to $\boldsymbol{V}$.

5.2. Idempotent ranges, minimal sets and traces. Tame congruence theory exploits information contained in polynomial algebras induced on certain sets associated with congruences or pairs of congruences, and especially certain sets called *minimal sets*, *bodies* and *traces* associated with a minimal nonzero congruence.

DEFINITION 5.3. Fix a finite algebra $\boldsymbol{A}$ and a minimal nonzero congruence α of $\boldsymbol{A}$.

(1) An *idempotent* of $\boldsymbol{A}$ is a unary polynomial function $e(x)$ of $\boldsymbol{A}$ such that $\boldsymbol{A} \models \forall x : e(e(x)) = e(x)$.
(2) An *idempotent range* of $\boldsymbol{A}$ is the range of any idempotent of $\boldsymbol{A}$.
(3) $\mathcal{N}_\alpha$ is the set of those idempotent ranges of $\boldsymbol{A}$ whose intersection with at least one α-block has more than one element.
(4) An *α-minimal set* is any member of $\mathcal{N}_\alpha$ which is minimal among all members of $\mathcal{N}_\alpha$ with respect to inclusion.
(5) An *α-trace* is any intersection of an α-minimal set with an α-block, provided the intersection has at least two elements.
(6) An *α-body* is the union of all the α-traces contained in some α-minimal set.

EXAMPLE 5.4. Consider the alternating group $\boldsymbol{A}_4$. It has exactly three normal subgroups: A_4, $\{1\}$, and V. Hence it has exactly three congruences: 1, 0 and the congruence α whose blocks are the cosets of V. $\boldsymbol{A}_4$ has several hundred idempotents and idempotent ranges. Let us explore a tiny fraction of them. Fix an element $a \in A_4$ of order 3. Define

$$e_1(x) = x^4 \qquad U_1 = e_1(A_4)$$
$$e_2(x) = a(a(ax^4)^4)^4 \qquad U_2 = e_2(A_4)$$
$$e_3(x) = a(a^{-1}x)^3 \qquad U_3 = e_3(A_4).$$

Since each element of V has order 2 or 1 while each element of $A_4 \setminus V = aV \cup a^2V$ has order 3, it is easy to see that $e_1(x) = x$ for all $x \in \{1\} \cup aV \cup a^2V$

while $e_1(x) = 1$ for $x \in V \setminus \{1\}$. In particular, e_1 is idempotent, $U_1 = \{1\} \cup aV \cup a^2V$, and U_1 is an idempotent range.

Using the above description of e_1 it can be deduced that e_2 maps V to 1, aV to a and a^2V to a^2. Hence e_2 is an idempotent, $U_2 = \{1, a, a^2\}$, and U_2 is an idempotent range. (In fact, every transversal of the cosets of V is an idempotent range.)

Again by considering orders of elements, it is easy to see that $e_3(x) = x$ for $x \in aV$ while $e_3(x) = a$ for $x \in V \cup a^2V$. Hence e_3 is idempotent, $U_3 = aV$, and U_3 is an idempotent range. (Similarly, V and a^2V are idempotent ranges.)

$U_2 \notin \mathcal{N}_\alpha$ since U_2 does not meet any α-block in more than one element. By contrast, both U_1 and U_3 are in $\mathcal{N}_\alpha$. U_1 is not an α-minimal set because it properly contains another member of $\mathcal{N}_\alpha$, for example U_3. Finally, it can be shown (better: a computer can show) that U_2 is an α-minimal set. In fact, the cosets of V are precisely the α-minimal sets of $\mathbf{A}_4$.

Since each α-minimal set in this example is contained entirely inside an α-block, the α-traces and α-bodies are identical to the α-minimal sets, i.e., the cosets of V.

The example just considered is rather simpler than what one should expect. In general, α-traces are *not* entire α-blocks, but subsets of α-blocks. α-traces are *not* identical to α-minimal sets, because α-minimal sets will *not* be contained entirely inside a single α-block. An α-minimal set may be strictly larger than its α-body, and an α-body may consist of more than one α-trace.

5.3. The five-fold classification. The notions of α-minimal set and α-trace may seem at first sight to be rather ad hoc. The first hint of their importance appears in the following theorem of P.P. Pálfy [33].

THEOREM 5.5. *Let $\mathbf{A}$ be a finite algebra, α a minimal nonzero congruence of $\mathbf{A}$, and N an α-trace. Then the induced polynomial algebra $\mathbf{A}|_N$ is polynomially equivalent to one of:*

(1) *A primitive permutation group construed as a G-set (as in Example 5.1), or a 2-element set with no functions.*
(2) *A 1-dimensional vector space over a finite field.*
(3) *A 2-element boolean algebra.*
(4) *A 2-element lattice $\langle 2; \wedge, \vee\rangle$.*
(5) *A 2-element semilattice $\langle 2; \wedge\rangle$.*

The second hint of the importance of α-traces is the following theorem of Hobby and McKenzie.

THEOREM 5.6. [13, Corollary 5.2(2)] *Let $\mathbf{A}$ be a finite algebra and α a minimal nonzero congruence of $\mathbf{A}$. The polynomial algebras induced on any two α-traces are isomorphic.*

The last sentence needs amplification, since it is problematic how to compare the languages of different induced polynomial algebras. The proper interpretation is that *there exists an identification of their languages with respect to which* the induced polynomial algebras are isomorphic.

Taken together, the two previous theorems assign to each minimal nonzero congruence α of a finite algebra an invariant, namely, the unique (up to isomorphism) primitive permutation group as a G-set, 1-dimensional vector space, or 2-element boolean algebra, lattice, semilattice or set with no functions which is isomorphic to the polynomial algebra induced on any α-trace. We declare the *type* of α to be the numerical label (1–5) of the item in Theorem 5.5 which contains this invariant. For example, the unique minimal congruence of the alternating group A_4 considered in Example 5.4 is "Type 2." In general, for every finite algebra A, the minimal nonzero congruences of A are each assigned a type from $\{1, 2, 3, 4, 5\}$. This is analogous to the abelian/nonabelian dichotomy for congruences described in Section 4, except that (i) the classification has five flavors rather than two; (ii) it applies only to finite algebras; and (iii) it applies only to minimal nonzero congruences of those algebras.

Theorem 5.6 remains true if "α-trace" is replaced with "α-body" or "α-minimal set." For reasons that will not be fully explained in this article, understanding the structure of the polynomial algebras induced on α-bodies and α-minimal sets is fundamental to applications the theory. However, these polynomial algebras are *not* classified up to isomorphism in all cases; in particular, there is no analogue of Theorem 5.5 for α-minimal sets.

5.4. Globalization (Part 1). The type of a minimal nonzero congruence of a finite algebra describes the *local* structure of the algebra within blocks of the congruence. To be useful, this local structure must reflect the global structure of the algebra. One mechanism that ties the local and global structures together is the "geometry" of α-traces and α-minimal sets with respect to polynomial functions. The following result is a rudimentary statement of the situation.

THEOREM 5.7. [13, Theorem 2.8, Example 2.9(2)] *Let A be a finite algebra and α a minimal nonzero congruence of A.*

(1) *Connectedness. Each α-block having more than one element is the connected union of α-traces.*
(2) *Isomorphism. For any two α-traces N, N' there exist unary polynomial functions f, g of A such that $f|_N$ is a bijection from N to N' and $g|_{N'} = (f|_N)^{-1}$. Moreover, there is an identification of the languages of $A|_N$ and $A|_{N'}$ with respect to which $f|_N$ and $g|_{N'}$ are mutually inverse isomorphisms.*
(3) *Density. Any two distinct elements a, b of an α-block can be mapped by some unary polynomial function f of A to distinct elements of an α-trace. Moreover, f can be taken to be an idempotent of A whose range is an α-minimal set.*

Using slightly more information than this, one can prove the following.

THEOREM 5.8. [13, Theorem 5.5, 5.6] *Let $\mathbf{A}$ be a finite algebra and α a minimal nonzero congruence. Then α is abelian if and only if its type is* 1 *or* 2.

PROOF SKETCH. If the type of α is 3, 4 or 5, and N is an α-trace, then $\mathbf{A}|_N$ is a 2-element boolean algebra, lattice or semilattice and hence is nonabelian. Since the functions of $\mathbf{A}|_N$ are derived from those of $\mathbf{A}$, the failure of $\mathbf{A}|_N$ to be abelian also witnesses the failure of α to be an abelian congruence of $\mathbf{A}$.

Conversely, assume the type of α is 1 or 2. The geometry of α-traces would in this case permit a failure of the abelian condition for α to be localized in an α-trace N and hence in the induced algebra $\mathbf{A}|_N$. But $\mathbf{A}|_N$ is either a vector space, a G-set, or a set with no functions, all of which are abelian. Hence α must be abelian. ⊣

5.5. Generalization to non-minimal congruences. There is a simple way to extend the above classification to non-minimal congruences. Suppose $\mathbf{A}$ is a finite algebra and let δ, α be congruences of $\mathbf{A}$ with $\delta < \alpha$ and such that α is minimal among all congruences strictly greater than δ. Such a pair (δ, α) is called a *cover pair* (or *prime quotient*). Given the cover pair (δ, α), the larger congruence α corresponds in the standard way to a minimal nonzero congruence $\overline{\alpha}$ of $\mathbf{A}/\delta$. We can then assign to the cover pair (δ, α) the type 1–5 of $\overline{\alpha}$ in $\mathbf{A}/\delta$. In this way every cover pair in $\operatorname{Con} \mathbf{A}$ is assigned one of five types.

Many applications of the theory require information about cover pairs (δ, α) that is not available in the quotient algebra $\mathbf{A}/\delta$. Additional information of interest to us is captured in the polynomial algebras induced on certain subsets of A which, when collapsed by δ, become minimal sets, bodies or traces for $\overline{\alpha}$ in $\mathbf{A}/\delta$. We define a (δ, α)-*minimal set* to be a subset U of A which (i) is the range of an idempotent of $\mathbf{A}$, (ii) becomes an $\overline{\alpha}$-minimal set of $\mathbf{A}/\delta$ when collapsed by δ, and (iii) is minimal under inclusion with respect to properties (i) and (ii). A (δ, α)-*trace* is an intersection of a (δ, α)-minimal set with an α-block, provided the intersection meets at least two distinct δ-blocks. A (δ, α)-body is the union of all (δ, α)-traces contained in a (δ, α)-minimal set.

The polynomial algebras induced on (δ, α)-traces by $\mathbf{A}$ are completely understood in the type 3 and 4 cases only; in the remaining three cases, they are largely understood, and of course are completely understood modulo δ (as their quotients modulo δ are the polynomial algebras induced on $\overline{\alpha}$-traces by $\mathbf{A}/\delta$ and these are described in Theorem 5.5). Less is known of the polynomial algebras induced on (δ, α)-bodies and (δ, α)-minimal sets. When applications demand it, one must occasionally make a detailed study of these induced algebras.

5.6. Globalization (Part 2). In addition to the geometry of traces and minimal sets, there are two more elements of the theory that link the local and global structures of a finite algebra. These elements relate two of the most basic notions of universal algebra – congruences and equations – between $\boldsymbol{A}$ and polynomial algebras induced on minimal sets, bodies and traces by $\boldsymbol{A}$.

LEMMA 5.9. [13, Lemma 2.3] *Let $\boldsymbol{A}$ be an algebra, U the range of an idempotent of $\boldsymbol{A}$, and S any nonempty subset of A.*

(1) *The map $\theta \mapsto \theta|_S$ is a $\wedge$-preserving map from* $\operatorname{Con} \boldsymbol{A}$ *into* $\operatorname{Con} \boldsymbol{A}|_S$.
(2) *The map $\theta \mapsto \theta|_U$ is a $\wedge$- and $\vee$-preserving map from* $\operatorname{Con} \boldsymbol{A}$ *onto* $\operatorname{Con} \boldsymbol{A}|_U$.

LEMMA 5.10. [13, Proof of Lemma 8.3] *Let $\boldsymbol{A}$ be a finite algebra and let $\mathcal{L}_e(\boldsymbol{A})$ be the reduct of the language of $\boldsymbol{A}|_A$ which consists of those functions which satisfy $\boldsymbol{A}|_A \models \forall x : f(x, \ldots, x) = x$. Then for every set $T \subseteq A$ which is either a (δ, α)-minimal set or (δ, α)-trace for some cover pair (δ, α) of $\boldsymbol{A}$, there is an identification of $\mathcal{L}_e(\boldsymbol{A})$ with a reduct of the language of $\boldsymbol{A}|_T$ such that every plain $\mathcal{L}_e(\boldsymbol{A})$-equation, that is, an equation of the form $f(x_{i_1}, \ldots, x_{i_n}) = g(x_{j_1}, \ldots, x_{j_m})$ with $f, g \in \mathcal{L}_e(\boldsymbol{A})$, if universally true in $\boldsymbol{A}|_A$, is also universally true in $\boldsymbol{A}|_T$.*

To illustrate the application of the previous two lemmas, we sketch the proof of two technical results that will be cited in the next section. Note the interplay between algebras $\boldsymbol{B}$ and their induced polynomial algebras $\boldsymbol{B}|_S$ in the proofs.

THEOREM 5.11. [13, Theorem 8.5] *Let $\boldsymbol{A}$ be a finite algebra. Then $\boldsymbol{HSP}(\boldsymbol{A})$ is congruence modular if and only if for every finite $\boldsymbol{B} \in \boldsymbol{HSP}(\boldsymbol{A})$ and every cover pair (δ, α) of $\boldsymbol{B}$,*

(1) $\operatorname{typ}(\delta, \alpha) \notin \{1, 5\}$.
(2) *(δ, α)-bodies coincide with (δ, α)-minimal sets.*

SKETCH OF PROOF. Assume first that $\boldsymbol{HSP}(\boldsymbol{A})$ is congruence modular. Then this fact is witnessed by the existence of a system of plain $\mathcal{L}_e(\boldsymbol{A})$-equations of a certain kind (known as *Day identities*) holding universally in $\boldsymbol{A}|_A$. Moreover, the functions occurring in the Day identities can be defined by terms in the language of $\boldsymbol{A}$ without parameters. Hence the universal validity of these Day identities is inherited by $\boldsymbol{B}|_B$ for every $\boldsymbol{B} \in \boldsymbol{HSP}(\boldsymbol{A})$ and consequently (using Lemma 5.10) by $\boldsymbol{B}|_T$ whenever $\boldsymbol{B}$ is finite and T is a minimal set or trace for a cover pair of $\boldsymbol{B}$.

Now let $\boldsymbol{B}$ be a finite member of $\boldsymbol{HSP}(\boldsymbol{A})$, (δ, α) a cover pair of $\boldsymbol{B}$, U a (δ, α)-minimal set and N a (δ, α)-trace. We may factor by δ and thus assume $\delta = 0$. Then $\boldsymbol{B}|_N$ is completely known (Theorem 5.5) and cannot model Day identities in case its type is 1 or 5. A careful analysis shows that $\boldsymbol{B}|_U$ cannot model Day identities if U does not coincide with its body. This proves items (1) and (2).

Conversely, assume that for every finite $\boldsymbol{B} \in \boldsymbol{HSP}(\boldsymbol{A})$ and every cover pair (δ, α) of $\boldsymbol{B}$, items (1) and (2) above hold. It suffices to prove that $\operatorname{Con} \boldsymbol{B}$ is a modular lattice for each *finite* member $\boldsymbol{B}$ of $\boldsymbol{HSP}(\boldsymbol{A})$. For each cover pair (δ, α) of such a $\boldsymbol{B}$ pick a (δ, α)-minimal set $U = U_{(\delta,\alpha)}$ and let $I = \{U_{(\delta,\alpha)} : (\delta, \alpha) \text{ a cover pair of } \boldsymbol{B}\}$. By items (1) and (2) and detailed

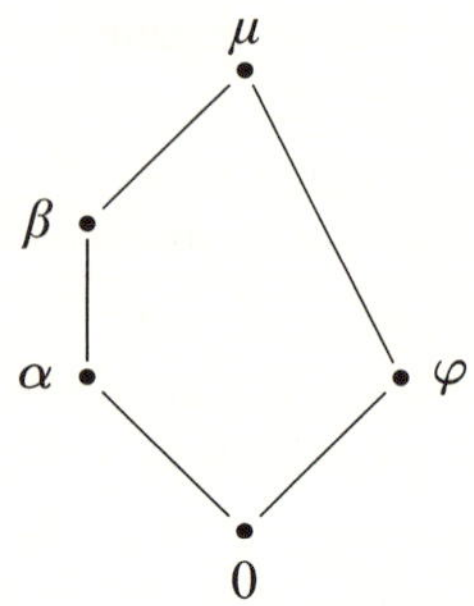

FIGURE 1. Pentagon sublattice.

knowledge of polynomial algebras induced on bodies of types 2–4, it can be shown that Con $\boldsymbol{B}|_U$ is a modular lattice for each $U \in I$.

It can be deduced from Lemma 5.9 that the family of maps $\{\theta \mapsto \theta|_U : U \in I\}$ is a representation of the lattice Con $\boldsymbol{B}$ by the modular lattices Con $\boldsymbol{B}|_U$ $(U \in I)$. This implies Con $\boldsymbol{B}$ itself is a modular lattice, as required. ⊣

LEMMA 5.12. [13, Lemma 6.5] *Suppose $\boldsymbol{A}$ is a finite algebra and* $0, \alpha, \beta, \theta, \mu$ *are distinct congruences forming a pentagon sublattice of* Con $\boldsymbol{A}$ *as pictured in Figure* 1. *Assume* (α, β) *and* $(0, \varphi)$ *are cover pairs. If* $\text{typ}(\alpha, \beta) \neq 1$ *then* $\text{typ}(0, \varphi) \in \{3, 4, 5\}$.

SKETCH OF PROOF. Let U be an (α, β)-minimal set, B its body, and $T = U \setminus B$. The map $\theta \mapsto \theta|_B$ is a $\wedge$-preserving map from Con $\boldsymbol{A}$ to Con $\boldsymbol{A}|_B$ by Lemma 5.9, from which it follows that either $\varphi|_B = 0|_B$ or $\{\alpha|_B, \beta|_B, \varphi|_B\}$ generate a pentagon sublattice of Con $\boldsymbol{A}|_B$. Using detailed knowledge of $\boldsymbol{A}|_B$ in types 2–5, one can show that the latter case is possible only if $\text{typ}(\alpha, \beta) = 1$.

So assume $\varphi|_B = 0|_B$. Because $\alpha|_B < \beta|_B$ and $\alpha|_T = \beta|_T$ (by definition of (α, β)-bodies) while $\beta_U \leq \alpha|_U \vee \beta|_U$ (by Lemma 5.9) and $\varphi|_B = 0|_B$ (by assumption), it follows that some block of $\varphi|_U$ is contained neither entirely in B nor entirely in T. Assume $\text{typ}(0, \varphi) \notin \{3, 4, 5\}$. Then φ is abelian by Theorem 5.8, and this is inherited by $\varphi|_U$. Now a careful analysis of $\boldsymbol{A}|_U$ in types 2–5 shows that in those cases it cannot have a congruence $\varphi|_U$ with these two properties. Hence $\text{typ}(\alpha, \beta) = 1$ if $\text{typ}(0, \varphi) \notin \{3, 4, 5\}$ in this case. ⊣

5.7. A sample application. One of the early successes of tame congruence theory was the following application by Hobby and McKenzie to the RS conjectures.

THEOREM 5.13. [13, Theorem 9.18, 10.4] *Let ε be any lattice equation which is nontrivial, i.e., fails to be true in some lattice. Then Theorem* 4.2 *is true with "ε" replacing "modularity." That is, the "Gap" and "Recursive" RS conjectures hold for those $\boldsymbol{A}$ for which* Con $\boldsymbol{B}$ *universally satisfies ε for every $\boldsymbol{B} \in \boldsymbol{HSP}(\boldsymbol{A})$.*

SKETCH OF PROOF. One can effectively determine from $\boldsymbol{A}$ whether $\boldsymbol{HSP}(\boldsymbol{A})$ is congruence modular. By Theorem 4.2, we may therefore assume that $\boldsymbol{HSP}(\boldsymbol{A})$ is not congruence modular. What will be shown in this case is that there is no bound to the cardinalities of the irreducibles in $\boldsymbol{HSP}(\boldsymbol{A})$.

The hypothesis that Con $\boldsymbol{B}$ satisfies ε for every $\boldsymbol{B} \in \boldsymbol{HSP}(\boldsymbol{A})$ yields a system of plain $\mathcal{L}_e(\boldsymbol{A})$-equations holding universally in $\boldsymbol{A}|_A$ but which does not hold universally in any primitive permutation group as a G-set, the 2-element set with no functions, or the 2-element semilattice. As in the proof of Theorem 5.11, it follows that no cover pair for a finite algebra in $\boldsymbol{HSP}(\boldsymbol{A})$ can have type 1 or 5.

Because $\boldsymbol{HSP}(\boldsymbol{A})$ is not congruence modular, there is a finite algebra $\boldsymbol{B} \in \boldsymbol{HSP}(\boldsymbol{A})$ for which Con $\boldsymbol{B}$ has a pentagon sublattice as in Figure 1. By finiteness of Con $\boldsymbol{B}$ we may assume that (α, β) is a cover pair. Moreover, we may assume that either $(0, \varphi)$ is a cover pair or the pentagon sublattice can be extended to a larger sublattice whose shape is known by the technical name "D_2." A lemma similar to Lemma 5.12 state that whenever Con $\boldsymbol{B}$ has a D_2 sublattice, a particular cover pair in the convex closure of the D_2 sublattice must have type 1 or 5. Since we have established that $\boldsymbol{HSP}(\boldsymbol{A})$ omits types 1 and 5, it follows that we are in the former case, i.e., $(0, \varphi)$ can be assumed to be a cover pair. Then Lemma 5.12 implies that $\operatorname{typ}(0, \varphi) = 3$ or 4.

The argument now breaks into cases according to the type of (α, β). If $\operatorname{typ}(\alpha, \beta) = 2$ then in $\boldsymbol{B}$ we essentially have a 2-element lattice or boolean algebra (defined on a $(0, \varphi)$-trace) "independent from" a vector space (defined on an (α, β)-trace). A uniform construction similar in spirit to the Freese-McKenzie construction in the congruence modular case produces irreducibles in $\boldsymbol{HSP}(\boldsymbol{A})$ of every infinite cardinality. On the other hand, if $\operatorname{typ}(\alpha, \beta) = 3$ or 4 then the underlying geometry together with the "independence" of two polynomial algebras induced on type 3 or 4 (α, β)-minimal sets form the basis for a separate construction. ⊣

Note that the role of the assumption about the lattice equation ε in the preceding proof was exactly to eliminate the presence of types 1 and 5 in $\boldsymbol{HSP}(\boldsymbol{A})$. Thus we have the following corollary of the proof.

COROLLARY 5.14. *The "Gap" and "Recursive" RS conjectures hold for all finite algebras $\boldsymbol{A}$ for which $\boldsymbol{HSP}(\boldsymbol{A})$ omits types* 1 *and* 5.

§6. Universal algebra since 1985: a sampler.

The deepest results in universal algebra in the last twenty years have depended almost without exception on the tools made available by tame congruence theory and modular commutator theory. This section contains a few of the highlights.

6.1. Decidable first-order theories and good structure. The *Decidable Equational Class Problem*, in its simplest version, asks the following two questions:

(1) For which finite algebras $\boldsymbol{A}$ in a finite language is the first-order theory of $\boldsymbol{HSP}(\boldsymbol{A})$ recursive?
(2) For which finite algebras $\boldsymbol{A}$ does the first-order theory of $\boldsymbol{HSP}(\boldsymbol{A})$ fail to interpret the theory of graphs?

In the late 1980's, McKenzie and M. Valeriote [31], building on similar work of S. Burris and McKenzie [4] in the congruence modular case, completely solved (2) and reduced (1) to the following question:

(1)′ For which finite rings R is the first-order theory of R-modules recursive?

Their solution uses tame congruence theoretic arguments that fill a book to prove the following.

THEOREM 6.1. [31] *If $\boldsymbol{A}$ is a finite algebra, then the first-order theory of $\boldsymbol{HSP}(\boldsymbol{A})$ fails to interpret the theory of graphs if and only if $\boldsymbol{HSP}(\boldsymbol{A})$ itself decomposes in a particularly strong way as a "product"*

$$\boldsymbol{HSP}(\boldsymbol{A}) = \mathcal{V}_1 \otimes \mathcal{V}_2 \otimes \mathcal{V}_3$$

where

(1) *For $i = 1, 2, 3$, $\mathcal{V}_i$ is an equational subclass of $\boldsymbol{HSP}(\boldsymbol{A})$ in which only type i occurs.*
(2) *Every member $\boldsymbol{B} \in \boldsymbol{HSP}(\boldsymbol{A})$ decomposes uniquely as a direct product $\boldsymbol{B}_1 \times \boldsymbol{B}_2 \times \boldsymbol{B}_3$ with $\boldsymbol{B}_i \in \mathcal{V}_i$.*
(3) *Every member of $\mathcal{V}_1 \cup \mathcal{V}_2$ is abelian.*
(4) *$\mathcal{V}_3$ is a finitely generated discriminator variety (a well-studied category of congruence distributive equational classes which includes, as typical examples, the class of boolean algebras and more generally, for each $n \geq 2$, the class $\mathcal{CR}_n$ of (commutative) rings satisfying $\forall x : \; x^n = x$; such classes are noteworthy for the fact that their members have topological representations over Stone spaces and recursive first-order theories).*
(5) *$\mathcal{V}_2$ is congruence modular and is definitionally equivalent in a strong sense to the class of R-modules for some finite ring R.*
(6) *$\mathcal{V}_1$ is definitionally equivalent in a strong sense to an equational class of k-sorted algebras all of whose functions have arity 1 and whose category of unary term functions satisfies a simple condition.*

If in addition the language of $\boldsymbol{A}$ is finite, then the first-order theory of $\boldsymbol{HSP}(\boldsymbol{A})$ is recursive if and only if all of the above items are true and moreover the first-order theory of R-modules is recursive, where R is the finite ring associated with $\mathcal{V}_2$.

Using ideas contained in the proof of the previous theorem, B. Hart, S. Starchenko and Valeriote proved the following special case of Vaught's Conjecture (extended to incomplete theories).

THEOREM 6.2. [11] *Any equational class $\mathcal{V}$ in a countable language has either countably many or continuum many countable models. Moreover, if the number* *of countable models is countable, then* $\mathcal{V} = \mathcal{V}_1 \otimes \mathcal{V}_2$ *with* $\mathcal{V}_1$ *and* $\mathcal{V}_2$ *satisfying descriptions related to those in Theorem* 6.1.

In fact, Hart, Starchenko and Valeriote prove such a decomposition assuming only that every member of $\mathcal{V}$ is superstable. Though they are not employing tame congruence theory, their techniques were inspired by the combinatorial arguments of tame congruence theory, replacing finiteness with the assumption of superstability.

An interesting variation of the decidable equational class problem is the *finitely decidable equational class problem*, which poses the original pair of questions (recursive, noninterpretability of graphs) for the theory of the class of *finite* members of $\boldsymbol{HSP}(\boldsymbol{A})$. This variation was vigorously studied in the 1990's, culminating in the solution, due to J. Jeong [15] and P.M. Idziak [14], in the case that $\boldsymbol{HSP}(\boldsymbol{A})$ omits type 1 (the recursive question again being reduced to the case of R-modules for finite rings R). The solution gives characterizations of good structure for $\boldsymbol{HSP}(\boldsymbol{A})$ weaker and somewhat more complicated than the conditions given in Theorem 6.1; as in the proof of Theorem 6.1, the conditions are proved necessary by intricate tame congruence theoretic arguments, and sufficient by an appropriate Feferman-Vaught-style analysis.

Finally, J. Berman and Idziak have defined the notion of a finitely generated equational class $\boldsymbol{HSP}(\boldsymbol{A})$ having *few finite models*; by this they mean that the number of k-generated members of $\boldsymbol{HSP}(\boldsymbol{A})$ up to isomorphism is bounded by $2^{p(k)}$ for some polynomial $p(k)$. Using arguments inspired by and going beyond those in the study of the finitely decidable equational class problem, the monograph [3] of Berman and Idziak characterizes those $\boldsymbol{HSP}(\boldsymbol{A})$ which have few finite models and omit type 1.

6.2. Refutation of the RS conjectures. Before 1993, the best completely general result concerning the RS conjecture was the following theorem of McKenzie and Shelah, amplifying on an earlier result of W. Taylor.

THEOREM 6.3. [30] *Let* $\mathcal{V}$ *be an equational class in a countable language. Then the cardinalities of the irreducible members of* $\mathcal{V}$:

(1) *are not bounded by any cardinal, or*
(2) *include all infinite cardinals less than or equal to* $2^{\aleph_0}$ *but no cardinal larger than* $2^{\aleph_0}$, *or*
(3) *include* $\aleph_0$ *but no cardinal larger than* $\aleph_0$, *or*
(4) *include arbitrarily large finite cardinals but no infinite cardinal, or*
(5) *are bounded by a finite cardinal.*

The RS "Gap" conjecture asserts that only the first and last options can arise if $\mathcal{V} = \boldsymbol{HSP}(\boldsymbol{A})$ for some finite algebra $\boldsymbol{A}$.

In the fall of 1993, while attempting to prove the relative version of the RS "Gap" conjecture for irreducible algebras with least nonzero congruence of type 5, McKenzie discovered a technique by which he could do roughly the following. Given a universal Horn theory T of directed graphs extending the universal Horn theory of some finite relational structure $\boldsymbol{M}$, McKenzie could build a finite algebra $\boldsymbol{A}$ with the property that all but finitely many of the irreducible members of $\boldsymbol{HSP}(\boldsymbol{A})$ (i) have a least nonzero congruence of type 5, and (ii) are bi-interpretable with those models of T which have a *universally reachable vertex*, that is, a vertex which is reachable by paths from every vertex other than itself. Moreover, the language of $\boldsymbol{A}$ can be taken to be finite if T is finitely axiomatizable relative to the universal Horn theory of $\boldsymbol{M}$. If T is taken to be the theory of directed graphs asserting that the in- and out-degrees of every vertex are at most 1, then McKenzie showed that a suitable finite structure $\boldsymbol{M}$ exists, hence $\boldsymbol{A}$ can be built and $\boldsymbol{HSP}(\boldsymbol{A})$ has a unique countably infinite irreducible member but no uncountable irreducible members. This example refuted the RS "Gap" conjecture. By extensions of the technique to universal Horn theories of more complicated structures, McKenzie proved:

THEOREM 6.4. [26] *For each $k = 1, \dots, 5$ there exists a finite algebra $\boldsymbol{A}_k$ for which the cardinalities of the irreducible members of $\boldsymbol{HSP}(\boldsymbol{A}_k)$ are described by option (k) of Theorem* 6.3.

Except in the case $k = 4$, McKenzie also showed that it is possible to choose the algebra $\boldsymbol{A}_k$ witnessing the above theorem to have a finite language. The question of whether $\boldsymbol{A}_4$ can be chosen with a finite language has been open since 1971 and is known as the *restricted Quackenbush problem*.

McKenzie also saw that, by the same technique, he could construct a universal Horn theory, finitely axiomatizable relative to the universal Horn theory of some finite structure, whose models having the analogue of a universally reachable vertex encode fragments of computations of Turing machines. In this way he proved the following response to the RS "Recursive" conjecture.

THEOREM 6.5. [28] *For each $k = 1, \dots, 5$ let C_k denote the class of (integer codes of) finite algebras $\boldsymbol{A}$ in a finite language such that the cardinalities of the irreducibles in $\boldsymbol{HSP}(\boldsymbol{A})$ are described by option (k) of Theorem* 6.3. *Then C_1 and C_5 are recursively inseparable.*

6.3. Tarski's finite basis problem and other undecidable problems. Recall from section 1:

Finite Basis Problem: For which finite algebras $\boldsymbol{A}$ with a finite language is $\boldsymbol{HSP}(\boldsymbol{A})$ finitely axiomatizable?

Tarski's Finite Basis Problem: Is there an algorithm which, given a finite algebra $\boldsymbol{A}$ in a finite language, decides whether the class $\boldsymbol{HSP}(\boldsymbol{A})$ is finitely axiomatizable?

In 1993 McKenzie proved that the answer to Tarski's question is "no." Here is how he did it. To prove an early version of Theorem 6.5, McKenzie gave an effective procedure which associates to each Turing machine $\mathcal{T}$ a finite algebra $\boldsymbol{A}(\mathcal{T})$ in such a way that:

(1) If $\mathcal{T}$ halts when started on a blank initial tape, then there is a finite bound to the cardinalities of the irreducible members of $\boldsymbol{HSP}(\boldsymbol{A}(\mathcal{T}))$.
(2) If $\mathcal{T}$ does not halt when started on a blank initial tape, then $\boldsymbol{HSP}(\boldsymbol{A}(\mathcal{T}))$ has countably infinite irreducible members.

Moreover, the language of $\boldsymbol{A}(\mathcal{T})$ is finite, and one of the countably infinite irreducibles in $\boldsymbol{HSP}(\boldsymbol{A}(\mathcal{T}))$ in the nonhalting case has properties known to imply that its presence in $\boldsymbol{HSP}(\boldsymbol{A}(\mathcal{T}))$ guarantees that $\boldsymbol{HSP}(\boldsymbol{A}(\mathcal{T}))$ is *not* finitely axiomatizable. Motivated by this, McKenzie gave another construction $\boldsymbol{F}(\mathcal{T})$ for which $\boldsymbol{HSP}(\boldsymbol{F}(\mathcal{T}))$ had the same countably infinite irreducible member in the nonhalting case and which he could show was finitely axiomatizable in the halting case. Thus McKenzie had reduced the halting problem to the finite basis problem, thereby answering Tarski's problem negatively.

THEOREM 6.6. [27] *The answer to Tarski's finite basis problem is 'no': there is no algorithm which, given a finite algebra **A** in a finite language, determines whether $\boldsymbol{HSP}(\boldsymbol{A})$ is finitely axiomatizable.*

Since their publication, McKenzie's Theorems 6.5 and 6.6 have led to other undecidability results proved by his methods, including:

THEOREM 6.7 (McKenzie, J. Wood [32]). *For each $i \in \{2, 3, 4, 5\}$ there is no algorithm which, given a finite algebra **A** in a finite language, determines whether type i occurs in $\boldsymbol{HSP}(\boldsymbol{A})$.*

THEOREM 6.8 (Willard [38]). *There is no algorithm which, given a finite algebra **A** in a finite language, determines whether the first-order theory of $\boldsymbol{HSP}(\boldsymbol{A})$ has a model companion.*

6.4. Park's conjecture and positive finite axiomatizability results. Related to the finite basis problem is the following conjecture of R. Park from 1976.

PARK'S CONJECTURE 6.9. Let $\boldsymbol{A}$ be a finite algebra in a finite language. If there is a finite bound to the cardinalities of the irreducible members of $\boldsymbol{HSP}(\boldsymbol{A})$, then $\boldsymbol{HSP}(\boldsymbol{A})$ is finitely axiomatizable.

The impetus for this conjecture was primarily the celebrated finite basis theorem of K. Baker, which confirmed the conjecture in case $\boldsymbol{HSP}(\boldsymbol{A})$ is congruence distributive.

THEOREM 6.10. [1] *Park's conjecture is true whenever $\boldsymbol{HSP}(\boldsymbol{A})$ is congruence distributive.*

(Note: by Theorem 3.3, if $\boldsymbol{HSP}(\boldsymbol{A})$ is congruence distributive then the cardinalities of the irreducibles in $\boldsymbol{HSP}(\boldsymbol{A})$ automatically have a finite bound.)

McKenzie subsequently verified Park's conjecture in the congruence modular case.

THEOREM 6.11. [25] *Park's conjecture is true whenever* $\mathbf{HSP}(A)$ *is congruence modular.*

Interest in the conjecture grew significantly when McKenzie refuted the RS conjectures in 1993. From the properties stated above it is evident that his construction $A(\mathcal{T})$ would also solve Tarski's problem if Park's conjecture could be verified for $A(\mathcal{T})$ in the halting case. This was eventually done by me and led to further verifications of Park's conjecture.

THEOREM 6.12. [37] *Park's conjecture is true whenever* $\mathbf{HSP}(A)$ *omits the abelian types* 1 *and* 2.

This last theorem has itself been generalized in two directions. On the one hand, Baker, G. McNulty and J. Wang [2] have succeeded in weakening the hypothesis that there be a finite bound on the cardinalities of the irreducible elements in $\mathbf{HSP}(A)$. In a different direction, M. Maróti and McKenzie [24] have generalized the result to a significant class of universal Horn classes generated by finitely many finite algebras.

6.5. The RS problem, nilpotency, and variations of abelianness. McKenzie's refutation of the RS "Gap" and "Recursive" conjectures, rather than ending efforts to understand the nature of irreducibles in finitely generated equational classes, instead served to sharpen the focus of the small group of specialists engaged in this analysis. The following somewhat ill-defined problem and its accompanying concrete subproblem describe in part the ongoing aim of these specialists.

RS PROBLEM 6.13.

(1) (Structure) Characterize those finitely generated equational classes $\mathbf{HSP}(A)$ (A finite) for which there is a cardinal bound to the cardinalities of its irreducible members.
(2) (co-R.E.) Prove that the property above is a co-r.e. property of A.

By the way, it is now time to reveal the secret of the acronym "RS." It stands for "residually small," jargon for "having a cardinal bound to the cardinalities of its irreducible members."

Freese's and McKenzie's proof of Theorem 4.2 completely solved both parts of the RS Problem in the congruence modular case. The first paragraph of the proof of Theorem 5.13, together with the observation preceding Corollary 5.14, imply that the problem is solved in case types 1 and 5 are omitted.

THEOREM 6.14. *The RS "Structure" problem is solved and the "co-R.E." problem is confirmed for all finite algebras* A *for which* $\mathbf{HSP}(A)$ *omits types* 1 *and* 5.

McKenzie refuted the RS conjectures with finitely generated equational classes which omitted the abelian types 1 and 2. Notwithstanding this fact, McKenzie had previously solved the RS *problems* in that case. Moreover, if one relativizes the RS problem to the consideration of only those irreducibles whose least nonzero congruence is nonabelian, then McKenzie had completely solved this relativized problem.

THEOREM 6.15. [29] *When the RS problem is relativized to those irreducible members of* $\boldsymbol{HSP}(\boldsymbol{A})$ *with least nonzero congruence of types* 3–5, *the "Structure" problem is solved and the "co-R.E." problem is confirmed.*

Subsequently, K. Kearnes in unpublished work has solved the RS problems relativized to type 2.

THEOREM 6.16. [17] *When the RS problem is relativized to those irreducible members of* $\boldsymbol{HSP}(\boldsymbol{A})$ *with least nonzero congruence of type* 2, *the "Structure" problem is solved and the "co-R.E." problem is confirmed.*

Therefore in a sense the only work remaining is the study of irreducibles whose least nonzero congruence has type 1. This statement obscures the magnitude of the problem, however; such irreducibles live outside "classical" mathematics, and their understanding seems to require entirely new tools. The situation is easiest to describe when considering equational classes $\boldsymbol{HSP}(\boldsymbol{A})$ which admit only the abelian types 1 and 2. Finite algebras in such classes are *solvable* in a sense analogous to the notion of solvable groups. Also by analogy it is possible to define the concept of one congruence *centralizing* another, so that a congruence is abelian if and only if it centralizes itself. With this notion of centrality one can define usual notions of nilpotency, but unlike the situation in groups the notions are not equivalent, and in general are not well-behaved or understood. In 1993 Kearnes published a ground-breaking paper [16] that uncovered deep consequences of these notions of nilpotency in finite algebras, proving, for example, that every homomorphic image of a *finite* abelian algebra is nilpotent in all senses.

Building on Kearnes' earlier work, Kearnes, E. Kiss and Valeriote in [20] developed a new theory of higher-dimensional traces, or *multitraces*, for cover pairs of types 1–3, and exploited this theory in [21] to resolve the RS problems in the case that every member of $\boldsymbol{HSP}(\boldsymbol{A})$ is abelian. In fact they prove:

THEOREM 6.17. *The RS "Structure" problem is solved and the RS conjectures are confirmed for all finite algebras* $\boldsymbol{A}$ *for which every member of* $\boldsymbol{HSP}(\boldsymbol{A})$ *is abelian.*

Tame congruence theory contained from the beginning a second notion of abelianness, called *strong abelianness*, which separates G-sets from R-modules (the former are, the latter are not). Kearnes and Kiss in [18] introduced a strong centrality relation which captures strong abelianness, and teased from it two further notions of centrality (*weak* and *rectangular*), which in turn give rise to their own notions of abelianness. As well, they solve the RS problems for finite strongly nilpotent algebras:

THEOREM 6.18. *The RS "Structure" problem is solved and the RS conjectures are confirmed for all finite strongly nilpotent algebras* $\boldsymbol{A}$.

Kearnes and Kiss then considered the more general case of nilpotent algebras. In [19] the authors define the notion of the *twin group* of a $(0, \alpha)$-trace and relate its properties to weak abelianness and nilpotence. In particular, they prove that if $\boldsymbol{A}$ is a finite nilpotent algebra, then a necessary consequence of $\boldsymbol{HSP}(\boldsymbol{A})$ being residually small is that the twin groups which arise must be abelian and hence every member of $\boldsymbol{HSP}(\boldsymbol{A})$ must be weakly abelian. As a corollary they are able to solve the RS "Structure" problem and verify the RS conjectures for a special kind of nilpotent algebra which arises naturally in tame congruence theory. And here rests the current state of the RS problems.

§7. Further reading. For an introduction to universal algebra see the classic text of S. Burris and H.P. Sankappanavar [5], which is out of print but can be downloaded for free (see the reference). Modular commutator theory is thoroughly covered in [8]; a slimmer introduction is [9]. The authoritative text for tame congruence theory is [13], which can be downloaded from the American Mathematical Society. The reader might wish to consult the introductions to (parts of) tame congruence theory given by [6] and [22]. The chapters [23, 35, 36] from [12] give more detailed surveys of parts of this article.

REFERENCES

[1] K. A. BAKER, *Finite equational bases for finite algebras in a congruence-distributive equational class*, ***Advances in Mathematics***, vol. 24 (1977), no. 3, pp. 207–243.

[2] K. A. BAKER, G. F. MCNULTY, and J. WANG, *An extension of Willard's finite basis theorem: congruence meet-semidistributive varieties of finite critical depth*, ***Algebra Universalis***, vol. 52 (2004), no. 2-3, pp. 289–302.

[3] J. BERMAN and P. M. IDZIAK, *Generative complexity in algebra*, ***Memoirs of the American Mathematical Society***, vol. 175 (2005), no. 828, pp. viii+159.

[4] S. BURRIS and R. MCKENZIE, *Decidability and Boolean representations*, ***Memoirs of the American Mathematical Society***, vol. 32 (1981), no. 246, pp. viii+106.

[5] S. BURRIS and H. P. SANKAPPANAVAR, ***A Course in Universal Algebra***, Graduate Texts in Mathematics, vol. 78, Springer-Verlag, New York, 1981, Free download at `www.math.uwaterloo.ca/~snburris/`.

[6] M. CLASEN and M. VALERIOTE, *Tame congruence theory*, ***Lectures on Algebraic Model Theory*** (B. Hart and M. Valeriote, editors), Fields Institute Monographs, vol. 15, American Mathematical Society, Providence, RI, 2002, pp. 67–111.

[7] R. FREESE and R. MCKENZIE, *Residually small varieties with modular congruence lattices*, ***Transactions of the American Mathematical Society***, vol. 264 (1981), no. 2, pp. 419–430.

[8] ———, ***Commutator Theory for Congruence Modular Varieties***, London Mathematical Society Lecture Note Series, vol. 125, Cambridge University Press, Cambridge, 1987.

[9] H. P. GUMM, *Geometrical methods in congruence modular algebras*, ***Memoirs of the American Mathematical Society***, vol. 45 (1983), no. 286, pp. viii+79.

[10] J. HAGEMANN and C. HERRMANN, *A concrete ideal multiplication for algebraic systems and its relation to congruence distributivity*, ***Archives of Mathematics***, vol. 32 (1979), no. 3, pp. 234–245.

[11] B. Hart, S. Starchenko, and M. Valeriote, *Vaught's conjecture for varieties*, ***Transactions of the American Mathematical Society***, vol. 342 (1994), no. 1, pp. 173–196.

[12] B. T. Hart, A. H. Lachlan, and M. A. Valeriote (editors), *Algebraic model theory*, NATO Advanced Science Institutes Series C: Mathematical and Physical Sciences, vol. 496, Dordrecht, Kluwer Academic Publisher, 1997.

[13] D. Hobby and R. McKenzie, ***The Structure of Finite Algebras***, Contemporary Mathematics, vol. 76, American Mathematical Society, Providence, RI, 1988, `www.ams.org/online_bks/conm76/`.

[14] P. M. Idziak, *A characterization of finitely decidable congruence modular varieties*, ***Transactions of the American Mathematical Society***, vol. 349 (1997), no. 3, pp. 903–934.

[15] J. Jeong, *Finitely decidable congruence modular varieties*, ***Transactions of the American Mathematical Society***, vol. 339 (1993), no. 2, pp. 623–642.

[16] K. A. Kearnes, *An order-theoretic property of the commutator*, ***International Journal of Algebra and Computation***, vol. 3 (1993), no. 4, pp. 491–533.

[17] ———, *Residual smallness relativized to congruence types*, preprint.

[18] K. A. Kearnes and E. W. Kiss, *Finite algebras of finite complexity*, ***Discrete Mathematics***, vol. 207 (1999), no. 1-3, pp. 89–135.

[19] ———, *Residual smallness and weak centrality*, ***International Journal of Algebra and Computation***, vol. 13 (2003), no. 1, pp. 35–59.

[20] K. A. Kearnes, E. W. Kiss, and M. A. Valeriote, *Minimal sets and varieties*, ***Transactions of the American Mathematical Society***, vol. 350 (1998), no. 1, pp. 1–41.

[21] ———, *A geometric consequence of residual smallness*, ***Annals of Pure and Applied Logic***, vol. 99 (1999), no. 1-3, pp. 137–169.

[22] E. W. Kiss, *An easy way to minimal algebras*, ***International Journal of Algebra and Computation***, vol. 7 (1997), no. 1, pp. 55–75.

[23] ———, *An introduction to tame congruence theory*, ***Algebraic Model Theory (Toronto, ON, 1996)***, NATO Advanced Science Institutes Series C: Mathematical and Physical Sciences, vol. 496, Kluwer Academic Publisher, Dordrecht, 1997, pp. 119–143.

[24] M. Maróti and R. McKenzie, *Finite basis problems and results for quasivarieties*, ***Studia Logica. An International Journal for Symbolic Logic***, vol. 78 (2004), no. 1-2, pp. 293–320.

[25] R. McKenzie, *Finite equational bases for congruence modular varieties*, ***Algebra Universalis***, vol. 24 (1987), no. 3, pp. 224–250.

[26] ———, *The residual bounds of finite algebras*, ***International Journal of Algebra and Computation***, vol. 6 (1996), no. 1, pp. 1–28.

[27] ———, *Tarski's finite basis problem is undecidable*, ***International Journal of Algebra and Computation***, vol. 6 (1996), no. 1, pp. 49–104.

[28] ———, *Recursive inseparability for residual bounds of finite algebras*, ***The Journal of Symbolic Logic***, vol. 65 (2000), no. 4, pp. 1863–1880.

[29] ———, *Residual smallness relativized to congruence types. I*, ***Journal of Pure and Applied Algebra***, vol. 163 (2001), no. 2, pp. 209–220.

[30] R. McKenzie and S. Shelah, *The cardinals of simple models for universal theories*, ***Proceedings of the Tarski Symposium (Proc. Sympos. Pure Math., Vol. XXV, Univ. California, Berkeley, Calif., 1971)***, American Mathematical Society, Providence, R.I., 1974, pp. 53–74.

[31] R. McKenzie and M. Valeriote, ***The Structure of Decidable Locally Finite Varieties***, Progress in Mathematics, vol. 79, Birkhäuser, Boston, MA, 1989.

[32] R. McKenzie and J. Wood, *The type set of a variety is not computable*, ***International Journal of Algebra and Computation***, vol. 11 (2001), no. 1, pp. 89–130.

[33] P. P. Pálfy, *Unary polynomials in algebras. I*, ***Algebra Universalis***, vol. 18 (1984), no. 3, pp. 262–273.

[34] J. D. H. Smith, ***Mal'cev Varieties***, Lecture Notes in Mathematics, vol. 554, Springer-Verlag, Berlin, 1976.

[35] M. A. VALERIOTE, *Decidable equational classes*, ***Algebraic Model Theory (Toronto, ON, 1996)***, NATO Advanced Science Institutes Series C: Mathematical and Physical Sciences, vol. 496, Kluwer Academic Publisher, Dordrecht, 1997, pp. 207–221.

[36] R. WILLARD, *Three lectures on the RS problem*, ***Algebraic Model Theory (Toronto, ON, 1996)***, NATO Advanced Science Institutes Series C: Mathematical and Physical Sciences, vol. 496, Kluwer Academic Publisher, Dordrecht, 1997, pp. 231–254.

[37] ———, *A finite basis theorem for residually finite, congruence meet-semidistributive varieties*, ***The Journal of Symbolic Logic***, vol. 65 (2000), no. 1, pp. 187–200.

[38] ———, *Determining whether* $V(A)$ *has a model companion is undecidable*, ***International Journal of Algebra and Computation***, vol. 14 (2004), no. 3, pp. 325–355.

DEPARTMENT OF PURE MATHEMATICS
UNIVERSITY OF WATERLOO, WATERLOO, ONTARIO
N2L 3G1, CANADA
E-mail: rdwillar@uwaterloo.ca

LECTURE NOTES IN LOGIC

1. *Recursion Theory.* J. R. Shoenfield. (1993, reprinted 2001; 84 pp.)

2. *Logic Colloquium '90; Proceedings of the Annual European Summer Meeting of the Association for Symbolic Logic, held in Helsinki, Finland, July 15–22, 1990.* Eds. J. Oikkonen and J. Väänänen. (1993, reprinted 2001; 305 pp.)

3. *Fine Structure and Iteration Trees.* W. Mitchell and J. Steel. (1994; 130 pp.)

4. *Descriptive Set Theory and Forcing: How to Prove Theorems about Borel Sets the Hard Way.* A. W. Miller. (1995; 130 pp.)

5. *Model Theory of Fields.* D. Marker, M. Messmer, and A. Pillay. (First edition, 1996; 154 pp. Second edition, 2006; 155 pp.)

6. *Gödel '96; Logical Foundations of Mathematics, Computer Science and Physics; Kurt Gödel's Legacy. Brno, Czech Republic, August 1996, Proceedings.* Ed. P. Hajek. (1996, reprinted 2001; 322 pp.)

7. *A General Algebraic Semantics for Sentential Objects.* J. M. Font and R. Jansana. (1996; 135 pp.)

8. *The Core Model Iterability Problem.* J. Steel. (1997; 112 pp.)

9. *Bounded Variable Logics and Counting.* M. Otto. (1997; 183 pp.)

10. *Aspects of Incompleteness.* P. Lindstrom. (First edition, 1997; 133 pp. Second edition, 2003; 163 pp.)

11. *Logic Colloquium '95; Proceedings of the Annual European Summer Meeting of the Association for Symbolic Logic, held in Haifa, Israel, August 9–18, 1995.* Eds. J. A. Makowsky and E. V. Ravve. (1998; 364 pp.)

12. *Logic Colloquium '96; Proceedings of the Colloquium held in San Sebastian, Spain, July 9–15, 1996.* Eds. J. M. Larrazabal, D. Lascar, and G. Mints. (1998; 268 pp.)

13. *Logic Colloquium '98; Proceedings of the Annual European Summer Meeting of the Association for Symbolic Logic, held in Prague, Czech Republic, August 9–15, 1998.* Eds. S. R. Buss, P. Hájek, and P. Pudlák. (2000; 541 pp.)

14. *Model Theory of Stochastic Processes.* S. Fajardo and H. J. Keisler. (2002; 136 pp.)

15. *Reflections on the Foundations of Mathematics; Essays in Honor of Solomon Feferman.* Eds. W. Seig, R. Sommer, and C. Talcott. (2002; 444 pp.)

16. *Inexhaustibility; A Non-exhaustive Treatment.* T. Franzén. (2004; 255 pp.)

17. *Logic Colloquium '99; Proceedings of the Annual European Summer Meeting of the Association for Symbolic Logic, held in Utrecht, Netherlands, August 1–6, 1999.* Eds. J. van Eijck, V. van Oostrom, and A. Visser. (2004; 208 pp.)

18. *The Notre Dame Lectures.* Ed. P. Cholak. (2005; 185 pp.)

19. *Logic Colloquium 2000; Proceedings of the Annual European Summer Meeting of the Association for Symbolic Logic, held in Paris, France, July 23–31, 2000.* Eds. R. Cori, A. Razborov, S. Todorčević, and C. Wood. (2005; 408 pp.)

20. *Logic Colloquium '01; Proceedings of the Annual European Summer Meeting of the Association for Symbolic Logic, held in Vienna, Austria, August 1–6, 2001.* Eds. M. Baaz, S. Friedman, and J. Krajíček. (2005; 486 pp.)

21. *Reverse Mathematics 2001.* Ed. S. Simpson. (2005; 401 pp.)

22. *Intensionality.* Ed. R. Kahle. (2005; 265 pp.)

23. *Logicism Renewed: Logical Foundations for Mathematics and Computer Science.* P. Gilmore. (2005; 230 pp.)

24. *Logic Colloquium '03: Proceedings of the Annual European Summer Meeting of the Association for Symbolic Logic, held in Helsinki, Finland, August 14–20, 2003.* Eds. V. Stoltenberg-Hansen and J. Väänänen. (2006; 407 pp.)

25. *Nonstandard Methods and Applications in Mathematics.* Eds. N. J. Cutland, M. Di Nasso, and D. Ross. (2006; 248 pp.)

26. *Logic in Tehran: Proceedings of the Workshop and Conference on Logic, Algebra, and Arithmetic, held October 18–22, 2003.* Eds. A. Enayat, I. Kalantari, and M. Moniri. (2006; 341 pp.)

27. *Logic Colloquium '02: Proceedings of the Annual European Summer Meeting of the Association for Symbolic Logic and the Colloquium Logicum, held in Münster, Germany, August 3–11, 2002.* Eds. Z. Chatzidakis, P. Koepke, and W. Pohlers. (2006; 359 pp.)

28. *Logic Colloquium '05: Proceedings of the Annual European Summer Meeting of the Association for Symbolic Logic, held in Athens, Greece, July 28–August 3, 2005.* Eds. C. Dimitracopoulos, L. Newelski, D. Normann, and J. Steel. (2007; 272 pp.)

29. *Logic Colloquium '04: Proceedings of the Annual European Summer Meeting of the Association for Symbolic Logic, held in Torino, Italy, July 25–31, 2004.* Eds. A. Andretta, K. Kearnes, and D. Zambella. (2007; 220 pp.)